electronic

processes in

materials

Leonid V. Azároff

PROFESSOR OF PHYSICS AND DIRECTOR
INSTITUTE OF MATERIALS SCIENCE
UNIVERSITY OF CONNECTICUT

James J. Brophy

DIRECTOR OF TECHNICAL DEVELOPMENT
IIT RESEARCH INSTITUTE

McGraw-Hill Book Company

New York   **San Francisco**   **Toronto**   **London**

electronic

processes in

materials

ELECTRONIC PROCESSES IN MATERIALS

Dedicated to
Carmen and Muriel

Preface

In recent years, there has been a growing awareness of the need to study the relation between the structure and properties of materials. This has led to the gradual adoption of the term *materials science* to describe such a study and has given the impression that such an approach to materials is of very recent vintage. Nothing could be further from the truth.

As a matter of fact, materials science properly begins when man first seeks to understand what makes one material behave differently from another. It is a matter of record that Empedocles, Democritus, and other Greek philosophers speculated over two thousand years ago that certain materials are composed of atoms that have "hooks" on them that enable them to stick to each other firmly, while others are composed of more slippery atoms that can easily move past each other. A more scientific approach was adopted by metallurgists and mineralogists in the nineteenth century to relate many distinguishable properties of crystalline metals and minerals either to chemical composition or to structural features visible with the aid of a microscope. The discovery of x-ray diffraction in 1913 and its confirmation of the periodic nature of crystals made it possible to evolve the much later theoretical concepts of materials and to establish the completely new sciences of solid-state chemistry and physics. Historically, metallurgists assimilated these twentieth-century developments into their field whereas mineralogists were slower to do this, so that the science of ceramics, fostered jointly by metallurgy and mineralogy, was born. As the amount of activity in these areas kept increasing, it became abundantly clear that the same processes were operative in all materials, including, for example, plastics, and that a great deal could be gained by attempting a more unified approach to their study. This is the modern-day version of materials science, which does not include, but does make use of, recent developments in solid-state chemistry and physics in order to relate the properties of materials to their structure on a submicroscopic as well as a macroscopic scale. Concurrently, it is concerned with applying any newly gained perspectives to the improvement of the specific properties that materials can possess and to their wiser utilization.

In writing this book, the authors have been motivated by this growing interest in a unified approach to materials. Since the properties of materials are chiefly determined by their atomic composition and the nature of the atomic aggregation, it is reasonable to say that they are controlled by the crystal structure in crystalline materials. The interplay of structure and properties, however, can be expressed in essentially two ways. One is to relate the structural features to the observed properties in a phenomenological way. The other is to make use of mathematical theories developed mainly with the aid of statistical mechanics and quantum mechanics to describe the behavior of materials under the influence of various forces. This book attempts to combine both approaches to discuss those properties of materials that are chiefly determined by electronic processes in materials. A companion volume is in preparation in which the properties of materials best described by thermodynamic processes will be considered. Until this second volume is completed, the reader is urged to consult *Introduction to solids*,† which, in a necessarily brief account, discusses most of the properties of materials not considered in this volume.

This textbook presupposes a familiarity with the elements of chemistry, physics, and calculus. Although some discussions make use of differential equations, their manipulation is clearly indicated, so that the mathematics serve primarily to decrease the number of words required to transmit a specific idea. After an introductory discussion of geometrical and x-ray crystallography, the fundamental concepts of quantum mechanics and statistical mechanics are explained and their application to bonding and other phenomena is illustrated. The rest of the book is concerned with the theoretical developments in our understanding of solids and their application to the elucidation of the electric, magnetic, optic, and thermal processes in materials. The relation of these properties to crystal structure is pointed out, and the interplay between theory and practice is often illustrated by a discussion of some of the more noteworthy devices that have been developed. In view of the dominant role that semiconductor technology has played in the evolution of mate-

† Leonid V. Azároff, *Introduction to solids* (McGraw-Hill Book Company, Inc., New York, 1960).

rials science, the fundamental principles underlying the operation of semiconductor devices are explained and related to commonly used semiconductor materials. No attempt has been made to discuss all materials or all conceivable devices; rather, the main classes of each have been explored at moderate depth. Exercises have been included at the end of each chapter in order to permit the reader to test his grasp of the subject matter and to stimulate further development of some of the ideas mentioned in the text. The answers to some of the problems are also provided so that the reader can evaluate his own progress. The literature lists at the end of each chapter have been divided into two groups, one referring to sources of supplementary information, the other to somewhat more advanced discussions of the same topics. References to individual contributions are not included because it is the objective of this book to provide a general synthesis rather than a discussion of specific details.

The units used throughout this book are mks units because these units are most appropriate for describing electric phenomena. It has become the practice to use mixed units in semiconductor technology because the measured quantities are usually quite small. Since this often leads to confusion, it is hoped that this book will help place the matter of units on a more consistent basis. Unfortunately, mks units are not the most natural units possible for the discussion of magnetic phenomena; nevertheless, this also can be done as discussed in the text. The equivalences of mks and cgs designations are given in Appendix 2. Another innovation in this book is that the symbols recently recommended by the American Institute of Physics, in cooperation with the International Union of Pure and Applied Physics, are used throughout. The most significant change is that capital letters, usually representing the first letter of a unit named in honor of its discoverer, are used in place of the previously used abbreviations. A partial listing of the most commonly used units and their symbols is given in Appendix 1, and each symbol is identified in the text when it first appears.

The authors are deeply indebted to many of their colleagues, who, through their publications, have made most of the information presented in this book available. In addition, a number of them consented to have their illustrations reproduced in this book. We gratefully acknowledge the permissions granted by the authors and previous publishers of the following illustrations:

Chapter 8: Fig. 16, J. E. Hill and K. M. van Vliet, *J. Appl. Phys.*, vol. 29 (1958), p. 177; Fig. 25, T. H. Geballe, *Phys. Rev.*, vol. 98 (1955), p. 940.

Chapter 9: Fig. 14, R. F. Brebrick and W. W. Scanlon, *Phys. Rev.*, vol. 96 (1954), p. 598.

Chapter 10: Fig. 18, M. Keilson (ed.), *Electronic Progress* (Raytheon Manufacturing Co., Lexington, Mass., 1957).

Chapter 11: Figs. 7 and 9, G. A. Haas and J. T. Jensen, Jr., *J. Appl. Phys.*, vol. 31 (1960).

We also express our thanks to those of our colleagues who kindly read and criticized this manuscript. In particular, we wish to thank Professor L. I. Grossweiner for carefully reading the entire manuscript. Our thanks go also to Mrs. Marion Vogt for preparing the final typescript and to those unsung heroines, our wives, who selflessly encouraged us throughout its preparation.

Leonid V. Azároff

James J. Brophy

Contents

1. structure of crystals

Introduction to crystallography

All materials are composed of atoms or molecules. What distinguishes crystals from noncrystals is the way that the atoms are arrayed inside the material. In crystals the atomic array is *periodic*; that is, a representative unit, or motif, is repeated at regular intervals along any and all directions in the crystal. In noncrystals, the same atomic groups are arrayed more randomly. This is true of all noncrystalline materials, whether in the form of gases, liquids, or solids. In fact, one can distinguish noncrystalline solids from liquids only by their relative viscosity. By comparison, the periodicity of the atomic array in crystals is reflected in all their properties. Before proceeding with a discussion of the properties of materials, therefore, it is first necessary to consider the different periodic arrays that can exist in three dimensions. The study of these different possibilities and the laws that govern them is called *pattern theory*, or *geometrical crystallography*. It should be noted that periodicity is not an exclusive property of crystals but occurs quite commonly in nature. As the reader becomes familiar with the ways a motif can be repeated periodically in space, it will prove interesting to examine critically such common periodic arrays as tiles in a floor, patterns in a dress or a necktie, and so forth.

It is possible to consider the permissible ways of periodically repeating a motif without having to consider the intimate details of the repeated unit. For convenience, therefore, the figure seven will be used in the illustrations to symbolize a fundamental group of atoms or molecules in a crystal. There are four principal means for repeating such a motif in space:

1. *Translation* by an amount t, shown in Fig. 1

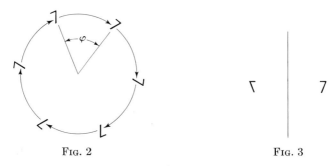

Fig. 1

2. *Rotation* through an angle φ, shown in Fig. 2

3. *Reflection* across a line (in two dimensions) or plane (in three dimensions), shown in Fig. 3

4. *Inversion* through a point, shown in Fig. 4.

Note that repetition by a translation or by a rotation leaves the motif unchanged whereas a reflection or an inversion changes the character of

Fig. 2   Fig. 3

the motif from a *right-handed* to a *left-handed* one. Note also that a translation repeats the motif an infinite number of times along a given direction while the other operations repeat the motif a finite number of times. The ways in which three mutually noncoplanar translations can combine to produce an infinite three-dimensional periodic array are considered next. The finite operations of repetition are called *symmetry elements* and will be considered in a subsequent section.

Periodicity in crystals. The single translation in Fig. 1 above produces an infinite linear array of the repeated object. If such a translation, t_1, is combined with another, noncollinear translation, t_2, then a two-dimensional array (Fig. 5A) obtains as follows: The entire linear array due to translation t_1 (Fig. 1) is repeated an infinite number of times by the second translation

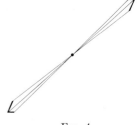

Fig. 4

t_2. Another way of looking at this is to say that the linear array due to t_2 is repeated by t_1. Since the nature of the repeated objects in Fig. 5A does not affect the translation periodicity, it is conventional to represent this periodicity by replacing each object in the array with a point. The resulting collection of points shown in Fig. 5B is called a *lattice*, in this case, a two-dimensional or *plane lattice*. It should be remembered that a point is an imaginary, infinitesimal spot in space, and consequently *a lattice of points is imaginary* also. On the other hand, the array of sevens in Fig. 5A is real. It is *not* a lattice of sevens, because a lattice is an imaginary concept; instead it is correctly designated a *lattice array* of sevens.

It is possible to add a third translation to the plane lattice in Fig. 5B or to the lattice array in Fig. 5A. In each case, the third translation

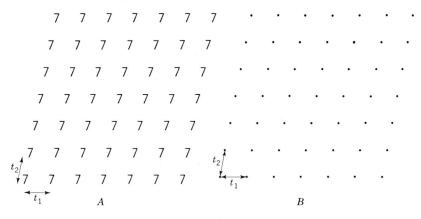

Fig. 5

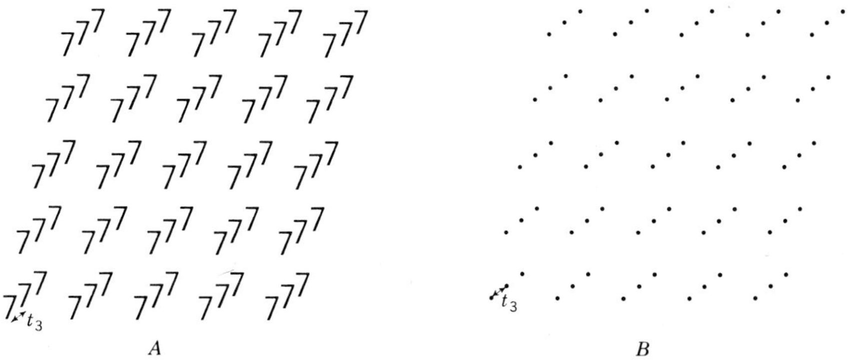

Fig. 6

repeats the entire plane at equal intervals t_3. This third translation thus produces a *space lattice* (Fig. 6B) or a *three-dimensional lattice array* (Fig. 6A). Because it is easier to draw a plane lattice than a space lattice on a sheet of paper, wherever possible plane lattices are used in this book. The extension of the principles illustrated to three dimensions follows directly from the above discussion.

Because a lattice array of groups of atoms is merely a periodic repetition of the fundamental grouping, it is not really necessary to consider the entire array. Instead, it is sufficient to consider the fundamental unit that is being repeated. This is normally done by selecting a set of translations that define a *unit cell* in the lattice. The entire lattice can then be visualized as an infinite collection of such unit cells repeated by the translations that form the edges of the chosen cell. The way that a unit cell can be chosen in a plane lattice is illustrated in Fig. 7. If pairs of translations such as t_1, t_2 or t_3, t_4 are chosen, they are said to define a *primitive cell*, so called because it contains only one lattice point, as can be easily seen by slightly displacing the origin of the shaded cells in Fig. 7 from a lattice point. If, instead, translations like t_5 and t_6 are chosen, the unit cell contains more than one lattice point and it is called a non-primitive or *multiple cell*. As shown later, it is sometimes advantageous

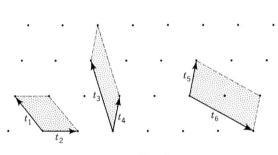

Fig. 7

to select a multiple or a *centered cell* as the unit cell of a lattice, even though it contains a larger number of lattice points, and therefore atoms, than a primitive cell does.

In three dimensions, the three mutually noncoplanar translations selected as unit-cell edges are called the *crystallographic axes* $\mathbf{a}$, $\mathbf{b}$, $\mathbf{c}$, and the angles between them, α, β, γ. The relative disposition of these axes and angles is shown in Fig. 8. Use can be made of these axes to define any other translation direction in the lattice by forming an appropriate vector sum. For example, the translations $\mathbf{t}_1$ and $\mathbf{t}_2$ shown in Fig. 9 can be expressed

$$\mathbf{t}_1 = 2\mathbf{a} + 1\mathbf{b}$$

and
$$\mathbf{t}_2 = 1\mathbf{a} + 4\mathbf{b}. \tag{1}$$

Note that any translation parallel to $\mathbf{t}_2$ in this lattice is an identical translation $\mathbf{t}_2$ regardless of its point of origin.

A translation direction in a space lattice can be represented by the vector sum

$$\mathbf{t} = u\mathbf{a} + v\mathbf{b} + w\mathbf{c}. \tag{2}$$

Actually, it is not necessary to write the entire expression in (2) once **a**, **b**, and **c** have been selected, since the only thing that changes for different directions is the set of coefficients u, v, and w.   It is convenient,

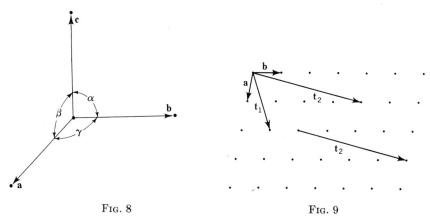

Fig. 8   Fig. 9

therefore, to use the shorthand notation $[uvw]$ to designate a translation direction in the lattice.

Representation of planes.   It is frequently necessary to consider planes passing through a space lattice.   In order to uniformalize the designation of a chosen plane, the following procedure has been adopted:

1. Determine the intercepts of the plane along **a**, **b**, and **c**.

2. Invert the intercepts; that is, write the numbers as their reciprocals.

3. If fractions result, multiply them by their lowest common denominator.

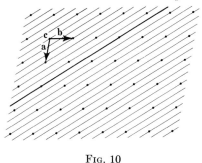

Fig. 10

For example, consider the planes drawn in Fig. 10, which shows a view of a space lattice along **c** and a set of planes that are parallel to **c** so that they are also seen in an edge view.   (Note that an identical plane passes through each lattice point as a result of the periodic nature of a lattice.) Consider the plane shown by a heavy line in Fig. 10:

Indexing procedure	a	b	c	
1. Determine intercepts	2	3	∞	
2. Note their reciprocals	$\dfrac{1}{2}$	$\dfrac{1}{3}$	$\dfrac{1}{\infty} = 0$	(3)
3. Clear fractions	3	2	0	

It is left to the reader to show that the same three integers result regardless of which parallel plane in Fig. 10 is chosen for this analysis (Exercise 1).

The resulting three integers are called the *Miller indices* of a plane and are conventionally enclosed in parentheses: (*hkl*). The meaning of these indices is that a set of parallel planes (*hkl*) cuts the a axis into h parts, the b axis into k parts, and the c axis into l parts. As an illustration, a (100), (110), and (122) plane is shown in Fig. 11.

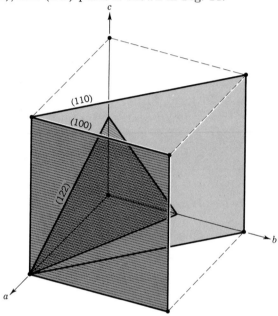

FIG. 11. The (100), (110), and (122) planes in a unit cell. Actually, identical parallel planes pass through each and every lattice point, but only the plane nearest to the origin is shown for clarity.

A special case of indexing arises when a lattice can be described by a unit cell having two equal axes inclined at 120 deg and a third axis that is orthogonal to the plane of these two axes (Fig. 12). As can be seen in Fig. 13, the plane of the two equal axes contains a third axis that is equal in length to the other two. It will be shown later in this chapter that this type of lattice occurs in the *hexagonal crystal system*, in which case the three coplanar axes are equivalent by symmetry. There is some advantage to displaying this symmetry equivalence in the indices. If the four hexagonal axes $\mathbf{a}_1$, $\mathbf{a}_2$, $\mathbf{a}_3$, $\mathbf{c}$ are used, then the corresponding *hexagonal* indices are (*hkil*). It is easy to show that the relationship between the three equivalent axes is

$$\mathbf{a}_1 + \mathbf{a}_2 = -\mathbf{a}_3 \tag{4}$$

and between the corresponding indices,

$$h + k = -i. \tag{5}$$

A negative index is written with a bar over it; that is, $(11\bar{2}1)$ means $(1, 1, -2, 1)$. Since the relationship in Eq. (5) is easily remembered, the i is sometimes replaced by a dot when the hexagonal indices $(hk \cdot l)$ are used.

A plane in the lattice can also be repeated by other operations of symmetry, such as a rotation or a reflection. The collection of all planes

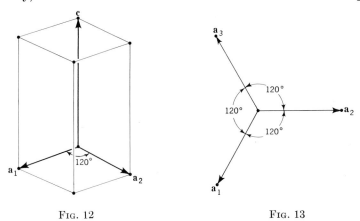

FIG. 12   FIG. 13

so repeated is designated $((hkl))$† and is said to constitute a *form*. The terminal faces of many crystals occurring in nature belong to one or more forms. Because most properties of crystals depend on direction, another relation between planes is often of interest. Whenever two or more planes have one direction in common, that is, they intersect along a common line, they are said to belong to a *zone*. Their intersection direction is called the *zone axis*. The indices of the zone axis $[uvw]$ and any plane (hkl) belonging to this zone must obey the relation

$$uh + vk + wl = 0. \tag{6}$$

For example, any plane $(h0l)$ belongs to the zone whose zone axis is $[010]$. The zone axis $[uvw]$ of any two intersecting planes $(h_1k_1l_1)$ and $(h_2k_2l_2)$ can be determined as follows:

$$\begin{aligned} u &= k_1l_2 - k_2l_1 \\ v &= l_1h_2 - l_2h_1 \\ w &= h_1k_2 - h_2k_1. \end{aligned} \tag{7}$$

† The equivalent designation for a set of symmetry-related translation directions is $[[uvw]]$.

Symmetry elements.   As already indicated, it is possible to repeat an object by a rotation, a reflection, or an inversion.   Combinations of two such operations into a single symmetry operation are also possible. If a symmetry element repeats a right-handed motif as another right-handed one, the pair is said to be *congruent*.   If a right hand becomes a left hand after repetition, the pair is said to be *enantiomorphous*.   A sequence of rotations about an imaginary line or axis producing a congruent set is called a *proper* rotation, and the symmetry element, a proper-rotation axis.   Reference to Fig. 2 shows that the so-called *throw* of a rotation axis φ must equal some integral submultiple of 360 deg, so that the number of repetitions n is finite.

$$\varphi = \frac{360 \text{ deg}}{n} \qquad \text{where } n = 1, 2, 3, \ldots \ . \tag{8}$$

The rotation axis whose throw φ is determined by (8) is called an n-fold axis.   Although, in general, n can be any integer, it must be remembered that any symmetry element present in a crystal must be consistent with the fundamental requirement of periodicity.

To see what restrictions a lattice imposes, consider the periodic repetition of n-fold axes A_n by a translation t as shown in Fig. 14.   Each

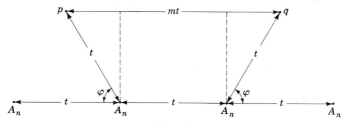

Fig. 14

rotation axis, in turn, has the effect of rotating all space surrounding it n times.   Such a rotation about two axes in the lattice row is shown in Fig. 14.   The two new lattice points generated at p and q are equidistant from the lattice row by construction.   Thus the line joining p to q is parallel to t and, in fact, must equal some integral multiple of t, say, mt. According to Fig. 14,

$$mt = t + 2t \cos \varphi \qquad m = 0, \pm 1, \pm 2, \pm 3, \ldots \tag{9}$$

and $\pm m$ is used, depending on whether the translation is measured to the right or left.   Dividing through by t and rearranging the terms in (9),

$$\cos \varphi = \frac{m - 1}{2} = \frac{N}{2} \qquad N = 0, \pm 1, \pm 2, \pm 3, \ldots \tag{10}$$

since $m - 1$ is also an integer, say, N.

It is now possible to determine what values φ can have in a lattice by solving (10) for all values of N. As shown in Table 1, only five rotation

Table 1
Determination of rotation axes allowed in a lattice

N	$\cos \varphi$	φ, deg	n
-2	-1	180	2
-1	$-\frac{1}{2}$	120	3
0	0	90	4
$+1$	$+\frac{1}{2}$	60	6
$+2$	$+1$	360 or 0	1

axes can exist in a lattice because solutions of (10) are not possible for $N > 2$. The axes are designated by the value that n has in Table 1; that is, a two-fold axis is called a 2, a three-fold axis is called a 3, and so forth. The five allowed rotation axes are depicted in Fig. 15.

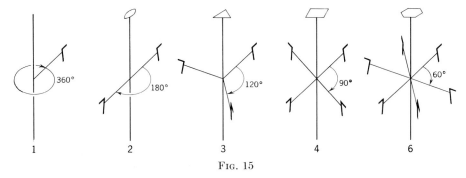

FIG. 15

As suggested earlier, it is possible to combine a proper rotation with a reflection into a single hybridized operation of repetition. The resulting set of repeated motifs are enantiomorphous to each other, and the operation is called an *improper* rotation. As an example of an improper rotation, consider the operation of *rotoreflection* shown in Fig. 16. The operation of two-fold rotation is combined with a reflection to give a two-fold rotoreflection axis $\tilde{2}$, pronounced "two tilde." It is important to realize that $\tilde{2}$ is neither a 2 nor a reflection, but a hybrid of both operations; that is, a rotoreflection is a new symmetry element encompassing the joint oper-

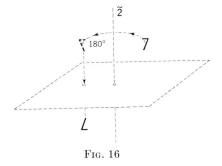

FIG. 16

ations of a rotation and a reflection. Note that the resulting enantiomorphous pair of sevens in Fig. 16 is identical with the pair shown
in Fig. 4. The symbol for an inversion through a center is $\bar{1}$ (pronounced "one bar"). Thus it follows that a $\tilde{2} = \bar{1}$, and is called an
inversion center, or *symmetry center*. It is similarly possible to combine
a rotation with an inversion into a new operation called a *rotoinversion*
axis. It can be shown, however, that the five resulting rotoinversion
axes are equivalent to the five possible rotoreflection axes, in pairs.

Finally, it should be noted that it is possible to combine a rotation with
a translation parallel to the rotation axis and a reflection with a translation parallel to the reflection plane, or *mirror plane m*, to produce new
hybridized symmetry operations. The translation components in each

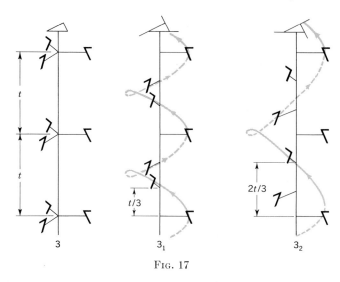

FIG. 17

case must be submultiples of the lattice translation in that direction,
otherwise a new symmetry operation does not result. When a subtranslation is combined with a rotation axis, a *screw axis* is produced.
For example, consider the combinations of a three-fold axis with subtranslations $t/3$ and $2t/3$ shown in Fig. 17. Starting at the bottom of
each axis, a rotation of the seven by 120 deg in a clockwise direction
(looking up along the axis) is combined with a subtranslation $t/3$ in the
central screw axis labeled 3_1. This operation is repeated two more times
until the translation equivalent seven is reached. The name screw axis is
derived from the spiral path followed in this process. One proceeds
similarly in the screw axis labeled 3_2, except that each subtranslation now
is $2t/3$. After three rotations of 120 deg and subtranslations, by an
amount $2t/3$, a translation equivalent seven is again reached. This

seven, however, lies two translations t above the initial seven. Since everything in the lattice must be repeated by the translation t, the other sevens shown in Fig. 17 are produced. Note that 3_2 is not unlike 3_1, except that successive sevens can be reached by a right-handed spiral in one and by a left-handed spiral in the other. The two kinds of screw axes, therefore, are enantiomorphous to each other.

When a reflection plane is combined with a subtranslation into a single operation of symmetry, a *glide plane* is produced. The crystallographic direction along which the subtranslation or glide component lies is used to designate the glide-plane type. Thus an a glide has a subtranslation of $a/2$ parallel to the plane and the a axis, a b glide has a glide component of $b/2$, and so forth.

Symmetry groups, The symmetry elements that can exist in crystals have been described above. They do not occur as isolated individuals in crystals, however, but form symmetry groups that are periodically repeated by the translations of a lattice. Since the environment of

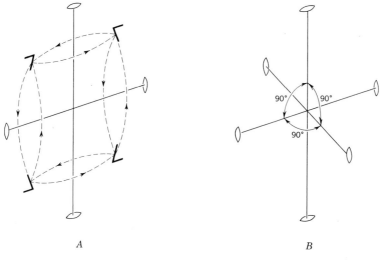

A B

F<small>IG</small>. 18

each lattice point must be exactly the same as that of any other lattice point, consider first the way in which the various symmetry elements can combine at a point. Suppose two two-fold axes intersect at 90 deg to each other. The way that each repeats a seven is shown in Fig. 18A. Note that the sevens in the resulting array are related to each other in pairs by still a third two-fold axis at right angles to the first two. Thus it can be seen that a combination of two two-fold axes at 90 deg produces a third two-fold axis mutually oriented as shown in Fig. 18B.

As another example, consider two mutually orthogonal mirror planes intersecting each other, shown in an edge view in Fig. 19*A*. The way that they repeat a seven is also indicated. Again note that a two-fold axis lying along the intersection (Fig. 19*B*) relates opposite pairs of sevens. It can be shown quite similarly that the combination of any two symmetry elements at a point automatically produces a third symmetry element passing through the same point. The combination of

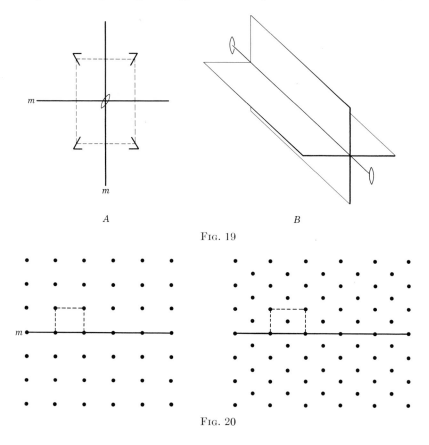

A   B

Fɪɢ. 19

Fɪɢ. 20

symmetry elements at a point is called a *point group*. Further examples of point groups and their derivation are considered in the exercises at the end of this chapter. It should be noted here that, in all, 32 point groups are possible in crystals.

Before proceeding to see the way that point groups can be combined with lattices in crystals, it is necessary to consider the various lattice types that can occur. Of primary interest are the restrictions placed on a lattice by the presence of certain kinds of symmetry elements. For example, suppose a mirror plane is to be combined with a lattice. As can be seen in Fig. 20, this is possible only if the lattice points lie along rows that are normal to the mirror plane. Note that the orthogonal unit cell in one lattice shown in Fig. 20 is primitive and in the other it is a multiple or centered cell. Other symmetry elements impose similar restrictions on a lattice. Thus a 4 requires that the lattice points lying in planes normal to the axis have a square array, and so forth. In fact, it has been shown by Bravais that 14 kinds of space lattices are necessary to accommodate the 32 point groups that can occur in crystals, so that these 14 space lattices are commonly called *Bravais lattices*.

When all possible combinations of point groups with the Bravais lattices are considered, 230 *space groups* result. The way that these can be classified into six *crystal systems* is described next.

Classification of crystals. Crystals occurring in nature often reflect the symmetry of their internal atomic arrangements in the symmetry relating their terminal faces. Thus it is possible to use this symmetry to

Table 2
Crystal systems

System	Minimal symmetry	Unit cell
Triclinic	1 (or $\bar{1}$)	$a \neq b \neq c$ $\alpha \neq \beta \neq \gamma$
Monoclinic	2 (or m)	$a \neq b \neq c$ $\alpha = \beta = 90 \deg \neq \gamma$
Orthorhombic	222 (or mmm)	$a \neq b \neq c$ $\alpha = \beta = \gamma = 90 \deg$
Tetragonal	4	$a = b \neq c$ $\alpha = \beta = \gamma = 90 \deg$
Isometric (cubic)	Four 3's	$a = b = c$ $\alpha = \beta = \gamma = 90 \deg$
Hexagonal	6 or 3	$a = b \neq c$ $\alpha = \beta = 90 \deg$ $\gamma = 120 \deg$

classify crystals, and this was done long before their internal arrangements were known. The symmetry exhibited by a crystal appears to be centered at its center, so that the point group of a crystal can be used to designate its *crystal class*. Historically, the crystal classes were collected into six crystal systems according to the symmetry elements that they had in common. The crystal systems and the minimum symmetry that the point groups belonging to each system must share are listed in Table 2. The last column in this table shows the relative dimensions of the unit-cell edges and angles of a space lattice that can accommodate the minimal symmetry in the second column. The corresponding Bravais lattices

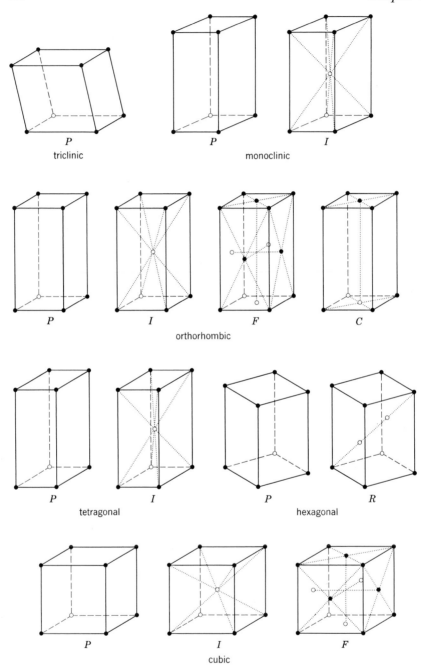

P   *P*   *I*

triclinic   monoclinic

P   *I*   *F*   *C*

orthorhombic

P   *I*   *P*   *R*

tetragonal   hexagonal

P   *I*   *F*

cubic

FIG. 21

are shown in Fig. 21. The letter P designates a primitive lattice, I a body-centered lattice, F an all-face-centered lattice, C a one-face-centered lattice, and R a special type of centered lattice that occurs only in the hexagonal system. The primitive cell in this lattice has the shape of a rhombohedron, and once it was common practice to refer crystals having such a lattice to a rhombohedral subsystem of the hexagonal system.† Nowadays it is considered more correct to select the multiple unit cell shown in Fig. 21 when describing such crystals.

A special symbolism has been developed to describe the combination of lattice-type and symmetry elements present in a crystal. The necessary information is collected in shorthand form in a *space-group symbol*, which first lists the lattice type, followed by the principal symmetry elements lying either parallel to or normal to the unique directions in a unit cell. For example, in the tetragonal and hexagonal systems, the three unique directions are c, a, and $[110]$. Thus the space-group symbol $I\,\dfrac{4}{m}\,cm$ means that the lattice is body-centered (I), that a 4 is parallel to the c axis and perpendicular to a mirror plane $\left(\dfrac{4}{m}\right)$, that a c glide is parallel to the a axis, and that a mirror plane is parallel to $[110]$. Since the space group contains only one four-fold axis, reference to Table 2 shows that it belongs to the tetragonal system. As might be expected from the previous discussions of symmetry groups in crystals, the space-group symbol does not enumerate all the symmetry elements actually present. The rules for deriving the other symmetry operations generated by those given in the symbol are discussed in some of the books listed at the end of this chapter. A full description of all 230 space groups can be found in *International tables for x-ray crystallography*, volume 1.‡

Equivalent positions in a unit cell. It is possible, of course, to combine a point group with a plane lattice to produce a *plane group*. As an example, consider the combination of a 4 with a primitive plane lattice denoted p (Fig. 22). Note that the unit cell of this lattice has the shape of a square; that is, $a = b$ because of the four-fold symmetry. A point located within this cell is repeated four times by the 4, and the resulting additional symmetry that relates these points is also indicated in Fig. 22. The coordinates of the points xy are normally expressed as

---

† The minimal symmetry that the rhombohedral lattice can accommodate is a 3. It is sometimes incorrectly referred to as a *trigonal* lattice for this reason. Actually, a 3 can be accommodated by either of the two hexagonal lattices in Fig. 21. Although it was once proposed that all crystals possessing one three-fold axis be grouped into a trigonal system, this practice has been abandoned.

‡ Norman F. M. Henry and Kathleen Lonsdale, *International tables for x-ray crystallography*, vol. 1 (The Kynoch Press, Birmingham, 1952).

fractions of the unit-cell edges; that is, x is the actual distance along the a axis, from the origin to the point, divided by the length of a. Similarly, the fraction y is the actual distance along b divided by the length of b. This designation has the advantage that the coordinates of all points need be determined only once for a particular plane group. One complete translation along a is $a/a = 1$, and so forth. Proceeding in a clockwise manner, the coordinates of the next point in Fig. 22 are the fraction y along the a axis and $1 - x = 1 + \bar{x} = \bar{x}$ along b. Since it is

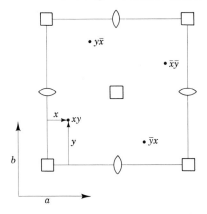

customary to list the coordinate along a first and along b next, the coordinates of this point are $y\bar{x}$. The coordinates of the other symmetry-related points are also indicated in Fig. 22. Such a set of symmetry-related points are said to belong to an *equipoint* set. Since four points belong to this set, its *rank* is said to be four.

FIG. 22.   Plane group $p4$.

Next consider the two points at $\frac{1}{2}0$ and $0\frac{1}{2}$. These are the locations of the two-fold axes in Fig. 22. Since each of these points is "shared" by two adjacent cells, only two such points "belong" to the unit cell shown and the rank of this equipoint set is two. Finally, it is easy to see that the ranks of the equipoints at 00 and $\frac{1}{2}\frac{1}{2}$ are, respectively, one. This information is collected in Table 3. As can be seen therein, there are four equipoint sets possible in $p4$. Three of them are *special positions* in the unit cell, whereas the coordinates of the fourth set can take on any values lying between 0 and 1. Such a set is said to contain the *general positions* in the unit cell.

Table 3
Equipoints of plane group $p4$

Rank of equipoint	Symmetry of location	Coordinates of equivalent points
1	4	00
1	4	$\frac{1}{2}\frac{1}{2}$
2	2	$0\frac{1}{2}, \frac{1}{2}0$
4	1	$xy, y\bar{x}, \bar{x}\bar{y}, \bar{y}x$

It is easy to extend the above discussion to the three-dimensional space group $P4$. The equipoints of this space group are obtained by adding the appropriate z coordinate to the equipoints of $p4$, as shown in Table 4.

Since the symmetry of $P4$ does not affect z, this coordinate is quite general for all equipoints. Note that Table 4 contains an additional column headed *Wyckoff notation*. This notation was proposed in one of the first tabulations of such data, and it has become common practice to refer to a position in a unit cell by its rank and letter.

Table 4
Equipoints of space group $P4$

Rank of equipoint	Wyckoff notation	Symmetry of location	Coordinates of equivalent points
1	a	4	$00z$
1	b	4	$\frac{1}{2}\,\frac{1}{2}z$
2	c	2	$0\frac{1}{2}z,\ \frac{1}{2}0z$
4	d	1	$xyz,\ y\bar{x}z,\ \bar{x}\bar{y}z,\ \bar{y}xz$

The closest packings of spheres

Hexagonal and cubic closest packings. To a very close approximation, the atoms in crystals can be likened to hard impenetrable spheres. It is instructive, therefore, to consider some of the ways that spheres can be packed together in three-dimensional arrays. As a start, consider the arrangement of like spheres in a plane. An array in which the maximum available space is occupied is shown in Fig. 23A. This array is called a

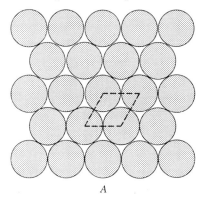

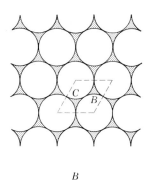

A *B*

Fig. 23

hexagonal closest-packed layer, or simply a closest-packed layer, and it can be shown that the circles in Fig. 23A occupy 90.7 per cent of the available space (Exercise 10). The unit cell of the plane lattice of this array is shown by broken lines in Fig. 23A. Note that each unit cell contains two kinds of void spaces, labeled B and C in Fig. 23B. Note also that the two kinds of triangular voids in this array differ in that the apices of the voids *B* point up whereas those labeled *C* point down.   If the origin of the unit cell is placed at either of these two voids, it is clearly seen that all voids of one kind are related by the same lattice as the circles in Fig. 23.   This means that three alternative placements of spheres can be distinguished, namely, *A*, *B*, or *C*, each of which results in a closest-packed layer of spheres.

The above discussion has a very real importance when the stacking of hexagonal closest-packed layers above each other is considered.   If a layer is named *A*, *B*, or *C*, accordingly as the spheres in that layer occupy *A*, *B*, or *C* sites, then a twofold choice exists in placing one closest-packed layer above another.   Let the first layer be an *A* layer.

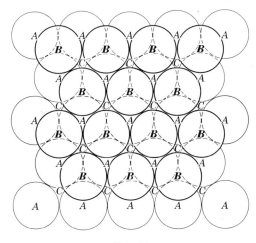

Fig. 24

The next layer can be either a *B* layer or a *C* layer.   Say that it is a *B* layer as shown in Fig. 24.   The next layer above can now be either an *A* layer or a *C* layer, and so forth.   From this discussion it follows that the stacking sequence of closest-packed layers can be designated by representing the layers in a sequence by the letters *A*, *B*, *C*.

It is evident from the above discussion that the total number of different possible ways of stacking closest-packed layers is infinite.   It turns out, however, that two stacking sequences occur most commonly in crystals.   In one, alternate layers are identical, so that the stacking can be represented $\cdots ABAB \cdots$.   This results in the *hexagonal closest packing* shown in Fig. 25.   A unit cell in the primitive hexagonal lattice of this array is shown in Fig. 26.   Note that the lattice is truly primitive since only the spheres in *A*-type layers are related by translations.   Thus the unit cell contains one lattice point but two atoms, at 000 and $\frac{1}{3}\frac{2}{3}\frac{1}{2}$,

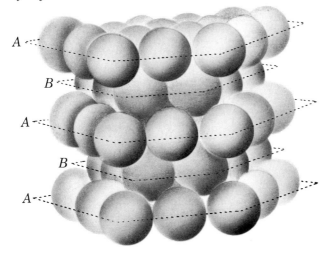

FIG. 25

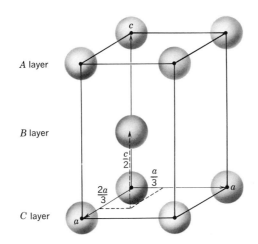

FIG. 26. Primitive unit cell in a hexagonal closest packing. In addition to the spheres at the lattice points, chosen to lie in an A layer, there is a sphere from a B layer inside the unit cell. Thus there are two spheres per lattice point in this cell. The occluded sphere from the B layer is not at a lattice point because it is not joined by a translation to the spheres in the A layer. (The spheres are shown reduced in size for clarity.)

respectively. This is an example, therefore, of an array whose structure and lattice are *not* the same. If the closest-packed layers are stacked in the sequence $\cdots ABC \cdots$, the closest packing shown in Fig. 27 results. A unit cell in this packing has been separated in Fig. 28, which clearly shows that this is a face-centered cubic array commonly called a

cubic closest packing.   Note that each atom occupies a lattice point in this
array, so that its structure and lattice are similar.

Body-centered cubic packing.   The above two arrays are called
closest packings because they are examples of the way equal spheres can
be packed to occupy the available space most efficiently (Exercise 11).

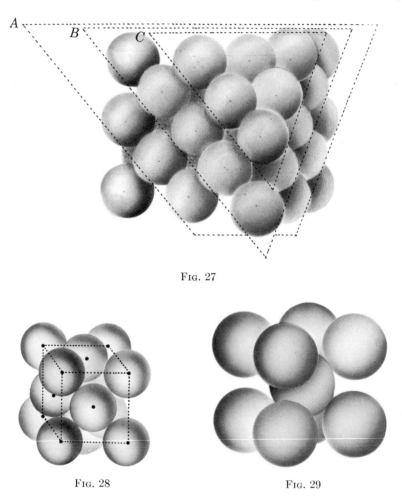

FIG. 27

FIG. 28   FIG. 29

Each sphere in such a packing has 12 nearest neighbors with which it is in
contact.   By comparison, Fig. 29 shows a packing in which each sphere
has only 8 nearest neighbors.   Such a body-centered cubic array is called
a close packing, but not a closest packing, to distinguish it from the arrays
discussed in the preceding section.   It is also a fairly common structure
type and is found to occur in many metal crystals.

Voids in closest packings. The kinds of voids existing in closest packings and their distribution in space is of equal importance to the array of the spheres themselves. This is so because smaller spheres can be placed between these closest-packed spheres to produce various arrays. It turns out that the atomic arrays of most inorganic crystals are of this type. The larger atoms form a closest packing, while the smaller atoms

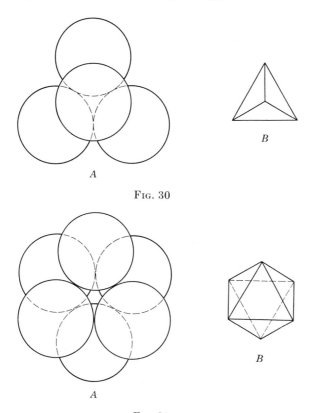

Fig. 30

Fig. 31

are distributed among the available voids. It is of practical importance, therefore, to consider these voids in some detail.

Two kinds of voids occur in all closest packings. If the triangular void in a closest-packed layer (Fig. 23A) has a sphere directly over it, there results a void with four spheres around it, as shown in Fig. 30A. The four spheres are arranged on the corners of a tetrahedron (Fig. 30B), and such a void is called a *tetrahedral void*. On the other hand, if a triangular void pointing up in one closest-packed layer is covered by a triangular void pointing down in an adjacent layer, then a void surrounded by six

spheres results (Fig. 31*A*).   These six spheres are arranged on the corners
of an octahedron (Fig. 31*B*), and such a void is called an *octahedral void*.
It can be shown, by examining Fig. 24 carefully, that these are the only
kinds of voids that can occur in a closest packing despite the fact that the
number of different closest packings possible is infinite.

The number of voids surrounding any sphere in a closest packing is
readily determined.   A sphere in a closest-packed layer *A* is surrounded
by six triangular voids of two kinds, *B* and *C*.   When the next closest-
packed layer above is added, say that it is a *B* layer, then the three *B* voids
become tetrahedral voids and the three *C* voids become octahedral voids.
If the added layer is a *C* layer, the *C* voids become tetrahedral voids and
the *B* voids become octahedral voids.   Similarly, the closest-packed layer
below the *A* layer gives rise to three tetrahedral and three octahedral
voids.   Furthermore, the particular sphere in layer *A* being considered
itself covers a triangular void in the closest-packed layer above and in

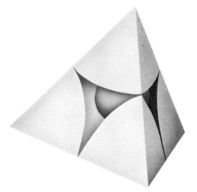

Fig. 32.

the layer below the sphere.   Thus two more tetrahedral voids surround
the sphere.   This results in $2 \times 3 + 1 + 1 = 8$ tetrahedral voids and
$2 \times 3 = 6$ octahedral voids surrounding the sphere.   Since the total
number of spheres and voids in a closest packing is very large, it is possible
to determine only the average number of voids of each kind belonging
to a sphere.   Each octahedral void is surrounded by six spheres, and each
sphere is surrounded by six octahedral voids.   The number of octahedral
voids belonging to one sphere is given by the ratio

$$\frac{\text{Number of octahedral voids around sphere}}{\text{Number of spheres around void}} = \frac{6}{6} = 1. \tag{11}$$

Each tetrahedral void is surrounded by four spheres, and each sphere is
surrounded by eight such voids.   The number of tetrahedral voids
belonging to one sphere is given by the ratio

$$\frac{\text{Number of tetrahedral voids around sphere}}{\text{Number of spheres around void}} = \frac{8}{4} = 2. \qquad (12)$$

The number of octahedral voids in a closest packing, therefore, is equal to the number of spheres. Similarly, the number of tetrahedral voids in a closest packing is twice the number of spheres, or twice the number of octahedral voids present.

The disposition of closest-packed spheres about a tetrahedral void is shown in Fig. 32, where the void is occupied by a smaller dark sphere. The radius of a sphere that just fits inside this void can be determined with the aid of Fig. 33*A*, which shows the tetrahedron inscribed inside a unit cube. The shaded (110) plane in this figure is considered separately in

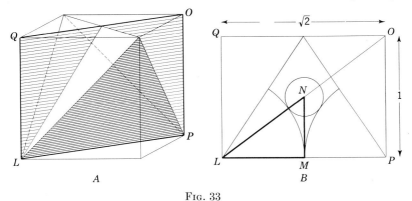

Fig. 33

Fig. 33*B*. It is easily seen that the right triangle *LNM* is similar to the right triangle *LOP*. Therefore

$$\frac{LM}{LN} = \frac{LP}{LO} = \sqrt{\frac{2}{3}}. \qquad (13)$$

Let *r* be the radius of the sphere inside the void, and *R* the radius of spheres in a closest packing. Then

$$\frac{LM}{LN} = \frac{R}{R + r} = \sqrt{\frac{2}{3}} \qquad (14)$$

and the radius of a sphere that just fits inside this void is obtained by rearranging the terms in (14).

$$r = \frac{\sqrt{3} - \sqrt{2}}{\sqrt{2}} R$$
$$= 0.225R. \qquad (15)$$

The disposition of closest-packed spheres about an octahedral void is shown in Fig. 34*A*. A diagonal plane passing through this octahedron is displayed in Fig. 34*B*. This cross section can be used to determine the radius of a sphere that just fits inside the octahedral void. It is left to Exercise 12 to prove that $r = 0.414R$.

Voids in body-centered cubic packing. Although not immediately apparent in Fig. 29, there are also two types of voids present in a body-

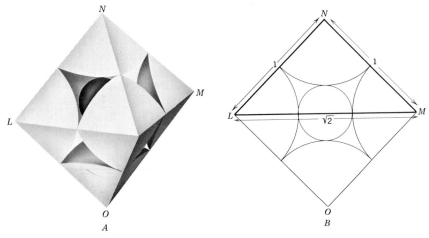

Fig. 34

centered cubic packing. In fact, they are deformed tetrahedral and octahedral voids, but there are three times as many octahedral voids as there are atoms in the packing and six times as many tetrahedral voids

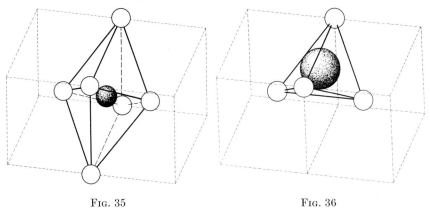

Fig. 35   Fig. 36

as atoms. Their relative disposition in the packing is illustrated in Figs. 35 and 36. The dimensions of the spheres forming the packing have been intentionally reduced in size so that the locations of the voids, represented by the larger dark spheres, are clearly visible. It is left to Exercise 13

621.381 A3 16 e

C. 1

to show that the radii of spheres that just fit inside these two voids are $r = 0.154R$, for the octahedral void, and $r = 0.291R$, for the tetrahedral void. Note that a sphere that just fits into a tetrahedral void is larger than the one in the octahedral void in this packing because of the distorted shapes of these voids. Nevertheless, the larger tetrahedral void cannot accommodate as large a sphere as can fit into an octahedral void in a closest packing of like spheres. The relative dimensions of these voids become important when actual atomic packings are considered, as discussed in the next section and elsewhere in this book.

Atomic packings in crystals

The atomic arrays, or crystal structures, of most elements are either the two closest packings or the body-centered cubic packing already discussed. Similarly, most inorganic compounds crystallize with structures comprised of closest-packed arrays of one kind of atoms, usually the larger atoms in the compound, with the smaller atoms dispersed among the available voids. Thus the crystal structures of chemically similar compounds often differ only in the way that the voids are occupied. Because the atoms occupying the voids are normally the relatively smaller metal atoms, Pauling has suggested that it is very convenient to think of a crystal structure as an array of the *coordination polyhedra* of these metal atoms. In closest packings, these are the already encountered octahedra (Fig. 34*A*) and tetrahedra (Fig. 32). In order to see the spatial array of the coordination polyhedra of voids in both closest packings, Fig. 37 shows an exploded view of a cubic closest packing and Fig. 38 of a hexagonal closest packing. There are several noteworthy features visible in these two illustrations. (Note that the closest-packed atoms are represented by points marking their centers at the corners of the polyhedra in Figs. 37 and 38.)

The cubic closest packing in Fig. 37 has been exploded along [100], not [111], which is the stacking direction of the closest-packed layers, so that the cubic symmetry is more clearly visible. The octahedra share edges with each other and faces with the tetrahedra. Similarly, the tetrahedra share edges with each other and faces with the octahedra. By comparison, the hexagonal closest packing in Fig. 38 has been exploded along [0001], the stacking direction. It is clearly seen that tetrahedra share corners with each other in a layer and pair up across a shared face with tetrahedra in the adjacent layer. The octahedra in this packing share edges with octahedra in the same layer but share opposite faces with octahedra in adjacent layers. The way that atoms can fit inside these polyhedra, therefore, must be different for the two kinds of packings. For example, it is possible for all the tetrahedral voids in a cubic

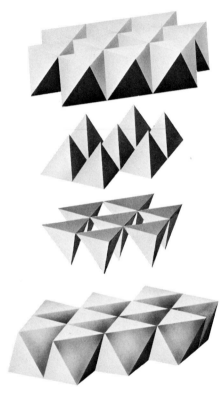

FIG. 37. Exploded view of a cubic closest packing of coordination polyhedra. In the collapsed view (Fig. 40) the tetrahedra are hidden because they just fit inside the octahedra of adjacent layers.

FIG. 38. Exploded view of a hexagonal closest packing. Note that the tetrahedra and octahedra in adjacent layers share faces across a mirror plane, whereas like polyhedra share edges only in a cubic closest packing.

closest packing to be occupied, whereas it is very unlikely that all the tetrahedral voids in a hexagonal closest packing can be filled. This is so because the atoms sitting at the centers of tetrahedral voids in a hexagonal closest packing would be very close to each other since the tetrahedra share a face. Some of the rules that govern these polyhedral arrays, that is, the ways in which metal atoms can be distributed among the available voids in a closest packing, are considered next.

Effect of atomic size. In view of the different sizes of the two kinds of voids in a closest packing, it follows that a particular atom can best fit into one or the other, depending on its size relative to that of the closest-packed atoms. Thus the number of atoms surrounding a central atom, called the *coordination number* (CN) of that atom, depends on the radius ratio of the two kinds of atoms. It is, of course, possible that the coordination number of an atom differs from four or six in a crystal structure that is not based on a closest packing. The most frequently encountered CN values are 3, 4, 6, 8, and 12. Three is the coordination number of atoms forming planar arrays, four and six have already been encountered, eight is the case for a cubic array, and twelve is the coordination of like atoms in a closest packing (Exercises 14 and 15).

To see what effect the radius ratio has in determining the coordination polyhedra that different atoms can have in a crystal, consider the case of an AB compound containing equal numbers of A and B atoms. First of all, it is clear that if an A atom has $CN = 6$, then $CN = 6$ for the B atom also. Thus the only coordination numbers that need be considered are the ratios $1:1$, $2:2$, $3:3$, $4:4$, $6:6$, $8:8$, and $12:12$. For the case $1:1$ and $2:2$, the two atoms can have any relative sizes and still be able to have these coordination numbers. For the other ratios this is no longer true. Consider the ratio $4:4$. It was shown in (15) that the smallest sphere that can fit into a tetrahedron formed by four touching atoms has $r/R = 0.225$. This means that if an A atom is to touch all four neigh-

Table 5
Radius-ratio limits in AB compounds for different coordination numbers

Coordination number	A's requirements from A's point of view (r_A/r_B)	B's requirements from B's point of view (r_B/r_A)	B's requirements from A's point of view (r_A/r_B)	Radius-ratio limits
1:1	$0-\infty$	$0-\infty$	$\infty-0$	$0-\infty-0$
2:2	$0-\infty$	$0-\infty$	$\infty-0$	$0-\infty-0$
3:3	$0.155-\infty$	$0.155-\infty$	$\infty-0.155$	$0.155-1/0.155$
4:4	$0.225-\infty$	$0.225-\infty$	$\infty-0.225$	$0.225-1/0.225$
6:6	$0.414-\infty$	$0.414-\infty$	$\infty-0.414$	$0.414-1/0.414$
8:8	$0.732-\infty$	$0.732-\infty$	$\infty-0.732$	$0.732-1/0.732$
12:12	1.0	1.0	1.0	1.0

boring B atoms, r_A/r_B can range from 0.225 to infinity. This has been called A's requirements from A's point of view in Table 5. Similarly, if a central B atom is to touch all its four neighboring A atoms, r_B/r_A can range from 0.225 to infinity. This has been called B's requirements from B's point of view in Table 5. On the other hand, the relative size of B, from A's point of view, must be such that r_A/r_B ranges from ∞ to 0.225. Finally, the requirements of both kinds of atoms can be met if the radius ratio is limited to the range 0.225–$1/0.225$. These values are recorded in Table 5 for all ratios of importance in AB compounds. The ranges of radius ratio for the different coordinations are also plotted in Fig. 39.

This means that relative coordination numbers of the atoms in AB compounds can be determined if their respective radii are known. The radii of the most abundant atoms are tabulated in Appendix 3. Using the values listed there for Na and Cl,

$$\frac{r_{Na}}{r_{Cl}} = \frac{0.97}{1.81} = 0.535. \tag{16}$$

According to Fig. 39 (and Table 5), the radius ratio is too small for 8:8, so that the coordination numbers of Na and Cl in common table salt, NaCl,

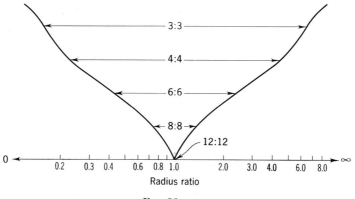

Fig. 39

are 6:6. Similarly, it can be shown that $r_{Zn}/r_S = 0.402$, so that the coordination numbers in ZnS must be in the ratio 4:4. Finally, the radius-ratio limits for AB_2, AB_3, A_2B_3, and other compounds can be similarly determined (Exercises 16 to 17).

Common crystal-structure types. As already noted, the crystal structures of most inorganic compounds can be considered as closest packings of the larger atoms, with the smaller ones distributed among the appropriate voids. For example, the chlorine atoms in NaCl can be assumed to form a closest packing. As already seen in (16), the sodium atoms must occupy the octahedral voids because their CN = 6. Since the number of octahedral voids in a closest packing just equals the number of closest-packed atoms, this means that all the octahedral voids are occupied. As can be seen in Fig. 38, however, the octahedra share opposite faces in a hexagonal closest packing, so that this structure is not

FIG. 40. Structure model of NaCl showing the packing of coordination polyhedra of sodium.

FIG. 41. Atomic model of the NaCl structure. The larger spheres represent closest-packed chlorine atoms. Note that the smaller sodium atoms occupy octahedral voids.

likely to be as stable as the cubic closest packing in Fig. 37. (If faces are shared by two coordination polyhedra, the like atoms at their centers are brought too near together and like atoms tend to repel each other.) Thus the most likely structure of NaCl has the polyhedra arrayed as shown in Fig. 40. This array has been confirmed experimentally, and an atomic model of the structure is shown for comparison in Fig. 41.

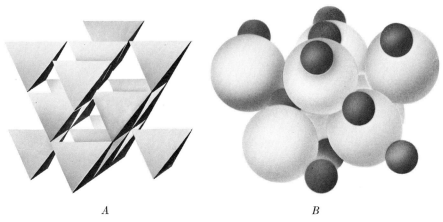

A   B

Fig. 42.   Sphalerite, or zinc blende, form of ZnS.

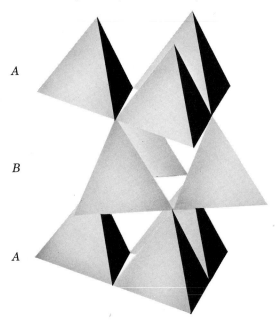

Fig. 43.   Wurtzite modification of ZnS.

As a second example, consider zinc sulfide.   The radius ratio of $r_{Zn}/r_S = 0.402$, so that the zinc atoms must occupy the tetrahedral voids in a closest packing of sulfur.   There are twice as many tetrahedral voids as there are atoms forming the closest packing, however, so that only one-half of the voids are actually occupied.   It can be shown that the tetrahedral voids in each closest packing are divided into two sets in which the tetrahedra share corners only.   One such set in a cubic closest packing is shown in Fig. 42*A*. This is the actual crystal structure of cubic zinc sulfide and occurs in the mineral sphalerite, sometimes called zinc blende. The atomic model of the sphalerite structure is shown next to it in Fig. 42*B*.

If one-half of the tetrahedral voids in a hexagonal closest packing are occupied, the array looks like Fig. 43. The tetrahedra in adjacent layers having apices pointing up (or down) are occupied; the others are empty. Thus the stacking sequence of the tetrahedral layers in this structure of ZnS, occurring in the mineral wurtzite, is $\cdots AB \cdots$. By comparison, the tetrahedral layers in the sphalerite structure (Fig. 42*A*) are stacked in the sequence $\cdots ABC \cdots$ along the [111] direction. The ability of a compound like ZnS to crystallize in two different structures is called *polymorphism* and is discussed further in the next section.

Variations in atomic packings. The actual crystal structures of a number of elements and simple compounds are described in some of the books listed at the end of this chapter. As already shown for ZnS, it may be possible for a compound to crystallize in more than one crystal structure. Such *polymorphous modifications* are possible whenever two or more arrays of the coordination polyhedra serving as the "building" blocks in the structure can occur. Conversely, it was shown above that the sodium octahedra in NaCl can pack in only one way, so that NaCl can normally exist in one structure only. It is possible, however, to alter the coordination polyhedra at very high temperatures or very high pressures. In such a case, the structural array of the deformed or altered polyhedra can assume different forms. Thus the existence, or *stability*, of a particular polymorphous modification depends on temperature (or pressure). In the case of elements, the terms *allotropy* and *allotropic modification* are sometimes used instead.

It is, of course, possible that two different compounds can crystallize with the same structural arrangement. If the two compounds have the same crystal structure but different chemical compositions, they are said to be *isostructural*, or *isotypic*. If their structures are the same and their chemical compositions are similar but not identical, so that they can form composite crystals called *solid solutions*, then they are said to be *isomorphous*.† For example, NiO, CoO, and PbS all have the structural arrangement shown in Fig. 41. The two oxides form solid solutions with each other but not with lead sulfide. Therefore NiO and CoO are isomorphous with each other but both are isotypic with PbS.

In forming solid solutions, the metal atoms of the two isomorphs usually distribute themselves randomly among the specified voids in the anion

---

† Strictly speaking, the term *isomorphous* means similar crystal forms (morphology). When two elements can substitute for each other in a crystal structure, they are best called *vicarious* elements.

close packing.   The replacement of the ions of one crystal by corresponding ions of the other can be likened to the solution of one crystal in the
other; hence the name solid solution.   If the two vicarious cations are
dimensionally alike and have similar properties, then they are completely
miscible; for example, NiO and CoO can be combined in any ratio of Ni
to Co.   On the other hand, KCl and NaCl are isomorphous, but the
difference in the atomic size of K and Na prevents more than a small
fraction of potassium atoms from entering the NaCl structure.   Minerals
are common examples of solid solutions because their structures usually
consist of closest-packed large atoms containing metal atoms in their
interstices.   Since the melts from which some minerals form contain a
large variety of metal constituents, it is quite common to find nonstoichiometric ratios of metal atoms in mineral crystals.   The similarity
between the structures and the properties of pure metals similarly

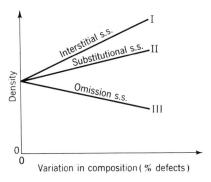

Fig. 44

accounts for the ease of formation of metallic solid solutions.   In fact,
most metals used in industry are solid solutions.
     There are several ways in which solid solutions can form.   Accordingly,
they are classified into the following:
     1. *Substitutional solid solution,* in which vicarious replacement of one
atom for another takes place.   Goldschmidt observed that substitutional
solid solutions can occur only if the radius of the larger atom does not
exceed the radius of the smaller atom by more than 15 per cent.
     2. *Interstitial solid solution,* in which limited amounts of solute atoms
occupy interstitial positions in the solvent crystal.   For example,
sphalerite can accommodate relatively large amounts of interstitial atoms
because of its "open" structure.   Similarly, certain metals can accommodate carbon atoms interstitially to form solid-solution carbides, or
nitrogen to form nitrides, and so on.
     3. *Omission solid solution,* in which the number of atoms of one kind is
slightly less than that required by stoichiometry.   This type of solid

solution can occur only if the remaining atoms can adjust their electrical charges to preserve charge neutrality. The name *defect structures* has been proposed to describe such crystals.

The three types of solid solutions can be readily distinguished if their composition and unit-cell size are known. The density of a crystal can be computed for a given composition and cell size. It is possible, therefore, to prepare a graph showing the variation in density with change in composition, as shown in Fig. 44. Curve I illustrates what happens in the case of interstitial solid solution; as more atoms fill the interstices, the density increases, so that the curve has a positive slope. The density change in a substitutional solid solution is shown by the other straight line (curve II) having a positive slope, for the case in which the solute atom is heavier than the atom for which it substitutes. Finally, curve III, calculated for an omission solid solution, has a negative slope because atoms are being removed from the crystal. Having prepared such a graph, it is possible to determine which type of solid solution exists in a crystal by measuring its density for a known composition and noting on which of the three curves the measured density value falls.

Suggestions for supplementary reading

Leonid V. Azároff, *Introduction to solids* (McGraw-Hill Book Company, Inc., New York, 1960).

N. V. Belov, *Struktura ionnich kristallov i metallitcheskich faz* (in Russian) (Akademiya Nauk S.S.S.R., Moscow, 1947).

Sir Lawrence Bragg, *The Crystalline state*, vol. 1, *A general survey* (G. Bell & Sons, Ltd., London, 1949).

M. J. Buerger, *Elementary crystallography* (John Wiley & Sons, Inc., New York, 1956).

F. C. Phillips, *An introduction to crystallography* (Longmans, Green & Co., Ltd., London, 1951).

Exercises

1. Select any two planes shown in Fig. 10 by light lines. Using the procedure outlined in (3), show that the Miller indices of both planes are (320).

2. Draw a cube-shaped unit cell like the one in Fig. 11. Next draw a (112) and (211) plane in this unit cell. Indicate the [112] direction. Also prove that a plane (hkl) is always perpendicular to the direction $[hkl]$ in a cubic lattice.

3. Make a sketch of 4_1 and 4_2; the subtranslations in these two four-fold screw axes are $t/4$ and $2t/4$, respectively.

4. Consider two mirror planes intersecting at 45 deg. Show that these two symmetry elements give rise to a four-fold axis lying at their intersection.

5. Demonstrate that whenever an evenfold axis, say, a 4, is perpendicular to a mirror plane, an inversion center appears at their point of intersection.

6. Show that the various operations of rotation and reflection require that there be at least five different kinds of plane lattices (a general, rectangular, diamond-shaped, square, and equilateral triangular plane lattice).

7. Show that it is possible to have a one-face-centered or an all-face-centered lattice in the orthorhombic system but not a two-face-centered lattice. What is the proper Bravais lattice in this case? (ANSWER: I.)

8. Draw a rectangular plane cell and pass mirror planes (reflection lines) through all the edges of the cell. Next place a seven in a general position xy and repeat it by these reflection lines. Show that new reflection lines occur halfway between the reflection lines originally drawn. This is the plane group *pmm*.

9. One of the uses of equipoint sets is in the calculation of interatomic distances. Consider a tetragonal crystal belonging to $P4$ whose unit-cell edges $a = 4.0$ Å and $c = 6.0$ Å. It contains two A atoms in equipoints $2c$ and four B atoms in $4d$. If the z coordinate of all six atoms is the same, what are the A-A, A-B, and B-B interatomic distances if $x = y = \frac{1}{4}$ for the equipoint set $4d$?

10. Show that the circles in Fig. 23A occupy 90.7 per cent of the available area "belonging" to each circle. (Divide the available area into adjacent hexagons by drawing tangent lines to each pair of touching circles.) Show that a square array of circles covers only 78.5 per cent of the available area.

11. Calculate the "packing efficiencies" of equal spheres located at the points of a primitive, a body-centered, and a face-centered cubic lattice. (Assume that the spheres are touching inside the unit cubes.)

12. Determine the relative radius of a sphere that just fits inside an octahedral void in a closest packing. Is its value the same for both kinds of closest packings?

13. In a body-centered cubic packing, show that the maximum radius of a sphere just fitting inside an irregular tetrahedral void is $0.291R$ and that inside an irregular octahedral void it is $0.154R$, where R is the radius of the spheres forming the packing.

14. Show that the radius ratio for a sphere inscribed by three larger spheres in a plane (triangle) is $r/R = 0.155$ and that the radius ratio for a sphere inscribed by eight larger spheres arranged at the corners of a cube about the central one is $r/R = 0.732$.

15. Show that it is not possible to have twelve-fold coordination of two different kinds of atoms in an AB compound. HINT: It is sufficient to prove that both kinds of atoms cannot have CN = 6 in a planar array.

16. The permissible CN values for an AB_2 compound are 2:1, 4:2, 6:3, and 8:4. What are the radius-ratio limits for these compounds? Using these results, what are the respective coordination numbers of Ca ($r = 0.99$ Å) and F ($r = 1.36$ Å) in CaF_2?

17. What is the most likely structure of Li_2S? Describe it as an occupation of available voids in either a cubic or a hexagonal closest packing of sulfur atoms ($r_{Li} = 0.60$ Å, $r_S = 1.84$ Å). This is an example of the so-called antifluorite structure, since in fluorite, CaF_2, the role of the metal atoms is reversed.

18. Zinc oxide is hexagonal, $a = 3.243$ Å, $c = 5.195$ Å, and each unit cell contains two formula weights in a unit cell. If the measured densities of two ZnO crystals are 5.470 and 5.60 g/cm³, determine which type of solid solution has occurred in each case. HINT: The ideal density of a crystal is given by the mass per unit cell divided by the unit-cell volume.

2. diffraction of x-rays

While studying electron optics in 1895, W. K. Röntgen observed that an invisible radiation produced inside his electron tube had the ability to penetrate objects that were opaque to visible light. Not knowing what the origin or properties of this new radiation were, he called it the x-radiation. Next he discovered that different materials absorbed this radiation in an unlike manner, so that the shadows cast on a photographic plate by more heavily absorbing parts of a material could be used to locate them inside the irradiated specimen. This process of utilizing x-rays is called radiography, and its applications in medicine and metallurgy are well known. Continued studies of the mysterious x-rays showed that they were a form of electromagnetic radiation having wavelengths in the neighborhood of 10^{-10} m. In 1912, Laue had the genius to realize that crystals, which were believed to consist of atoms having dimensions commensurate with x-ray wavelengths, could be used as diffraction gratings for this form of radiation. This suggestion was subsequently tested by two of his colleagues, Friedrich and Knipping, who demonstrated that an x-ray beam passing through a single crystal was indeed broken up into a collection of diffracted beams. When these results became known at Cambridge University in England, they aroused the simultaneous interest of two physicists, W. H. and W. L. Bragg, and a chemist, W. Barlow. W. L. Bragg and his father, W. H. Bragg, were already studying the properties of x-rays, whereas Barlow was speculating about the actual atomic arrangements that were most likely to occur inside crystalline materials. When Barlow learned that crystals can diffract x-rays, he persuaded the younger Bragg to test his belief that sylvite, KCl, and halite, NaCl, had similar structures based on cubic closest packings of chlorine atoms. In 1914, W. L. Bragg confirmed the postulated structures by comparing the intensities of the diffracted beams, as described in a later section, and thereby opened the way to the elucidation of the atomic arrangements in crystals. In the years that have followed this discovery, use has been made of crystal structures thus determined in explaining the properties and behavior of materials. It is not an overstatement to say that this discovery revolutionized the fields of chemistry, metallurgy, mineralogy, and solid-state physics and laid the foundation for what is called today materials science.

Before proceeding to discuss the nature of x-ray diffraction and its utilization, it should be noted that other kinds of radiation can be diffracted by crystals also. Following de Broglie's postulate in 1924 that electrons have wavelike properties, two American physicists, Davisson and Germer, showed in 1927 that electrons similarly can be diffracted by crystals. With the further evolution of atomic and nuclear physics, it was demonstrated that neutrons, deuterons, and other nuclear radiations could be diffracted also. Of these radiations, electrons and neutrons are used in crystal-structure determinations because they interact with atoms in crystals in slightly different ways from x-rays. The low penetrating power of electrons and the relative scarcity of nuclear reactors needed to produce sufficiently intense neutron beams have limited their application, however, so that x-ray diffraction remains the principal tool for studying the structure of materials. The basic diffraction principles are sufficiently similar so that the following discussion, although limited to the diffraction of x-rays, can be extended quite easily to include other radiations as well.

Elementary diffraction theory

Bragg law. The fundamental reason why crystals can diffract x-rays is that they consist of periodic arrays of atoms. The importance of

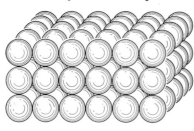

periodicity becomes apparent when the diffraction process is analyzed according to a procedure first suggested by Bragg. Suppose a crystal has the atomic array shown in Fig. 1. As already shown in Chap. 1, this array can be represented alternatively as a collection of parallel planes (*hkl*) periodically spaced a distance *d* apart. An edge view of these

Fig. 1

planes is shown in Fig. 2. When a wavefront of x-rays impinges on the crystal, the electrons in each atom scatter x-rays in all directions. Certain directions are of particular interest. Consider the incoming rays OE and $O'A$ inclined at the angle θ to the planes (hkl). Observe that the scattered rays AP and EP' also form the angle θ with (hkl). Since the total pathlengths of the rays $O'AP$ and OEP' are the same, these rays are said to scatter *in phase* with each other; that is, the waves of the individual rays arriving at PP' again form a common wavefront. This is the condition for scattering in phase by one plane in a crystal.

Next, consider the incoming ray $O'C$ and the scattered ray CP''. The total pathlength $O'CP''$ is greater than that of rays $O'AP$ and OEP' by an amount

$$\Delta = BCD$$
$$= 2BC. \tag{1}$$

Since $$BC = d \sin \theta \tag{2}$$

the path difference is $$\Delta = 2BC$$
$$= 2d \sin \theta. \tag{3}$$

If $O'CP''$ is to arrive at $PP'P''$ in phase with rays $O'AP$ and OEP', that is, if the two planes are to scatter in phase, then the path difference Δ

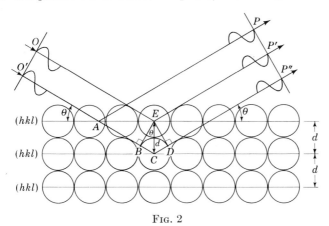

Fig. 2

must equal an integral number of wavelengths $n\lambda$, where $n = 0, 1, 2, 3, \ldots$. Thus the condition for in-phase diffraction by a set of parallel crystal planes is

$$n\lambda = 2d \sin \theta. \tag{4}$$

Note that the diffraction process can be likened to reflection of x-rays by a set of parallel planes (hkl), except that θ is the complement of the usual angle of incidence i. It is common practice, therefore, to interchange the words diffraction and reflection of x-rays.

Relation (4) is known as the *Bragg law*, and it shows that the diffraction intensities can build up only at certain values of θ, corresponding to a

specific value of n, λ, and d.   This is so because the wavelets scattered
from various parts of the crystal have a common wavefront only at these
angles.   Consequently, the amplitudes of all the individual wavelets add
up to give a resultant wave having the maximum amplitude possible, as
shown in Fig. 3A.   On the other hand, at other scattering angles, the
wavelets emanating from different parts of the crystal are not in phase
with each other.   When the amplitudes of the wavelets are summed at

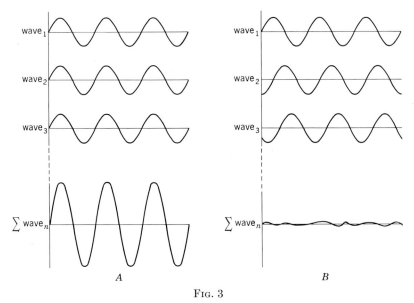

FIG. 3

these angles, some of the amplitudes are positive while others are negative, as shown in Fig. 3B.   The resultant wave, therefore, has an amplitude nearly equal to zero if a sufficiently large number of wavelets are
considered.

Recasting (4) to solve for θ,

$$\theta = \sin^{-1}\left(\frac{n\lambda}{2d}\right) \tag{5}$$

it is seen that rays diffracted by a crystal are given off in different directions corresponding to the different values of the interplanar spacing d.
It is possible to reverse this statement and say that from a knowledge of
the experimentally observed diffraction angles it is possible to determine
the interplanar spacings of a crystal.   From a list of such spacings it is
then possible to deduce the lattice of a crystal.   Thus one piece of
information that can be obtained from any diffraction experiment employing monochromatic (single-wavelength) radiation is the unit-cell size
and shape.

To see what influence the atomic arrangement in the crystal has on the observed diffracted beams, consider Fig. 4. This figure shows a crystal structure composed of two mutually displaced lattice arrays of atoms. At an angle θ satisfying Eq. (5) for both arrays, the atoms in each array scatter radiation in phase with other atoms in the same array. The total pathlength, however, is longer for rays scattered by one array than for the other. Consequently, each array contributes to the resultant wave, scattered by the whole crystal, waves that are not quite in phase with each other. This has the effect of reducing the intensity of the diffracted beam from what it would be if all the atoms in the crystal

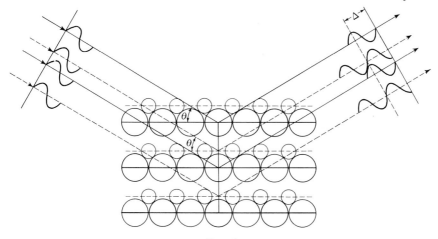

FIG. 4

structure scattered in phase. It follows, therefore, that the crystal structure modifies the diffraction intensities and, conversely, that a proper interpretation of the observed intensities can disclose the structure of a crystal.

Diffraction intensities. The way that the crystal structure determines the relative intensities of the diffracted beams can be seen by considering the determination of the structure of NaCl. This was first done by W. L. Bragg, who measured the intensities diffracted by successive orders of the planes (100), (110), and (111), shown in Fig. 5.† The significance of the observed intensity variations is that successive orders

---

† It should be noted that diffraction of successive orders n can be pictured as the diffraction by fictitious planes (nh,nk,nl) by rewriting (4)

$$\lambda = 2\frac{d}{n}\sin\theta = 2d_{nh,nk,nl}\sin\theta$$

since $d_{nh,nk,nl} = d_{hkl}/n$. These planes are said to be fictitious because they do not pass through lattice points the way that the real planes (hkl) do.

of 111 reflections are alternately weak and strong. This suggests that
the (111) planes are composed of interleaved layers of unlike atoms
because the atoms of one layered set diffract in phase with each other
but not with the atoms in the other set (Fig. 4). The structure model
proposed by Barlow is pictured on page 29 and clearly shows that alternate (111) planes are composed exclusively of sodium and chlorine atoms

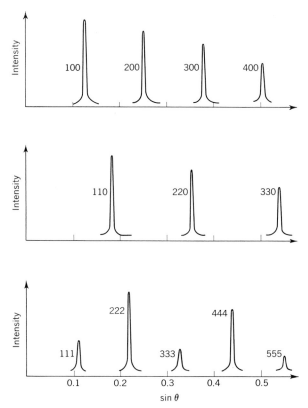

Fig. 5. Diffraction intensities (schematic) of the first few orders from the (100),
(110), and (200) planes of NaCl. As discussed in the text, the indices assigned
above to the 100 and 110 reflections are incorrect.

in direct agreement with Bragg's x-ray measurements. Note that
alternate (100) and (110) planes are composed of similar arrays of Na and
Cl atoms, so that an analogous intensity variation should not occur.

The x-ray scattering power of an atom is directly proportional to the
number of electrons composing it and can be expressed by a *scattering
factor f*. (The scattering power of an atom actually decreases as θ
increases because of the phasal relationship between wavelets emanating
from different points in the same atom. This is the reason why the

intensities of successive orders in Fig. 5 show a slight decline.) In considering the atomic array in a crystal, it is sufficient to add up the contributions from the atoms within a single unit cell. It is necessary to take into account their relative positions in the unit cell so that the pathlengths of the part of the beam scattered by each atom are properly related. This can be done by determining the *structure factor* F_{hkl} for any reflection by (hkl) planes according to

$$F_{hkl} = f_1 e^{2\pi i(hx_1+ky_1+lz_1)} + \cdots + f_n e^{2\pi i(hx_n+ky_n+lz_n)}$$
$$= \sum_{n=1}^{N} f_n e^{2\pi i(hx_n+ky_n+lz_n)} \tag{6}$$

where N is the total number of atoms in a unit cell. The exponential term expresses the relative phase of the radiation scattered by each atom n as a function of its position in the unit cell $x_n y_n z_n$. Thus if the distribution of atoms among available equipoints is known, structure factors can be determined for different hkl reflections.

The unit cell of NaCl contains four Na atoms at 000, $\frac{1}{2}\frac{1}{2}0$, $\frac{1}{2}0\frac{1}{2}$, $0\frac{1}{2}\frac{1}{2}$ and four Cl atoms at $\frac{1}{2}\frac{1}{2}\frac{1}{2}$, $\frac{1}{2}00$, $0\frac{1}{2}0$, $00\frac{1}{2}$, respectively. The structure factor for NaCl, therefore, is

$$F_{hkl} = f_{Na} \left[e^{2\pi i(0+0+0)} + e^{2\pi i\left(\frac{h}{2}+\frac{k}{2}+0\right)} + e^{2\pi i\left(\frac{h}{2}+0+\frac{l}{2}\right)} + e^{2\pi i\left(0+\frac{k}{2}+\frac{l}{2}\right)} \right]$$
$$+ f_{Cl} \left[e^{2\pi i\left(\frac{h}{2}+\frac{k}{2}+\frac{l}{2}\right)} + e^{2\pi i\left(\frac{h}{2}+0+0\right)} + e^{2\pi i\left(0+\frac{k}{2}+0\right)} + e^{2\pi i\left(0+0+\frac{l}{2}\right)} \right]$$
$$= [f_{Na} + f_{Cl}e^{\pi i(h+k+l)}][e^0 + e^{\pi i(h+k)} + e^{\pi i(h+l)} + e^{\pi i(k+l)}]. \tag{7}$$

Note that $e^{\pi i n}$ is equal to $+1$ if n is an even number including zero and -1 if n is an odd number. This means that if the Miller indices hkl are not all even or all odd, terms in the second parentheses in (7) equal zero since two exponential terms are $+1$ and two are -1. When the indices are not mixed, the terms in the second parentheses add up to four, so that the structure factor for NaCl can be written

$$F_{hkl} = 4[f_{Na} + f_{Cl}e^{\pi i(h+k+l)}]. \tag{8}$$

This can be further simplified by noting that

$$F_{hkl} = 4(f_{Na} + f_{Cl}) \qquad \text{when } h, k, l \text{ are all even}$$
and
$$F_{hkl} = 4(f_{Na} - f_{Cl}) \qquad \text{when } h, k, l \text{ are odd.} \tag{9}$$

The intensity of x-rays reflected by planes (hkl) in a crystal is actually proportional to the square of their structure factors. The expressions in (9) clearly show why reflections 111, 333, . . . have a lower intensity than the reflections 222, 444, . . . in the sodium chloride structure. Remember also that the reflections from planes having mixed indices cannot be observed because their structure factors are identically equal to zero.   Such reflections are called *space-group extinctions* and occur in all crystals having centered lattices.   Specifically, it can be shown that the extinction rule for an all-face-centered lattice requires that hkl be unmixed for a reflection to occur (Exercise 3).

The above statement appears to be in direct contradiction with the labeling of the reflections in Fig. 5.   The reason for this is that, when Bragg performed the first measurements, neither the values of λ nor of d were known, so that the correct value of n in Eq. (4) could not be determined.   The way that this problem was resolved is shown in the following analysis.   Substituting the $\sin \theta$ values for reflections from (100), (110), and (111) planes given in Fig. 5 into (4),

$$n_1\lambda = 2d_{100}(0.126), \qquad n_2\lambda = 2d_{110}(0.178), \qquad n_3\lambda = 2d_{111}(0.109). \quad (10)$$

From the known geometry of a cubic lattice, $d_{100} = \sqrt{2}d_{110}$, $d_{100} = \sqrt{3}d_{111}$, and $d_{110} = \sqrt{\frac{3}{2}}d_{111}$.   Forming these ratios between the relations in (10) eliminates λ and gives

$$\frac{d_{100}}{d_{110}} = \frac{(0.178)n_1}{(0.126)n_2} = \sqrt{2}$$

$$\frac{d_{100}}{d_{111}} = \frac{(0.109)n_1}{(0.126)n_3} = \sqrt{3} \qquad (11)$$

$$\frac{d_{110}}{d_{111}} = \frac{(0.109)n_2}{(0.178)n_3} = \sqrt{\frac{3}{2}}.$$

Now it is clearly evident that the last two ratios in (11) can be satisfied only if $n_1 = 2$, $n_2 = 2$, and $n_3 = 1$.   Thus it is clear that the indices shown in the upper two graphs in Fig. 5 should be doubled.   This supports the conclusion previously reached, namely, that the indices of allowed reflections from the NaCl structure must be all odd or all even.

Determination of unit-cell contents.   Pursuing the crystal-structure analysis of NaCl a step further, note that there are three unknown d values in (11) and only two independent equations relating them.   It is not possible, therefore, to determine their absolute values.   To do this, use can be made of the relation between the unit-cell volume υ and the density D of a crystal.

$$D = \frac{\text{mass of cell}}{\text{volume of cell}} = \frac{qA/N_0}{\upsilon} \qquad (12)$$

where q is number of formula weights per cell
A is atomic weight of formula unit, in atomic mass units (AMU)
N_0 is Avogadro's number (6.023×10^{26} molecules/kg-mole)
υ is unit-cell volume, in m^3.

The density of sodium chloride is 2.15, and there are four NaCl units per cell, each weighing 58.45 AMU.   Substituting these values in (12), the unit-cell volume is calculated to be 178×10^{-30} m³. The unit cell of NaCl is cubic, so that the length of its unit-cell edge

$$a = \mho^{\frac{1}{3}} = 5.63 \times 10^{-10} \text{ m.}$$

Since $a = d_{100} = 2d_{200}$ in cubic crystals, it is now possible to determine the absolute lengths of the interplanar spacings. Substituting these values in the Bragg equation (4), it is also now possible to determine the wavelength of the x-radiation used to measure the reflection intensities reproduced in Fig. 5. (It is convenient to introduce a new unit of length called an angstrom unit, 1 Å $= 10^{-10}$ m, because it turns out that atomic dimensions are of this order of magnitude.) For example, substituting in (4) the appropriate values for the 200 reflection,

$$\lambda = 2(5.63/2)(0.126) \text{ Å}$$
$$= 0.71 \text{ Å.} \tag{13}$$

Thus it was possible for Bragg to determine the actual wavelength of x-rays for the first time.

When the x-ray wavelength is known, (4) can be used directly to determine d from the measured reflection angle θ. As discussed in more detail later, a knowledge of the interplanar spacing values permits the determination of the lattice type and the unit-cell volume. This means that all quantities in (12) can be determined experimentally except q, the number of formula weights contained in a unit cell. This number can be predicted, however, because it must be an integer and it must be equal to one of the equipoint ranks of the appropriate space group. Use can be made of (12), therefore, to determine the accurate molecular weight of, say, a complicated organic molecule when its composition is known only approximately. The way that this can be done is further discussed in Exercises 5 and 6 at the end of this chapter.

Determination of atomic arrays. After the unit-cell dimensions of a crystal have been determined, it is possible to calculate the number of atoms that it contains as shown in the preceding section. It is next possible to make use of the observed diffraction intensities to determine the crystal structure, that is, the actual atomic array in a crystal. When the number of atoms contained in a unit cell is small, this can be done by trial-and-error procedures. Suppose that an AB_2 crystal belongs to the space group $P4$ and contains two formulas per unit cell. Making use of Table 4 in the preceding chapter, the two A atoms can occupy either the equipoints in $1a$ and $1b$ or in $2c$. The four B atoms, therefore, must be distributed among the equipoint set $4d$. (The A atoms cannot occupy $4d$ because a total of four like atoms are required by the symmetry.)

After calculating the structure-factor expressions for this crystal, the values of x, y, and z can be determined by seeking an agreement with the experimentally measured reflection intensities. If the unit cell contains a large number of atoms in general positions, such iterative procedures become difficult to carry out and more direct methods must be employed.

It has been shown by W. L. Bragg that the atomic distribution in a unit cell of a crystal can be represented by a Fourier series, called the *electron-density* function of a crystal, $\rho(xyz)$, where

$$\rho(xyz) = \frac{1}{\upsilon} \sum_{h} \sum_{k} \sum_{l} {}^{+\infty}_{-\infty} F_{hkl} e^{-2\pi(hx+ky+lz)}. \tag{14}$$

If the values of F_{hkl} are known, therefore, the electron density of a crystal can be calculated at each point of the unit cell xyz according to (14).

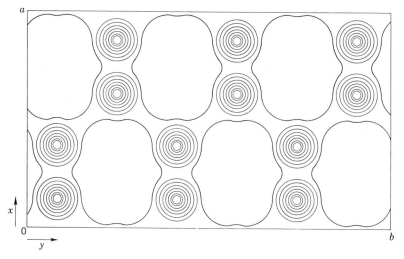

Fig. 6.   Electron density of Cu_2FeS_3 projected on the xy plane.

Since the greatest density occurs at atomic sites in the cell, it is possible to determine the exact crystal structure by this means. The electron density of the mineral cubanite, Cu_2FeS_3, is shown in Fig. 6 projected on the xy plane. The contours in Fig. 6 join points having the same electron-density values, and the x and y coordinates of the atoms are clearly visible. (Actually, the maxima in Fig. 6 represent pairs of atoms superimposed on each other.) The z coordinates of the atoms can be similarly determined, so that electron-density syntheses can be used to determine crystal structures very precisely.

There is one difficulty, however, in this procedure. Experimentally, one measures the intensity $I_{hkl} \propto |F_{hkl}|^2$, and although the proportionality constants are known, it is possible to determine the magnitude of F_{hkl} in (14), but not its phase relative to that of the direct beam. In the simplest case, the relative phase can be plus (in phase) or minus (out of phase). This means that it is not possible to combine correctly the different terms in (14), and generally, the electron density of a crystal cannot be synthesized unless the atomic coordinates are known beforehand, even approximately. [When the atomic positions are known, (6) can be used to calculate the phases.] A number of procedures for deducing these phases have been developed in the last fifteen years; however, they are very complicated, so that the determination of complex crystal structures is a major undertaking.

Reciprocal-lattice concept. The theory of x-ray diffraction was also analyzed by P. P. Ewald, who developed a particularly useful relation

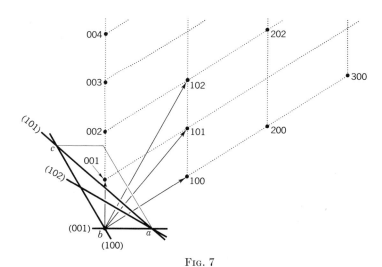

Fig. 7

between the diffracted x-ray beams. Instead of visualizing the crystal as a set of interpenetrating planes (hkl), each plane can be represented by its normal drawn from a common origin. The length of the normal is made proportional to $1/d_{hkl}$, so that its length and direction uniquely describe a set of parallel planes. It can be shown that the terminal points of all such possible normals form a lattice array. Without proving this rigorously, the way that such a lattice can be constructed is illustrated in Fig. 7, which shows the unit cell of a monoclinic crystal looking along its unique axis, here designated b. The cell edges lying in the plane of the drawing are accordingly a and c. The illustration also shows four ($h0l$) planes, seen in an edge view, namely, (100), (101), (102), and (001).

Since all these planes are parallel to b, their normals lie in the plane of the paper. To locate the points representing these planes, proceed as follows:

1. Draw the normal to each plane from a common origin.

2. Place a point on the normal at a distance from the origin equal to $1/d_{hkl}$.

The resulting collection of points preserves the important characteristics of the planes in the crystal. The direction to a point specifies the orientation of the planes it represents, and the distance to the point represents the interplanar spacing. An examination of Fig. 7 also shows that the points labeled 100, 101, and 102 lie along a straight line which is perpendicular to the crystal plane (001). Thus it becomes evident that a collection of all such points forms a lattice array. This is called the *reciprocal lattice*, because distances in this lattice are reciprocal to those in the crystal.

It is convenient to define a *reciprocal-lattice vector* $\boldsymbol{\sigma}_{hkl}$ whose magnitude is $1/d_{hkl}$ and whose direction is parallel to the normal to the (hkl) planes. Use can be made of these vectors to define a unit cell in the reciprocal lattice whose three edges are $\boldsymbol{\sigma}_{100}$, $\boldsymbol{\sigma}_{010}$, and $\boldsymbol{\sigma}_{001}$, so that its volume is directly reciprocal to the unit-cell volume υ in the crystal.

The reciprocal-lattice concept is very useful in interpreting x-ray diffraction experiments and in the discussion of electrical and other properties of crystals. This is so because it is easier to picture a collection of points in space than it is to visualize collections of interpenetrating planes in a crystal. To see its application to x-ray diffraction, rewrite the Bragg equation (4) to solve for the sine of the glancing angle θ_{hkl} in terms of the variables involved,

$$\sin\theta_{hkl} = \frac{\lambda/2}{d_{hkl}} = \frac{1/d_{hkl}}{2/\lambda}. \tag{15}$$

A direct geometrical interpretation of (15) is given in Fig. 8. Here θ is the angle between the diameter of a circle of radius $1/\lambda$ and the line drawn to the end of the line OP representing the reciprocal-lattice vector of length $1/d_{hkl}$. With this construction, a geometrical interpretation of x-ray diffraction can be given as follows:

1. Let AO be taken not only as a length, but also as the direction of the x-ray beam; then, since AP makes the angle θ with AO, AP has the same slope as the reflecting crystal plane.

2. OP is the normal to the reflecting plane ($\angle APO = 90$ deg), and hence it has the direction of the reciprocal-lattice vector $\boldsymbol{\sigma}_{hkl}$ drawn from the origin to the reciprocal-lattice point P. Its length is also $1/d_{hkl} = |\boldsymbol{\sigma}_{hkl}|$.

3. The above two conditions are indicated in Fig. 9. It is also clear from this figure that $\angle OSP = 2\angle OAP = 2\theta$; hence the vector from the

center of the circle S to the reciprocal-lattice point P_{hkl} represents the direction of the x-ray reflection.

The following can now be said about the meaning of Fig. 9:

1. The crystal can be pictured as located at the center of the circle S.

2. The point O, where the direct beam leaves the circle, is the origin of the reciprocal lattice, which is oriented so that every $\boldsymbol{\sigma}_{hkl}$ is normal to its plane (hkl).

3. Whenever a plane (hkl) of the crystal makes the angle θ with the direct-beam direction, the reciprocal-lattice point hkl at the end of $\boldsymbol{\sigma}_{hkl}$ lies

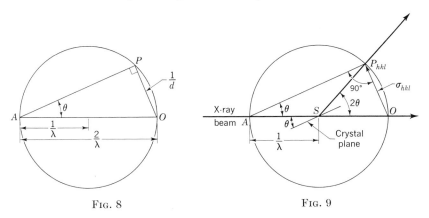

<table>
<tr><td>FIG. 8</td><td>FIG. 9</td></tr>
</table>

on the circumference of the circle, and the reflected x-ray beam passes through this point.

4. Diffraction can occur, therefore, only when a reciprocal-lattice point touches the circle.

In three dimensions, the circle in Fig. 9 becomes a sphere, commonly called the *sphere of reflection*, or *Ewald sphere*. To picture any diffraction experiment, consider the crystal inscribed inside a sphere of radius $1/\lambda$. Its reciprocal lattice is centered at the point where the incident x-ray beam leaves the Ewald sphere and bears the orientation shown in Fig. 9 to that of the crystal. To generate a diffracted beam, the crystal must be so positioned that a plane forms the proper angle θ_{hkl} with the incident beam, at which time the reciprocal lattice is correspondingly positioned so that the reciprocal-lattice point hkl just intersects the sphere of reflection.

The powder method

Experimental arrangement. Most materials do not normally occur in the form of single crystals. Instead, the steel forming an automobile body, the bricks forming the side of a house, and an ordinary pencil lead

are composed of many small crystals bearing random orientations relative
to each other.   For x-ray diffraction purposes they can be thought of as
random aggregates of many small crystallites not unlike sand grains on a
beach or sugar crystals in a grocer's bag.   The study of the crystallinity
of such aggregates is carried out by the *powder method* of x-ray crystallography.   Instead of one crystal, therefore, a collection of many
randomly oriented crystallites is placed in the path of a monochromatic
x-ray beam.   Consider the same set of planes (hkl) in each crystal.
Since these planes are assumed to have all possible orientations, their
reciprocal-lattice vectors have all possible orientations also, and the
reciprocal-lattice points of these planes form a sphere of radius σ_{hkl}.
The reciprocal lattice of a powder, therefore, consists of concentric

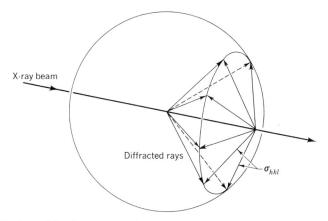

Fɪɢ. 10.   Reciprocal-lattice construction for one set of planes in a powder sample.
Some of the vectors σ_{hkl} satisfying the condition for x-ray diffraction (Fig. 9) are
indicated.

spheres.   The way that one such sphere intersects the sphere of reflection
is indicated in Fig. 10.   As can be seen in this drawing, the intersection
occurs along a circle and the x-rays diffracted by the planes (hkl) in each
crystal form cones that are concentric about the direct x-ray beam.   By
reference to Fig. 9, it can be seen that the half-opening angle of these
cones is 2θ.   Several such nested cones produced by different (hkl) planes
are shown in Fig. 11.
    The information sought in a diffraction experiment is the glancing
angle θ and the intensity of the diffracted beams.   Provided the orientation of the crystallites in the sample is truly random, the intensity
distribution throughout each cone is uniform.   It is sufficient, therefore,
to intercept the diffracted beams along any circle about the sample having
the x-ray beam as a diameter.   This can be done by using either an

ionization detector such as a Geiger counter or a photographic film. The conventional placement of a film strip is shown in Fig. 11, and a typical powder photograph in Fig. 12. The intensity of a reflection can be recorded directly by a Geiger counter, or it is deduced from the blackening of the photographic film. The Bragg angle θ can be measured quite easily on the film if the specimen-to-film distance is known. The necessary relations are illustrated in Fig. 13. The full-opening angle of

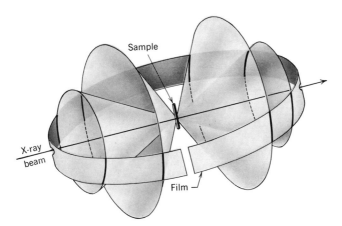

Sample

X-ray beam

Film

FIG. 11. Typical arrangement of film, cylindrical sample, and x-ray beam in a powder camera. The diffraction cones emanating from the sample intercept the film and are recorded as arc segments.

FIG. 12. Powder photograph of tungsten.

the diffraction cone 4θ is determined by measuring the distance S between two corresponding arcs on the powder photograph symmetrically displaced about the exit point of the direct beam.

$$4\theta = \frac{S}{R} \qquad \text{radians}$$

$$4\theta = \frac{S}{R}\left(\frac{180}{\pi}\right) \qquad \text{deg} \tag{16}$$

where R is the specimen-to-film distance, usually the radius of the camera housing the film (see also Exercise 8.) After θ is determined, Eq. (15) then can be used to calculate the d values for each reflection.

Determination of unit-cell dimensions. Once the d values are known, use can be made of the geometrical relation between the crystallographic axes, the Miller indices, and d_{hkl} to assign the appropriate indices to each reflection and to determine the unit-cell dimensions. For the cubic system, this relation has a particularly simple form,

$$d_{hkl} = (h^2 + k^2 + l^2)^{-\frac{1}{2}}a. \tag{17}$$

Since the possible values that the integers in parentheses can have are the same for all cubic crystals, a list of d's for one cubic crystal is the same as for any other cubic crystal, except that the entire list is multiplied by a different value of a in each case. This suggests that indexing of cubic crystals can be carried out most easily by constructing a chart in which

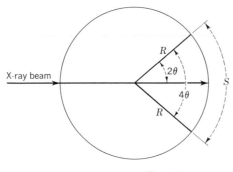

FIG. 13

the variation of d versus a is plotted for all possible combinations of h, k, l. Such a chart is shown in Fig. 14.

To use this chart, the measured d values are first marked on a strip of paper, using the same scale as in the chart. The point corresponding to $d = 0$ is placed on the vertical axis, and the paper strip is maintained in a horizontal position. The strip is then moved up and down on the chart until all the lines marked on the strip correspond to lines on the graph. In doing this, the largest d value can be expected to correspond to (100), (110), or (111). Once a match is obtained, the intersection of the top edge of the strip with the vertical axis of the graph marks the value of a for the crystal examined. Thus the unit-cell dimensions and the indices of the planes are determined at the same time. It is possible to do this by analytical trial-and-error methods also; however, the graphical procedure illustrated is more direct.

Relations like (17) exist for the other five crystal systems also. They become increasingly more complicated as the number of unknown unit-cell constants increases from two in the hexagonal and tetragonal systems to six in the triclinic system. Graphical and analytical methods have been developed for indexing powder photographs of crystals belonging to these systems; however, their discussion here is not warranted.

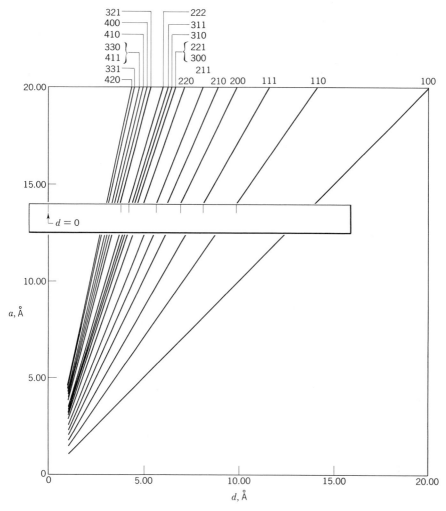

Fig. 14. Indexing chart for cubic crystals. A strip of paper on which the experimentally determined d values are noted is shown in the position where a match occurs.

Identification of unknown crystals. One of the features that emerges from the discussion of x-ray diffraction by crystals is that the Bragg angles determined by the lattice and the diffraction intensities determined by the atomic arrangements are different for each crystal. Although it is possible for two crystals to have the same lattice, or even the same composition, it is not possible for both to be the same unless they are identical crystals. The x-ray diffraction diagram of a crystal, therefore, is as unique as a fingerprint. It follows from this that x-ray diffraction methods can be used to identify unknown crystals by comparing their x-ray diffraction diagrams with a collection of known diagrams. Such a collection has been made and is called the *x-ray powder-data file.* The necessary information is tabulated on cards and in an index book. The diffraction diagrams are classified by noting the three most intense reflections and their *d* values in the upper left-hand corner of each card. To identify an unknown from its powder photograph, the *d* values of the three most intense reflections observed are used to find a match with a card in the file. The correctness of the identification is then verified by comparing all the observed reflections with those listed on the card. Mixtures of unknowns are similarly identified by an iterative procedure in which the reflections belonging to an identified crystal are eliminated from further consideration.

This method of identification has a number of decided advantages. It is first of all nondestructive, so that the same sample can be used for other purposes. Even very small samples can be analyzed because the irradiated area can have the dimensions of microns (10^{-6} m). The identification is unique in that it yields not only the chemical composition of the material, but also the state of aggregation. This is particularly important when the same combination of atoms can crystallize in several modifications. For example, TiO_2 crystallizes in three fairly common forms, only one of which is useful in paint manufacture. Furthermore, the analysis can be made quantitative by suitable admixtures of standards or by previous calibrations. Finally, such characteristics as crystallite size, shape, and degree of crystallinity also can be determined from a single x-ray exposure.

Single-crystal methods

Rotating-crystal method. Although most materials normally exist in the form of polycrystalline aggregates, certain applications, some of which are discussed in later chapters, require that single crystals be used. Some of the principal x-ray diffraction methods used to characterize these crystals are therefore briefly considered below. When dealing with a single crystal, some degree of freedom must be introduced into the experiment to allow different planes (*hkl*) to diffract x-rays. Reference to Fig. 9 suggests that this can be done for a monochromatic beam by rotating the crystal and its accompanying reciprocal-lattice construction so that different reciprocal-lattice points get an opportunity to intercept the sphere of reflection. It is customary to rotate the crystal about a direction that is normal to the direct beam, and the crystal is usually oriented so that one of its crystallographic axes is parallel to the rotation axis. Such an arrangement is shown in Fig. 15, in which a tetragonal crystal is oriented to rotate about its *c* axis. The corresponding reciprocal lattice has its origin at the point of emergence of the direct beam from the sphere of reflection. A portion of the reciprocal lattice is depicted as a collection of lattice nets parallel to the reciprocal-lattice vectors σ_{100} and σ_{010}. These nets are equally spaced along the rotation direction by an amount equal to σ_{001}. The indices of the points in each level have the same *l* index but different *h* and *k* indices. As the crystal is rotated,

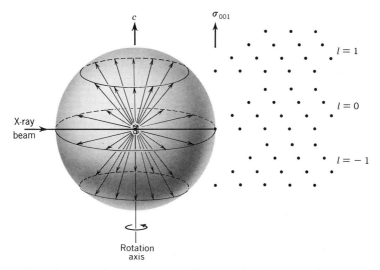

Fig. 15. Rotating-crystal arrangement. The crystal is shown at the center of the sphere of reflection. Part of the reciprocal lattice is also shown. Since the reciprocal-lattice origin lies at the point of emergence of the x-ray beam, it can be pictured as rotating about σ_{001} while the crystal rotates about *c*.

different points lying in the same layer intersect the sphere along a circle that is equidistant from its center. The diffracted beams passing through these intersections are generators of the same cone. Note that the cone is symmetrically disposed about the rotation axis of the crystal, at right angles to the direct beam. If a film is now cylindrically disposed about the crystal's rotation axis, it records the reflections corresponding to any reciprocal-lattice layer as individual spots lying along straight rows called *layer lines*.

A quick consideration of Fig. 15 shows that the height of a layer line above the equatorial line is directly proportional to the reciprocal-lattice vector $\sigma_{001} = 1/d_{001}$. It is thus possible to use the rotating-crystal method to determine the unit-cell dimensions of a single crystal. This procedure has the great advantage over the powder method that a single measurement gives the unit-cell constant directly. For crystals not belonging to the cubic system this means that two, or at most three, such photographs can be substituted for the far more tedious and time-consuming analytical indexing procedures.

This arrangement also has the advantage that reflections from individual (hkl) planes are resolved on the film. Thus the reflection intensity of individual spots can be measured and used, for example, to observe whether reflections having certain combinations of h, k, and l are missing. As previously discussed, such systematic absences can be used to establish the lattice type and form an important step in the determination of a crystal's space group. Actually, the single spots are not really representative of the intensity of individual reflections. This is so because certain reciprocal-lattice points such as $11l$, $1\bar{1}l$, $\bar{1}1l$, and $\bar{1}\bar{1}l$ are all equidistant from the rotation axis of the reciprocal lattice. Thus they all intersect the sphere of reflection at the same point. This situation is further aggravated by chance coincidences, so that some other scheme must be employed to resolve reflections from individual planes. A simple way of doing this is to limit the crystal's rotation to just a few degrees. This is called the *oscillation method*. It has a disadvantage over other methods described in the next section in that a very large number of films are required to measure reflections lying in different parts of the reciprocal lattice.

Moving-film methods. An alternative procedure for resolving individual reflections is to move the film concurrently with the rotation of the crystal. Thus each time a particular plane is in reflection position, the film has been displaced to a slightly different place and it is not possible for two reflections to superimpose on the photograph. In order to limit the number of reflections appearing on a single film and to aid in their identification, only the reflections corresponding to a single layer in the reciprocal lattice are allowed to reach the film. This is done by interposing a suitably shaped metal screen between the crystal and the film. This screen absorbs all diffracted beams except those belonging to the desired layer.

A number of different arrangements have been proposed for doing this. One of the more popular ones, the Buerger precession method, has the advantage that the reciprocal lattice is recorded without distortion. It is possible to assign indices to each reflection by direct inspection. The crystallographic axes and angles can be similarly measured directly from the photograph. A typical precession photograph is shown in Fig. 16A, next to a drawing of the corresponding reciprocal-lattice net in Fig. 16B.

The Laue method. There is another way to introduce the necessary degree of freedom required to permit the recording of reflections from different planes. Suppose the direct beam incident on a stationary crystal contains many different wavelength components. Each plane can then reflect those x-rays whose wavelength satisfies the Bragg law for the particular angle θ that it forms with the direct beam. This is the experimental arrangement that was used to test the feasibility of x-ray diffraction by crystals and is called the Laue method in honor of its discoverer. Although it simultaneously records the reflections from many planes in the crystal, it is difficult to use for the determination of unit-cell dimensions because the exact value of λ is not known for the different reflections. Nevertheless, methods for indexing the various reflections are available because their position on the film depends on the

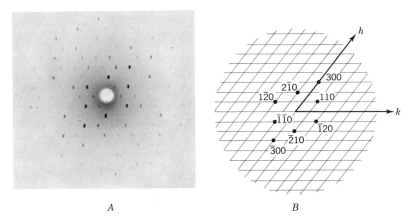

A B

FIG. 16. Precession-camera photograph of the reciprocal-lattice net of $Ca_2(Mn,Fe)$ $(PO_4)_2 \cdot 2H_2O$. (A) Actual photograph; (B) reciprocal-lattice net, with some of the reflection positions appropriately labeled.

orientation of the crystal relative to the incident-beam direction. Principal application of the Laue method nowadays is limited to the study of crystal orientation. It is possible to use very intense beams and direct-image-intesifier tubes so that crystals can be oriented by this method in a matter of minutes. It is very useful, for example, for the orientation of single crystals prior to cutting them into suitable shapes for device applications.

Preferred orientation studies

Wire texture. Although most common materials are polycrystalline, the orientation of the crystallites composing them is not always completely random. Their preferential alignment parallel to one or more crystallographic directions, called *texture*, may occur accidentally during formation or as a result of fabricating the material. For example, most materials deform plastically when subjected to external stresses. Suppose a metal is extruded through a small circular die to form a wire. The individual metal grains undergo slip along specific directions $[[uvw]]$ parallel to the extrusion direction, so that the resulting wire contains crystallites oriented not unlike the sketch in Fig. 17. Because the properties of a wire are related to the direction of preferred orientation

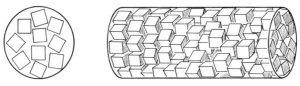

FIG. 17

along its axis, it is frequently necessary to determine what this crystallographic direction is. The obvious way to do this is to take an x-ray diffraction photograph with the x-ray beam directed at right angles to the wire axis. As already discussed, the reciprocal lattice of randomly oriented crystallites consists of concentric spheres. In the case of a wire, the normals to a set of planes (hkl) are constrained to lie at some fixed angle to the wire axis. (This is the real meaning of preferred orientation.) The corresponding reciprocal-lattice points are restricted to lie along rings about the wire-axis direction (Fig. 18). Whereas the reciprocal-

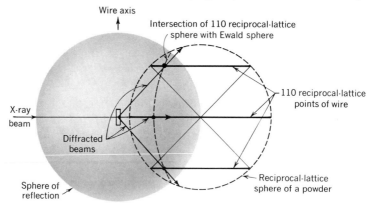

FIG. 18. Wire or fiber texture. The 110 reciprocal-lattice points are constrained to lie along the horizontal circles on the reciprocal-lattice sphere because of preferred orientation. Where these circles intersect the sphere of reflection, diffracted beams appear.

lattice sphere of a powder intersects the sphere of reflection in a circle (Fig. 10), the three rings intersect the Ewald sphere at six points (in front and behind the plane of the drawing). In actual wires, the preferred orientation is not as stringent as the illustration in Fig. 18 and the reciprocal-lattice construction consists of bands whose width ϕ is a measure of the angular deviation from perfect alignment (Fig. 19A). When a film is placed at right angles to the direct beam in Fig. 18, the reflections from various (hkl) planes appear as arcs (instead of full cones) (Fig. 19B), corresponding to the intersections of these bands with the sphere of reflection. Since the arcs lie along the diffraction rings already discussed in the powder method, their Miller indices can be readily determined. A simple analysis of the relevant geometrical relations between the wire axis and the photograph then permits an unambiguous determination of the preferred orientation direction of the crystallites in the wire.

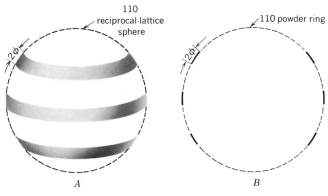

Fig. 19

Sheet texture. In the case of a wire, the crystallites align preferentially along a single direction, namely, the wire axis. Their orientation about the wire axis remains completely random. It is possible, however, to deform a polycrystalline material so that preferential orientation of two directions occurs simultaneously, for example, in the rolling of a metal sheet. The reciprocal-lattice construction for the resulting sheet texture consists of even more limited groupings of reciprocal-lattice points than those shown in Fig. 19A. In the limiting case, when all crystallites are perfectly aligned along two noncollinear directions, the material becomes a single crystal whose reciprocal lattice is a periodic array of discrete points. The crystallites do not align perfectly in actual sheets, however, so that sets of reciprocal-lattice points hkl are distributed not unlike the case depicted in Fig. 20.

Whether any of these reciprocal-lattice points will intersect the sphere of reflection now depends on the orientation of the sheet relative to the incident x-ray beam. In order to determine their actual distribution, therefore, it is necessary to obtain a number of diffraction diagrams at various inclinations of the sheet.   After the distribution has been determined, it is customary to present it graphically in a stereographic projection called a *pole figure*.   It should be noted that the density of reciprocal-lattice points in Figs. 19*A* and 20 is not necessarily uniform over the shaded areas shown.   Their density, of course, determines the number of x-rays diffracted at the specified angles.   Hence the intensity distribution along such arcs as are shown in Fig. 19*B* can be used as a measure of the number of planes having corresponding orientations in the sample. By this means it is possible to get a very accurate picture of the actual orientations of crystallites in any polycrystalline material.

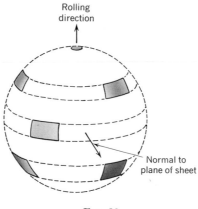

Fɪɢ. 20

Organic texture.   A very interesting form of aggregation is encountered in many organic materials comprised of large molecules.   Quite often these materials are not really crystalline, in the sense that their molecules are not periodically arranged in space.   On the other hand, the atoms forming each molecule are arranged in an orderly fashion, and this orderly array is common to all the molecules.   Moreover, the shape of the molecules may be influential in determining their packing in a solid, so that, although a true periodicity is lacking, the intermolecular distances tend to group very closely about some average value.   When such mean groupings are combined with fairly regular intramolecular arrays, the resulting material can be said to have a pseudo periodicity.   For example, the chainlike molecules comprising many organic fibers, ranging from nylon to human hairs, can be considered to be similar to elongated crystallites in a wire.   Because they tend to align themselves parallel to the fiber axis, an x-ray diffraction diagram of such fibers is not unlike the wire texture shown in Fig. 17*B*.   In fact, wire texture is frequently called *fiber texture* for that reason.   Similarly, it is possible for these organic molecules to align themselves in a way that resembles the sheet texture described in the preceding section. An example of such alignment can be found in polyethylene sheets. The degree of orientation in such sheets can be varied, say, by stretching the sheet, and these orientations can be readily determined by routine x-ray diffraction procedures. Such studies are useful, not only in elucidating the atomic or molecular arrays in materials, but also in relating these arrays to their properties. A number of examples of the effect of preferred orientation on the electrical, magnetic, and optical properties of a material are cited in subsequent chapters.

Suggestions for supplementary reading

Leonid V. Azároff and Martin J. Buerger, *The powder method in x-ray crystallography* (McGraw-Hill Book Company, Inc., New York, 1958).

Sir Lawrence Bragg, *The crystalline state*, vol. 1, *A general survey* (G. Bell & Sons, Ltd., London, 1949).

B. D. Cullity, *Elements of x-ray diffraction* (Addison-Wesley Publishing Company, Inc., Reading, Mass., 1956).

Suggestions for further reading

C. S. Barrett, *Structure of metals*, 2d ed. (McGraw-Hill Book Company, Inc., New York, 1952).

M. J. Buerger, *X-ray crystallography* (John Wiley & Sons, Inc., New York, 1942).

Dan McLachlan, Jr., *X-ray crystal structure* (McGraw-Hill Book Company, Inc., New York, 1957).

Exercises

1. Given that the unit-cell edge of a cubic crystal $a = 2.62$ Å, at what angles can the 100, 110, 111, 200, and 210 reflections of monochromatic x-rays be observed if $\lambda = 1.54$ Å?

2. Calculate the structure factor expressions for a body-centered cubic structure. (Assume one atom per lattice point of a body-centered cubic lattice.) What is the extinction rule for this lattice type, that is, for what combinations of h, k, and l does the structure factor equal zero? (ANSWER: $h + k + l$ must be even for $F \neq 0$).

3. By analogy to Exercise 2, determine the extinction rule for a face-centered lattice. Does it matter whether the lattice is cubic or othorhombic?

4. The structure of CsCl consists of one Cs atom at 000 and one Cl atom at $\frac{1}{2}\frac{1}{2}\frac{1}{2}$ in a cubic unit cell. Assuming that $f_{Cs} = 3f_{Cl}$, calculate the structure-factor expression for CsCl. By considering the extinction rules determined in Exercises 2 and 3, what is the lattice type of CsCl? HINT: What lattice types conform to cubic symmetry?

5. Zinc oxide has a hexagonal structure, $c = 5.2069$ Å, $a = 3.2492$ Å, and the density $D = 5.73 \times 10^3$ kg/m³. How many atoms of Zn and O does each unit cell contain? (ANSWER: 2ZnO per cell.)

6. The density of $(CH_3 \cdot C_6HSO_3)Mg \cdot nH_2O$ is 1.42 g/cm³, and the monoclinic unit cell has $a = 25.2$ Å, $b = 6.26$ Å, $c = 6.95$ Å, $\beta = 91°54'$. If the number of water molecules n lies between 4 and 8, what is the correct formula of magnesium p-toluene sulfonate hydrate? (HINT: The number of molecules per unit cell must be integral.) This is an example of the use of x-ray diffraction in the determination of correct molecular weights and compositions.

7. Suppose that an AB_2 crystal belongs to space group $P4$ and contains two formula weights per unit cell.

(a) Determine the structure-factor expressions for $hk0$ reflections, assuming both possible distributions of the A atoms discussed in the text.

(b) Suppose that the coordinates of the B atoms can be deduced from packing considerations. Which $hk0$ reflections are most sensitive to small changes in x and y values of these atoms? HINT: Which has the smallest contribution from A atoms?

(c) Suppose that the coordinates of the B atoms in $4d$ turn out to be $x = 0.125$ and $y = 0.125$ and that $f_A \simeq f_B$. Which $hk0$ reflections can be used to locate correctly the A atoms?

8. For convenience in manipulating Eq. (16), the radius of x-ray diffraction cameras is chosen in multiples or submultiples of the conversion constant from radians to degrees. If $S/2$, measured in millimeters on the film of a powder camera, corresponds directly to the Bragg angle in degrees, what is the radius of the camera? (ANSWER: 28.65 mm.)

9. If you want to determine the length of a for a hexagonal crystal from the layer-line spacing in a rotating-crystal photograph, about which direction of the crystal must you rotate it? HINT: Make a drawing of the hexagonal net of the reciprocal lattice and relate it to a.

10. Construct the reciprocal lattice of a two-dimensional "crystal" having $a = 1.24$ Å, $b = 2.48$ Å, and $\gamma = 120°$.

11. Making use of the extinction conditions determined in Exercises 2 and 3, what are the reciprocal lattices of a body-centered and a face-centered cubic crystal? HINT: When a reflection has zero intensity, the corresponding reciprocal-lattice point can be assumed to be absent. (ANSWER: FCC is the reciprocal of BCC.)

12. Å powder photograph of CsCl has been prepared using CuKα radiation ($\lambda = 1.54$ Å). If the Bragg angles measured for the first few observed diffraction lines are $10.72°$, $15.31°$, $18.88°$, $20.91°$, $24.69°$, and $27.24°$, what are the indices of the observed reflections and the value of a? HINT: CsCl is cubic. (ANSWER: 100, 110, 111, 200, 210, and 211).

13. A rotating-crystal photograph of a tetragonal crystal, $c = 4.62$ Å and $a = 3.08$ Å, is prepared with the crystal set to rotate about its four-fold axis using CuKα radiation ($\lambda = 1.54$ Å). At what angles (2θ) do the $hk0$ reciprocal-lattice points intersect the sphere of reflection? The $00l$ points?

14. If the radius of the cylindrical film used to record the photograph in Exercise 13 is 57.3 mm, what is the spacing between the recorded layer lines? (The cylindrical film is coaxial with the rotation axis of the crystal.)

3. quantum mechanics

Elements of theory

Up to the beginning of the twentieth century, the properties of all types of matter had been explained by means of a causal theory based on Newton's classical mechanics. An increasing number of observations, particularly those dealing with subatomic particles, however, could not be explained satisfactorily in this way. One noteworthy example was the behavior of electromagnetic radiation such as visible light. About 1675, Issac Newton proposed that light was composed of infinitesimal particles called corpuscles that traveled in straight lines at tremendous speeds. Although this theory was immediately criticized by Robert Hooke, who argued that light was composed of waves as originally suggested by Christian Huygens, Newton's prestige was so great that the corpuscular theory of light persisted into the nineteenth century. In 1804, however, Thomas Young performed an important experiment establishing the property of constructive and destructive interference by light beams emanating from two distinct sources. Such interference could be explained only by the wave theory of light. This viewpoint was further extended by the French scientist Augustin Fresnel to explain virtually all the phenomena of light observed by that time. Thus the wave theory of light gained in ascendancy until, a century later, the German physicist Max Planck issued his now famous radiation law, in 1901. It states that radiation is emitted in discrete bundles, or *quanta*, of energy,

$$E = nh\nu \tag{1}$$

where n is any integer, h is a universal constant equal to 6.63×10^{-34} J-sec (joule-second), and ν ($= c/\lambda$) is the frequency of the radiation equal to the velocity of light c divided by its wavelength λ.

The wave theory can be reconciled with the corpuscular theory by suitably combining waves of different frequencies to form *wave packets* like those illustrated in Fig. 1. Since the "presence" of a light quantum is measured by its intensity, which is proportional to the square of its amplitude, a wave packet like the one shown in Fig. 1*B* can be used to represent a quantum that is localized to a small region of space. It should

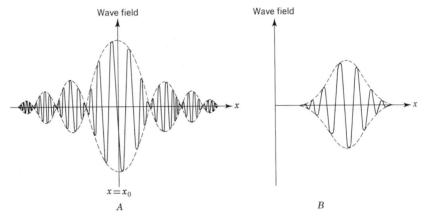

Fig. 1. One-dimensional wave packets. (*A*) Combination of waves having slightly different wavelengths and phases to form a wave packet that has a maximum amplitude near x_0 and decreases to zero as $|\pm x| > x_0$. (*B*) A wave packet that has a nonzero amplitude only in a very narrow range of x. Such wave packets can be used to represent short bursts of light produced by passing light through a small opening controlled by a rapidly closing and opening shutter.

be noted that the wave packet itself travels through space with a velocity called the *group velocity*, which is equal to the velocity of the particle as determined by classical mechanics, whereas the individual waves forming the packet have a velocity called the *phase velocity*, which is usually different. Fortunately, the subsequent discussions in this book do not require this distinction to be made, and the experimentally measurable group velocity only will be considered.

This dual nature of light was firmly established by Albert Einstein in 1905 when he proposed a correct explanation for the photoelectric effect (further discussed in Chap. 11). Einstein also extended the quantum theory of Planck to explain correctly the specific heat of solids, as discussed in Chap. 5. Two fundamental differences between the classical and quantum theories must be noted. According to classical mechanics, a bound particle can change its energy in a continuous manner, whereas, according to the quantum theory, such changes can occur only in discrete amounts, or quanta, equal to integral multiples of $h\nu$. The reason that many physical measurements fail to detect this discontinuous change is that Planck's constant h is such a small quantity. Thus, even at microwave frequencies ($\nu \simeq 10^{10}$ cps), $h\nu \simeq 10^{-25}$ J, an energy barely detectable by the most sensitive instruments now available. The second difference is that the motion of a particle is specified exactly in classical mechanics whereas quantum mechanics determines the *probability* of occurrence of any event. At first glance this may appear to be a severe limitation since experience with macroscopic bodies suggests that their motion can be predicted if their past history is known. It turns out that actual observations of phenomena involving subatomic particles are incapable of completely distinguishing the motion of individuals, only the collective behavior of the whole. This is not a limitation imposed by insufficiently accurate measuring apparatus, but rather a fundamental limitation of nature. As discussed in a subsequent section, the uncertainty attached to measurements involving submicroscopic particles actually sets a kind of limitation on the measuring devices themselves. Thus quantum mechanics is capable of explaining all observations most satisfactorily, and a more exact theory would be superfluous because its veracity could not be tested by experimental means.

Bohr atom. Early support for the quantum theory was given by spectrographic observations that the radiation emitted by specific atoms always has the same specific energy. Now, according to the laws of classical electrodynamics, an electron in an atom can radiate energy only while its velocity is changing. Since there is no reason according to classical mechanics why such a velocity change (acceleration) should not be continuous, it was not possible to reconcile experimental observations of only a limited number of energy values with classical theory. A way out of this dilemma was suggested by Niels Bohr, who postulated in 1913 two conditions concerning the structure of an atom:

1. For each atom, a discrete set of *energy states* exists in which the electrons can move without radiating energy. These states can be identified by a set of integers called *quantum numbers*.

2. Under suitable conditions, an electron can pass from one such state to another. Since the energy of each state is different, such a transition requires that the atom absorb or emit energy in order to satisfy the fundamental law of energy conservation.

The structure of an atom was established in 1911 by Ernest Rutherford, who showed that it consisted of a positively charged nucleus, having nearly the mass of the entire atom, surrounded by negatively charged electrons, each having a mass approximately equal to $1/2{,}000$ of the mass of the lightest atom, hydrogen. Rutherford pictured the nucleus at the

center of a miniature solar system, with the electrons traveling around it
as planets travel around the sun. An electron's orbit can be represented
by a thin cylindrical ring having the mass of the electron m. If the ring's
radius is r, then its moment of inertia is $I = mr^2$ and the angular momentum is $I\omega = mr^2\omega$. Quantitatively, Bohr proposed that an electron can
travel in any orbit provided that the electron's angular momentum is
equal to an integral multiple of $h/2\pi$.

$$I\omega = mr^2\omega = \frac{nh}{2\pi} \qquad n = 1, 2, 3, \ldots \qquad (2)$$

It follows from (2) that the angular velocity ω is given by

$$\omega = \frac{nh}{2\pi mr^2}. \qquad (3)$$

The electron is maintained in its orbit of radius r by a balance between
the electrostatic attractive force of the nucleus $Ze^2/4\pi\epsilon_0 r^2$ [Z is the atomic
number, Ze is the nuclear charge, $-e$ is the electronic charge, r their
separation, and ϵ_0 is a constant equal to 8.85×10^{-12} F/m (farad per
meter)] and the centrifugal force $m\omega^2 r$. This requires that

$$m\omega^2 r = \frac{Ze^2}{4\pi\epsilon_0 r^2} \qquad (4)$$

from which it follows that

$$\omega = \left(\frac{Ze^2}{4\pi\epsilon_0 mr^3} \right)^{\frac{1}{2}}. \qquad (5)$$

By equating the squares of Eqs. (3) and (5),

$$\omega^2 = \frac{n^2 h^2}{4\pi^2 m^2 r^4} = \frac{Ze^2}{4\pi\epsilon_0 mr^3}$$

and it is possible to determine the radius of the nth orbit:

$$r_n = \frac{n^2 h^2 \epsilon_0}{\pi m Z e^2}. \qquad (6)$$

Similarly, it is possible to calculate the energy corresponding to the nth
orbit. The potential energy V for an electrostatic force field is

$$V = -\frac{Ze^2}{4\pi\epsilon_0 r}$$

$$= -\frac{mZ^2 e^4}{4\epsilon_0^2 n^2 h^2} \qquad (7)$$

assuming that V goes to zero when the electron is at infinity. Its kinetic energy is

$$\begin{aligned} KE &= \tfrac{1}{2}mv^2 \\ &= \tfrac{1}{2}m\omega^2 r^2 \\ &= \frac{mZ^2e^4}{8\epsilon_0^2 n^2 h^2}. \end{aligned} \tag{8}$$

Finally, the total energy of an electron in the nth orbit is

$$\begin{aligned} E_n &= V + KE \\ &= -\frac{mZ^2e^4}{8\epsilon_0^2 n^2 h^2}. \end{aligned} \tag{9}$$

When an electron undergoes transitions from one state to another in accord with Bohr's second postulate, its energy changes by an amount equal to

$$\Delta E = E_{n_2} - E_{n_1} = nh\nu \tag{10}$$

where the E_n's are determined by Eq. (9). The integer n is the first and *principal quantum number* and determines the electron's energy. It also determines the distance separating the electron from the nucleus according to (6). Actually, this is not quite correct, because additional quantum numbers must be used to specify more exactly the electronic orbits, as discussed in a subsequent section. Nevertheless, it is frequently convenient to think of an atom in terms of the original Bohr model in which the location of an electron is limited to a radial distance r_n according to (6), so that the electrons move in essentially spherical shells around the nucleus. These shells are denoted K, L, M, N, etc., in order of increasing radius, and correspond to $n = 1, 2, 3, 4$, etc. Letters from the middle of the alphabet were chosen for these designations because the early investigators could not be sure that other shells, closer to the nucleus, did not exist.

The actual distribution of the electrons in atoms is considered in a later section. According to this model, the positively charged nucleus is surrounded by 2 electrons in the K shell, up to 8 electrons in the L shell, up to 18 electrons in the M shell, and so forth. The more distant electrons have higher energies, according to (9). Note, however, that E_n is a negative quantity because the potential energy of an electron (6) is chosen as zero when the electron is completely removed from the nucleus. Thus less work is required to remove an outer electron than an inner electron from the atom, because the energy of an outer electron is a smaller negative quantity and is closer to zero. The electrons that are most easily removed lie in the outermost shell of the atom and are called its *valence electrons*. The energy required to remove one valence electron from a neutral atom is called its *first ionization potential;* the energy required to remove the second valence electron is called the *second ionization potential;* and so on. The resulting positive *ion* has a higher energy than the neutral atom and is said to have a valence of $1+, 2+$, etc. As discussed in the next chapter, atoms prefer to have complete outer shells. When an atom lacks one or more electrons to complete its outermost shell, additional electrons may attach themselves to it. The resulting negatively charged ion is then said to have a valence of $1-$, $2-$, etc.

Introduction to wave mechanics. The dual nature of radiation having been established, it was left to de Broglie to make the bold suggestion in 1924 that small particles of matter such as electrons have the corresponding property of behaving like waves of wavelength

$$\lambda = \frac{h}{mv} \tag{11}$$

where h is Planck's constant, and m and v are the mass and velocity of the particle, respectively. This hypothesis was verified within three years by Davisson and Germer in the United States and by Thomson in England, who demonstrated that electrons are diffracted by crystals similarly to x-rays. It will be recalled from the discussion in Chap. 2 that diffraction, like the constructive and destructive interferences of light observed by Young, can be explained only by means of the wave theory. Although not specifically discussed there, x-rays also can be scattered without producing constructive interferences. Such incoherently scattered x-rays have a slightly longer wavelength (smaller energy) and their existence has been verified experimentally. This phenomenon was first explained by Compton, who postulated that the collision between x-rays and electrons is elastic, not unlike the collision between two billiard balls. The law of energy conservation requires that an x-ray quantum incident on an electron transfer some of its energy to the electron, so that the Compton effect illustrates the particle aspect of x-rays and electrons.

This ability to associate a wave nature with a particle like an electron suggests that the well-known wave equation can be utilized in characterizing its motion. How this is done is described in the next section. First, consider a more familiar application of the wave equation to the problem of a vibrating string of length L, clamped at both ends. If the string is set into motion by plucking it, its displacement U as a function of distance along the string x and time t is given by the solutions of

$$\frac{\partial^2 U(x,t)}{\partial x^2} = \frac{\mu}{T} \frac{\partial^2 U(x,t)}{\partial t^2} \tag{12}$$

where μ is the mass per unit length of the string, and T is the tension applied to the string. If the string is vibrating in such a manner that the nodes of the string are stationary in space and the maximum displacement at any point does not vary with time, that is, when *standing waves* are set up, then the equation giving the maximum displacement u as a function of x is

$$\frac{d^2u(x)}{dx^2} + \omega^2 \frac{\mu}{T} u(x) = 0 \tag{13}$$

where $\omega = 2\pi\nu$ is the angular frequency of vibration (Exercise 7). This equation has a general solution of the form

$$u(x) = A \cos \omega \sqrt{\frac{\mu}{T}} x + B \sin \omega \sqrt{\frac{\mu}{T}} x \tag{14}$$

as can be verified by direct substitution in (13).

The quantity $\sqrt{\mu/T}$ has the dimensions of seconds per meter, so that it equals the reciprocal of velocity v. It is possible to relate the arguments in (14), therefore, to the wavelength of a standing wave λ because $\nu/v = 1/\lambda$.

$$\omega \sqrt{\frac{\mu}{T}} = \frac{2\pi\nu}{v} = \frac{2\pi}{\lambda}. \tag{15}$$

Substituting (15) in (13),

$$\frac{d^2u}{dx^2} + \frac{4\pi^2}{\lambda^2} u = 0 \tag{16}$$

which is the time-independent wave equation whose general solution is

$$u = A \cos \frac{2\pi}{\lambda} x + B \sin \frac{2\pi}{\lambda} x. \tag{17}$$

The classical wave equation is used in many branches of physics, including acoustics, mechanics, optics, and others concerned with wave propagation. For the case of a vibrating string, the possible wavelengths of standing waves are determined by the length of the string L. The nodes of a standing wave divide the total length L into integral parts so that

$$\lambda = \frac{2L}{n} \qquad \text{where } n = 1, 2, 3, \ldots \tag{18}$$

since the distance between nodes is one-half a wavelength. By substituting (18) for λ in (17),

$$u = A \cos \pi \frac{n}{L} x + B \sin \pi \frac{n}{L} x \tag{19}$$

and it is possible to determine the particular solutions of the wave equation (16). This is done by making use of the *boundary conditions* that the displacement $u = 0$ when $x = 0$ or when $x = L$ because the string is clamped at both ends.

When $x = 0$:

$$u = 0 = A \cos (0) + B \sin (0)$$
$$= A. \tag{20}$$

When $x = L$:

$$u = 0 = B \sin \pi \frac{n}{L} L$$
$$= B \sin \pi n. \tag{21}$$

According to (20), A is zero; therefore, excluding the trivial solution when B is also zero, the only way that the right side of (21) can be zero is for $\sin \pi n$ to equal zero. This is the case when n is equal to an integer. Several possible solutions of (16) are plotted in Fig. 2 for $n = 1, 2$, and 3.

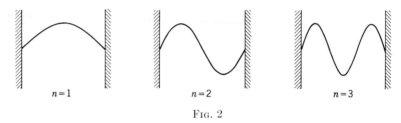

$$n = 1 \qquad\qquad n = 2 \qquad\qquad n = 3$$

Fig. 2

These are the familiar fundamental and higher-order harmonics. Note that B is an arbitrary constant determining the amplitude and is independent of the boundary conditions. An important conclusion to be drawn from the above example is that the solutions of boundary-condition problems expressed by wave equations like (16) are *quantized;* that is, they are different from zero only for a discrete set of values of n and are equal to zero for any other value of n.

Schrödinger theory. Following de Broglie's postulate of the wave nature of electrons, Schrödinger showed in 1926 that a wave equation can be used to characterize their behavior in agreement with the quantization requirements of Bohr's quantum theory. In fact, whenever the results of the Bohr theory agree with experimental observations, Schrödinger's theory gives an equivalent result. Moreover, Schrödinger's theory agrees with the experiment even when the Bohr theory does not. The Schrödinger theory is an exact theory in the sense that no approximations are inherent in its formulation. Its application to atomic and solid-state problems, however, gives results that are probabilistic in form, for reasons discussed below. The formulation adopted by Schrödinger constitutes a branch of mathematical physics called *wave mechanics.* An alternative approach developed by Born and Heisenberg employs matrices and leads to essentially identical results. It is now common practice to call this branch *quantum mechanics*, regardless of whether the wave equation or matrix formulation is used. Because the Schrödinger theory is mathematically simpler, it is considered in some detail below.

According to the de Broglie postulate, $\lambda = \dfrac{h}{mv}$ for an electron. Substituting this quantity for the wavelength in (16),

$$\frac{d^2\psi}{dx^2} + \frac{4\pi^2 m^2 v^2}{h^2}\, \psi = 0 \tag{22}$$

where the location of the electron along x is represented by the *wave function* $\psi(x)$ instead of the displacement variable u previously employed. Recalling that the kinetic energy of an electron is equal to the difference between its total energy E and its potential energy V,

$$\tfrac{1}{2}mv^2 = E - V \tag{23}$$

so that

$$m^2v^2 = 2m(E - V). \tag{24}$$

Substituting (24) in (22), the one-dimensional time-independent form of Schrödinger's equation results.

$$\frac{d^2\psi}{dx^2} + \frac{8\pi^2 m}{h^2}\,(E - V)\psi = 0. \tag{25}$$

In three dimensions $d^2\psi/dx^2$ must be replaced by partial derivatives with respect to the x, y, and z axes, so that the three-dimensional Schrödinger equation is

$$\left(\frac{\partial^2}{\partial x^2} + \frac{\partial^2}{\partial y^2} + \frac{\partial^2}{\partial x^2}\right)\psi + \frac{8\pi^2 m(E - V)}{h^2}\,\psi = 0. \tag{26}$$

The solutions of (26) are not unlike the solutions of (16) except that the wave function ψ is a function of three space coordinates x, y, z. In this sense, the wave function describes the position of an electron. There is one very important distinction, however. The displacement of a vibrating string can be readily observed, and it is meaningful, therefore, to describe its displacement precisely. The displacement or position of an electron, on the other hand, cannot be observed directly. The reason for this becomes obvious when the "tools" of observation are considered. In order to observe an electron, either light photons or x-ray photons or other electrons have to be used. These particles interact with the electron, changing its position immediately as the signal describing its previous position reaches the observer. Thus the information obtained from the experiment no longer correctly describes the state of the electron. It is not meaningful, therefore, to seek an equation that accurately describes the position of an electron, because the correctness of such an equation cannot be verified.

Although it is not meaningful to predict the exact position of an electron at a particular instant of time, it is possible to determine the most probable place where the electron will be. Max Born showed in 1926 that the quantity $|\psi|^2$ tells the probability of the presence of an electron in a unit volume of space at the time at which the wave function ψ is being considered. When a large number of identical experiments are performed, this so-called *probability density* accurately predicts what the spatial distribution of electrons will be. Since this wave function is a function of space coordinates, the probability density has different values as a function of space coordinates also. Moreover, it predicts quite accurately the most probable locations of an electron for each possible value of the energy E in (26). (These energy values are quantized just like the energy values assumed by Bohr.) In fact, the most probable values of the distance from the nucleus for an electron predicted by the probability density agree quite well with the radii of the Bohr orbits. Because the electrons are not limited to the occupation of prescribed positions in space or orbits according to quantum mechanics, an electron is said to occupy an *orbital* described by its orbital wave function.

Since the probability density $|\psi|^2$ expresses the most probable location of an electron in space, it follows that the integral of this quantity, integrated over all space, must be a constant which is independent of time. When *normalized*, that is, properly scaled, wave functions are used, this probability integral is equal to unity; that is,

$$\int_{-\infty}^{\infty} |\psi|^2 \, dv = 1. \tag{27}$$

The integral in (27) is one of the physical boundary conditions that the wave-function solutions of (26) must obey, not unlike the boundary conditions imposed on the vibrating string in the previous section. Other boundary conditions imposed on the wave functions are that they must be everywhere continuous and, mathematically speaking, well behaved. This means that $\psi(x)$ must be a continuously varying function of x and that its first derivative with respect to x, $d\psi/dx$, must be a continuous function of x also.

Uncertainty principle. The foregoing discussion indicates that the predictions of quantum mechanics are statistical in nature. At first glance this appears to be contradictory to everyday experience, in which it is usually possible to describe the position and motion of bodies exactly. As already stated, however, the experimental evidence of the behavior of electrons is inherently statistical in nature. An analogy can be drawn to an experiment in which it is attempted to locate the whereabouts of a firefly enclosed in an opaque box. This can be done, for example, by inserting a photographic plate inside the box. When the plate is subsequently developed, the darkest spots correspond to the positions where the firefly spent most of the time. It is not possible to tell from the photograph, however, exactly when the firefly was at any particular spot during the exposure. If an attempt is made to observe the firefly's whereabouts directly, say, by opening the lid of the box, then the firefly can escape from the box, thereby defeating the purpose of the experiment.

In addition to being unable to predict the exact location of an electron at every instant of time, it turns out that it similarly is not possible to specify the exact velocity of an electron when it occupies a particular point in space, nor its energy at a particular instant of time, and so forth. This limitation was expressed by Heisenberg in the *uncertainty principle,* which states that the product of the uncertainties in specifying two such conjugate characteristics can never be less than a constant value. This is not an artificial postulate required to make quantum mechanics agree with observations, but can be derived directly from the theory by making use of a mathematical relation known as the *Schwarz inequality.* The resulting analytical relations between the uncertainties in position Δx and momentum Δp have the form

$$\Delta x \times \Delta p \geq \frac{h}{2\pi}$$

or

$$\Delta E \times \Delta t \geq \frac{h}{2\pi} \tag{28}$$

where E and t are, respectively, energy and time, and h is Planck's constant. The meaning of the relations in (28) is that, as one attempts to specify one of the conjugate variables, say, the time t, more precisely, that is, as $\Delta t \to 0$, the uncertainty in the energy ΔE becomes correspondingly larger.

The uncertainties need not be very large, however, since their product can equal $\simeq 10^{-35}$ in mks units. The important consequence of the uncertainty principle is that quantum-mechanical calculations for subatomic particles yield a description of what is most probable rather than the usual picture one obtains from classical mechanics. It should be realized, of course, that even experiments involving large bodies have associated with them a certain degree of uncertainty, commonly termed the experimental accuracy of a measurement. According to classical mechanics, however, there is no reason, in principle, why a measurement cannot be very precise. To see when classical mechanics rather than quantum mechanics can be used, consider the form of the first relation in (28), after dividing both sides by the mass of the particles being studied.

Since $p = mv$,

$$\Delta x \times \Delta v \geq \frac{h}{2\pi m}. \tag{29}$$

It is clear from (29) that, as the mass of the particles increases, the uncertainty decreases. Thus, when a particle has a mass of one kilogram, or approximately that of 10^{30} electrons, then the uncertainty of its velocity and position can be specified with a correspondingly smaller uncertainty. In fact, the uncertainty becomes so small at this stage that it can be safely called a certainty instead. On the other hand, the uncertainty principle very definitely states that there is a limit to how precisely one can observe atomic phenomena.

Quantum numbers. A direct solution of the Schrödinger equation (25) is postponed to a subsequent section in which it is applied to the hydrogen atom. As shown there, the general solution is a wave function $\psi(xyz)$, which contains three constants called *quantum numbers*. One of these is designated n and can equal any integer from one to infinity. It determines the total energy of a electron, similarly to (9) in the Bohr theory, and is called the *principal* quantum number. The other two constants must be integers also, and their magnitudes are related to that of the principal quantum number n. The *azimuthal* quantum number l can have all integer values from $l = 0$ to $l = n - 1$, and it determines the electron's orbital angular momentum. Now an electron moving in such an orbital is a moving electric charge, so that, according to Ampère's rule, there is a magnetic field associated with it. In the presence of an external magnetic field, it turns out that the orbital can orient itself in $2l + 1$ different ways; that is, it can assume any one of the $2l + 1$ recognizably different orientations. That this is indeed the case was demonstrated in an experiment performed by Stern and Gerlach in 1922. They passed a stream of atoms through an inhomogeneous magnetic field and observed that the atoms were deflected in certain particular ways only, indicating that their magnetic moments were quantized. The specific orientation assumed by the orbitals is determined by the *magnetic* quantum number m, which can have all integer values from $m = -l$ to $m = +l$, including zero. Each possible solution of the Schrödinger equation, therefore, can be distinguished by means of the specific values that the three quantum numbers can have. The resulting orbitals are represented by the probability densities shown in Fig. 3 for several different values of l and m. As can be seen therein, the orbital corresponding to $l = 0$ is spherically shaped, and it is called the s orbital, following an early spectrographic designation. The three orbitals corresponding to $l = 1$ are called p orbitals; the five orbitals corresponding to $l = 2$ are called d orbitals. The relation between the Bohr atom shells,

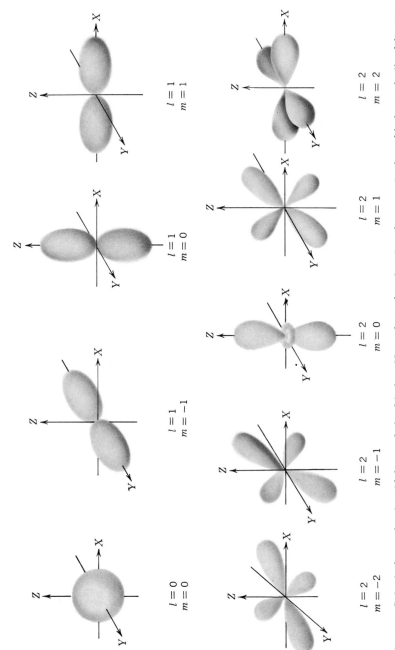

Fig. 3. Spherical s and spheroidal p and d orbitals. Note that when $l = 2$ and $m = -2$, the orbitals are inclined by 45 deg to X and Y; when $l = 2$ and $m = -1$, the orbitals are inclined by 45 deg to Y and Z; when $l = 2$ and $m = 1$, the orbitals are inclined by 45 deg to X and Z.

73

Table 1
The quantum numbers describing solutions of Schrödinger's equation

Bohr atom designation of shell	Principal quantum number n	Azimuthal quantum number l	Magnetic quantum number m	Spectrographic designation
K	1	0	0	1s (for "sharp")
L	2	0	0	2s
		1	0, ± 1	2p (for "principal")
M	3	0	0	3s
		1	0, ± 1	3p
		2	0, ± 1, ± 2	3d (for "diffuse")
N	4	0	0	4s
		1	0, ± 1	4p
		2	0, ± 1, ± 2	4d
		3	0, ± 1, ± 2, ± 3	4f (for "fundamental")

the three quantum numbers, and their spectrographic designations are summarized in Table 1.

In a hydrogen atom containing only one electron, all the wave functions having the same principal quantum number n have the same energy. When an atom contains more than one electron, however, the energy values are affected by interactions between the electrons. The main force acting on any electron is the electrostatic attraction of the positive nucleus at the atom's center. It can be shown that for $n \leq 3$, the energy of any electron is given by

$$E_n = -(13.6 \text{ eV}) \frac{(Z - Z_0)^2}{n^2} \tag{30}$$

where -13.6 eV (electron volts) corresponds to the energy of a 1s electron in a hydrogen atom, and Z_0 is a *shielding* constant which expresses the decrease in the nuclear attraction for a particular electron due to the shielding effect of the other electrons present. In an atom, the numerical value of Z_0 depends not only on n, but also on l, and varies from practically zero when $n = 1$ to almost Z for the highest occupied quantum state. The effect that this has on the energy values of the wave-function solutions is illustrated in Fig. 4. The relative positions, or *energy levels*, of the orbitals are indicated by dashes of length commensurate with the number of different wave functions that have the same energy. When more than one solution of the Schrödinger equation (for different values of m) is possible for the same energy value, such solutions are said to be *degenerate*. Thus the p orbitals are threefold degenerate, the d orbitals are fivefold degenerate, and so on.

The total number of permissible values that m can have is $2l + 1$, corresponding to the total number of ways that the magnetic vector due to the orbital angular momentum of the electron can differ in its orientation. If an atom is placed in a strong magnetic field, therefore, there is an odd number of ways that the magnetic vectors of electrons can align themselves with respect to the external field. Each different way of aligning in the magnetic field produces a small but finite change in the energy. This effect was first reported by Zeeman and bears his name. For certain atoms, notably alkali metals, however, even numbers of

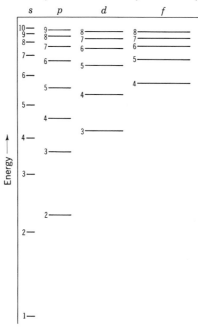

FIG. 4. Relative energy levels of atomic orbitals. The levels are labeled with appropriate values of n.

energy levels were observed. To explain this effect, an additional magnetic moment had to be postulated. It is assumed that the electric charge of an electron is not stationary but spins about an axis passing through the electron. The electron can spin only in essentially two unique directions, say, clockwise and counterclockwise. Thus the total number of ways to orient the magnetic vector of the spin momentum is equal to 2. Now the spin momentum, like the angular momentum due to the orbital motion of the electron, can give rise to $2m_s + 1$ levels. In order to limit the total number to 2, the magnitude of the so-called *spin* quantum number m_s must be equal to $\frac{1}{2}$. The direction of the electron spin vector is indicated by letting m_s equal $+\frac{1}{2}$ or $-\frac{1}{2}$, depending on whether the spin vector is parallel to the orbital momentum or not. Although the spin quantum number was originally introduced by Uhlenbeck and Goudsmit in an arbitrary way, a few years later it was shown by Pauli, and more elegantly by Dirac, that it could be incorporated directly into a more generalized form of the Schrödinger theory.

Pauli exclusion principle.   It is usually more convenient to speak of the electronic orbitals in terms of the four quantum numbers than in terms of the actual wave-function solutions of Schrödinger's equation. This is so because the values of the four quantum numbers actually specify what form the solution takes on.   An electron described by four such numbers has a definite energy and travels with a predictable average velocity in a predictable orbital.   A complete description of an atom can be given, therefore, by specifying the four quantum numbers of each electron and by listing the number of electrons of each kind.   A systematic procedure for doing this was made possible by Wolfgang Pauli in 1925 when he proposed the principle that *no two electrons in an atom can have exactly the same set of four quantum numbers*.   According to the *Pauli exclusion principle*, therefore, there can be two electrons present for each combination of n, l, and m, provided that each has a different value of m_s.   Thus hydrogen can have one $1s$ electron of either positive or negative spin, whereas helium ($Z = 2$) can have two $1s$ electrons only if their spins are oppositely directed.   In the absence of an external field, both $1s$ electrons have the same energy and are said to occupy the same energy level.   They can be distinguished, however, because they occupy two distinct *quantum states*, as can be verified by considering the four quantum numbers describing each state.   Accordingly, each s level contains two quantum states; each p level contains six quantum states; and so forth.

Applications of quantum mechanics

Hydrogen atom.   The application of Schrödinger's equation to actual atoms is illustrated by the simplest atom, hydrogen.   Since the hydrogen atom contains only one electron in addition to the nucleus, this is called a two-body problem and can be solved exactly.   The negatively charged electron of charge $-e$ moves in the field of the positively charged nucleus of charge $+e$, so that its potential energy in this central field is

$$V = -\frac{e^2}{4\pi\epsilon_0 r} \tag{31}$$

according to Coulomb's law, where r is the interparticle separation.   Substituting this value of V for the potential energy in (26),

$$\left(\frac{\partial^2}{\partial x^2} + \frac{\partial^2}{\partial y^2} + \frac{\partial^2}{\partial z^2}\right)\psi + \frac{8\pi^2 m}{h^2}\left(E + \frac{e^2}{4\pi\epsilon_0 r}\right)\psi = 0 \tag{32}$$

which is the Schrödinger equation for the hydrogen atom.

There are essentially two kinds of solutions of (32) possible. In the simplest kind, the wave function $\psi(r)$ depends on r only, so that it is spherically symmetric. Considering this class of solutions first, it is necessary to evaluate the operation of the first parentheses in (32) on $\psi(r)$. This means that the second partial derivative of $\psi(r)$ with respect to x, y, and z must be calculated. To do this, start by considering the first partial derivative with respect to x:

$$\frac{\partial \psi}{\partial x} = \frac{\partial \psi}{\partial r} \frac{\partial r}{\partial x}. \tag{33}$$

Remembering that $r = (x^2 + y^2 + z^2)^{\frac{1}{2}}$,

$$\frac{\partial r}{\partial x} = \frac{1}{2} \frac{2x}{(x^2 + y^2 + z^2)^{\frac{1}{2}}} = \frac{x}{r} \tag{34}$$

so that

$$\frac{\partial \psi}{\partial x} = \frac{x}{r} \frac{\partial \psi}{\partial r}. \tag{35}$$

Differentiating both sides of (35) again,

$$\frac{\partial^2 \psi}{\partial x^2} = \frac{1}{r} \frac{\partial \psi}{\partial r} - \frac{x^2}{r^3} \frac{\partial \psi}{\partial r} + \frac{x^2}{r^2} \frac{\partial^2 \psi}{\partial r^2}. \tag{36}$$

Identical expressions result for the other two variables:

$$\frac{\partial^2 \psi}{\partial y^2} = \frac{1}{r} \frac{\partial \psi}{\partial r} - \frac{y^2}{r^3} \frac{\partial \psi}{\partial r} + \frac{y^2}{r^2} \frac{\partial^2 \psi}{\partial r^2} \tag{37}$$

and

$$\frac{\partial^2 \psi}{\partial z^2} = \frac{1}{r} \frac{\partial \psi}{\partial r} - \frac{z^2}{r^3} \frac{\partial \psi}{\partial r} + \frac{z^2}{r^2} \frac{\partial^2 \psi}{\partial r^2}. \tag{38}$$

Adding the three differentials in (36), (37), and (38) together and remembering that $x^2 + y^2 + z^2 = r^2$,

$$\frac{\partial^2 \psi}{\partial x^2} + \frac{\partial^2 \psi}{\partial y^2} + \frac{\partial^2 \psi}{\partial z^2} = \left(\frac{3}{r} - \frac{r^2}{r^3} \right) \frac{\partial \psi}{\partial r} + \frac{r^2}{r^2} \frac{\partial^2 \psi}{\partial r^2}$$

$$= \frac{\partial^2 \psi}{\partial r^2} + \frac{2}{r} \frac{\partial \psi}{\partial r}. \tag{39}$$

Substituting (39) in the Schrödinger equation (32),

$$\frac{\partial^2 \psi}{\partial r^2} + \frac{2}{r} \frac{\partial \psi}{\partial r} + \frac{8\pi^2 m}{h^2} \left(E + \frac{e^2}{4\pi \epsilon_0 r} \right) \psi = 0. \tag{40}$$

The simplest solution of (40) that satisfies the requirement that the wave function be continuous and have a large value only near the nucleus (small r) has the form

$$\psi(r) = e^{-ar} \tag{41}$$

where a is a constant whose value must be determined. To do this, substitute (41) in (40).

$$a^2 e^{-ar} + \frac{2}{r}(-ae^{-ar}) + \frac{8\pi^2 m}{h^2}\left(E + \frac{e^2}{4\pi\epsilon_0 r}\right)e^{-ar} = 0. \qquad (42)$$

Factoring e^{-ar} out of Eq. (42) and rearranging the terms,

$$a^2 + \frac{8\pi^2 m}{h^2}E = \frac{2a}{r} - \frac{2\pi me^2}{\epsilon_0 h^2 r}. \qquad (43)$$

The left side of (43) does not depend on r, and yet the equality in (43) must be maintained for any value of r. The only way that this can be accomplished is if both sides in (43) are equal to zero. Thus, setting the right side of (43) equal to zero,

$$a = \frac{\pi me^2}{\epsilon_0 h^2}. \qquad (44)$$

Similarly, setting the left side of (43) equal to zero,

$$E = -\frac{a^2 h^2}{8\pi^2 m} = -\frac{me^4}{8\epsilon_0^2 h^2} \qquad (45)$$

as can be seen by substituting (44) for a in (45). Note that the energy in (45) is the same as that derived in the Bohr theory (9) for $Z = 1$ and $n = 1$. Consequently, this is the lowest energy level for an electron in hydrogen, called the *ground state* of hydrogen. Substituting for the constants in (45),

$$E = -\frac{(9.1 \times 10^{-31}\text{kg})(1.6 \times 10^{-19}\text{C})^4}{8 \times (8.85 \times 10^{-12}\text{ F/m})^2(6.63 \times 10^{-34}\text{ J-sec})^2} = -13.6\text{ eV} \quad (46)$$

(C is the abbreviation for coulomb).

According to the above analysis, the radially symmetric wave function in (41) is a solution of the Schrödinger equation for hydrogen provided a has the value specified in (44) and the energy E is given by (45). In order to determine the orbital corresponding to this wave function, it is necessary to calculate the probability $|\psi|^2\, d\upsilon$ of finding the electron in a unit volume $d\upsilon$. Since $\psi(r)$ is radially symmetric, the most convenient volume element to choose is a spherical shell about the nucleus at its center. The volume inscribed in a shell lying between r and $r + dr$ is $4\pi r^2\, dr$, so that the probability of finding the electron is proportional to

$$|\psi(r)|^2 r^2 = r^2 e^{-2ar}. \qquad (47)$$

This probability is shown plotted against r in Fig. 5. The maximum value of the probability can be determined by differentiating (47) and setting the derivative equal to zero.

$$\frac{d(r^2 e^{-2ar})}{dr} = 0 = 2re^{-2ar} - 2ar^2e^{-2ar}$$

and
$$r = \frac{1}{a} = \frac{\epsilon_0 h^2}{\pi m e^2} \tag{48}$$

according to (44). Note that the radial value of the maximum probability in (48) is equal to the radius r_1 of the first Bohr orbit for hydrogen according to (6). The wave function $\psi(r) = e^{-ar}$, therefore, represents an s state and has the spherical orbital shown in Fig. 3.

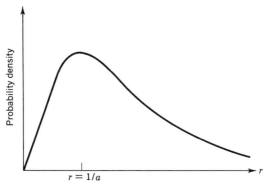

FIG. 5

A more general solution of (32) can be obtained by representing the wave function by a product of three functions $R(r)\ \Theta(\theta)\ \Phi(\phi)$, where r, θ, and ϕ are the usual three spherical coordinates. When this wave function $\psi(r,\theta,\phi)$ is substituted in the Schrödinger equation (32), it can be shown that it is possible to obtain three simultaneous differential equations in which each of the three variables is separated, similarly to the procedure used above. Without considering the solutions in detail, it turns out that completing the solutions of two of these simultaneous differential equations requires representing $R(r)$ and $\Theta(\theta)$ by power series. For example,

$$\Theta = \sin^m \theta(A_0 + A_1 \cos \theta + A_2 \cos^2 \theta + \cdots) \tag{49}$$

whose coefficients A_k are determined by a relation like

$$A_k = A_{k-2} \frac{(k + |m| - 1)(k + |m| - 2) - l(l + 1)}{k(k - 1)} \tag{50}$$

where k is a dummy variable denoting the coefficients of the series in (49), but l and m are the azimuthal and magnetic quantum numbers, respectively. The relation in (50) serves to define the numerical relations between the integer values that l and m can have. A detailed derivation of the relations between the quantum numbers is considered to be beyond the scope of this book; the results have been summarized in Table 1.

The two classes of solutions described above include a very large number of actual possible solutions. Thus a different set of solutions is possible for each energy E_n. The actual number of possible solutions depends on the value of the principal quantum number n, the number increasing as n increases. These degenerate solutions have the same energy in the hydrogen atom and are called *excited states* to distinguish them from the ground state corresponding to $n = 1$. When an atom contains more than one electron, the Schrödinger equation cannot be solved exactly because there is no exact solution possible for even the three-body problem in mechanics. The actual procedure used is to approximate the true atomic state by a central field like that of hydrogen, except that the charge on the nucleus is set equal to Ze. Next the wave functions of each electron moving in such a central field are computed separately, and the resulting probability densities are used to recompute the central field. This field is compared with the one initially assumed. If it is different, then this process is repeated, using the new field to obtain new wave functions, until a self-consistent field is obtained. This procedure was first proposed by Hartree, who has used the *self-consistent field approximation* to determine the charge distributions for many atoms with a satisfactory degree of accuracy. It should be noted that the field in the above discussion is substituted for V in (26). As shown in subsequent chapters, most solid-state problems requiring the Schrödinger equation for their solution are actually solved by selecting a suitable form of the potential field V. The approximations used in such solutions, therefore, are primarily concerned with selecting a potential field that is a very close representation of the actual field and yet has such a form that it is mathematically possible to obtain direct solutions of the Schrodinger equation.

X-ray spectra. As stated in the first section of this chapter, Bohr based his theory largely on the experimental observation that suitably excited atoms always emitted radiations that were characteristic of the emitting atom. A complete explanation of all the observed spectra and their origins, however, was not possible until quantum mechanics was developed. The detailed theory is not discussed here. Instead, the experimental results are explained in terms of the quantum-mechanical concepts already described, so that this section serves as a demonstration of the validity of these results. Specifically, the x-ray portion of the electromagnetic spectrum has been chosen for consideration in view of the discussion of x-ray diffraction in Chap. 2.

In order for an atom to undergo a transition to an excited state, it must gain energy from an external source. Subsequently, the atom can return to its ground state by emitting this energy in the form of a photon whose energy is

$$\Delta E = E_{\text{excited}} - E_{\text{ground}} = h\nu. \tag{51}$$

The values of $\nu(=c/\lambda)$ for x-rays range from 10^{15} to 10^{19} sec^{-1}. The wavelengths of x-rays most frequently used in x-ray diffraction studies have a much narrower range, 0.5 to 2.5 Å.

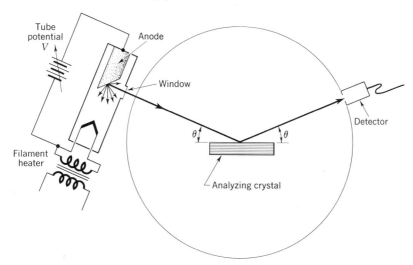

FIG. 6. X-ray spectrometer.

To see how an atom can be suitably excited to emit x-rays, consider the x-ray tube in Fig. 6 consisting of a hot filament and a metal block serving as the anode, placed in an evacuated glass envelope. Electrons emitted by the filament are accelerated toward the anode, commonly called the *target*, which absorbs the electrons and subsequently becomes a source of x-rays. The target emits x-rays in all directions, and those passing through an x-ray transparent window impinge on a crystal from which they are reflected to a detector according to Bragg's law,

$$\lambda = 2d_{hkl} \sin \theta. \tag{52}$$

In the x-ray spectrometer shown in Fig. 6, the interplanar distance d_{hkl} of the analyzer crystal is known, so that the wavelength of the radiation reflected at any angle θ can be determined with the aid of (52). By noting the intensity of the reflected beam at all possible Bragg angles, one thus obtains a complete spectral analysis of the x-rays emitted by the tube target. Such a spectrum, obtained at two different tube potentials, is shown in Fig. 7.

The acceleration, and hence the velocity and kinetic energy, of the electrons bombarding the target depend on the potential difference applied to the tube. Suppose that the tube target is a block of silver

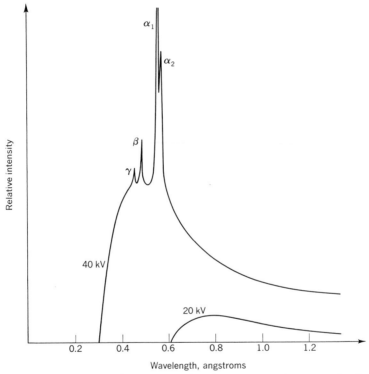

FIG. 7. X-ray spectrum of silver. Note that the spectrum of x-rays emitted with a 20-kV tube loading consists of a continuous distribution called the *continuous spectrum*. At the higher potential (upper curve) several *characteristic spectra* are seen superimposed on the continuous spectrum.

metal and the tube potential is 20 kV. The x-ray spectrum under these conditions is shown by the lower curve in Fig. 7. Next, suppose that the tube potential is increased to 40 kV. The x-ray spectrum then looks like the upper curve in Fig. 7. These spectra have several noteworthy features. At lower tube voltages, the spectrum has a uniformly varying intensity as a function of wavelength and is called the *continuous* spectrum. At higher voltages, peaks appear at specific wavelengths which are superimposed on the continuous spectrum. The wavelengths at which these peaks appear are different for each element used as a target, so that they are called *characteristic* spectra. Also note that the continuous spectrum has a sharp cutoff at the short-wavelength side, sometimes called the *short-wavelength limit,* and this limit moves to shorter wavelengths as the tube potential is increased. These features of the x-ray spectrum now can be explained by the quantum theory.

The continuous spectrum has the same appearance regardless of the metal used as a target in an x-ray tube. Thus the production of these x-rays must be a similar process in all materials. The necessary energy is supplied to the target by the electrons bombarding it. An electron of charge e accelerated by a potential V reaches the target with an energy equal to eV. One of three things can happen. Either it passes through the target without transferring any of its energy to the target atoms or it transfers part of its energy, or all its energy, to an atom. The first possibility is called electron scattering and is not unlike the case of x-ray scattering described in Chap. 2. The more likely situation, however, is one in which the electron interacts with an atom and loses energy of amount ΔE. This energy is then emitted by the electron in the form of an x-ray photon of frequency determined by $h\nu = \Delta E$. Since the incident electron can lose any amount of energy in such collisions, and generally collides with several atoms before coming to rest in the target, the emitted x-rays have all possible frequencies (wavelengths) and constitute the continuous spectrum. The maximum amount of energy that an electron can lose in a single collision cannot exceed its total kinetic energy, so that the largest frequency or shortest wavelength that the emitted x-rays can have is given by

$$eV = h\nu_{\max} = \frac{hc}{\lambda_{\mathrm{swl}}}. \tag{53}$$

The important point here is that the energy emitted is quantized, even though the quantum of energy transferred in each collision can vary from a maximum value of eV to a minimum value of zero. It follows from this that if the target is a very thin foil placed transversely to the electron beam, then the probability of an electron passing through it without interacting with an atom should increase. Moreover, the probability of multiple interactions should decrease even more, so that the x-ray spectrum from a thin target should show a maximum at the short-wavelength limit, corresponding to "head-on" collisions, and tail off rapidly at longer wavelengths. That this is indeed the case can be seen in Fig. 8. The continuous spectrum in Fig. 7, therefore, can be thought of as a superposition of many individual curves like Fig. 8. The maximum intensity corresponds to the most probable interaction, which is determined by the maximum in the electron energy-distribution curve. The position of this maximum, in turn, depends on the applied tube potential, so that the maximum in the continuous spectrum also shifts to shorter wavelengths with increasing potential.

There is still one more possible way in which an incident electron can be absorbed by an atom in the target. If the incident electron has sufficient energy, it can literally "knock out" one of the electrons bound to an atom. The energy required to knock out, say, a K electron is equal to the binding energy of that electron E_K. When the atom loses one of its two K electrons, it is said to be in a K *quantum state of the atom.* This is an excited state for the atom, so that it is quite likely that one of the other electrons in the atom "falls" into the vacancy in the K shell. If this electron is an L electron, then it leaves behind a vacancy in the L shell and the atom is said to undergo a transition to the L quantum state of the atom. If it is an M electron, the atom transfers to an M state,

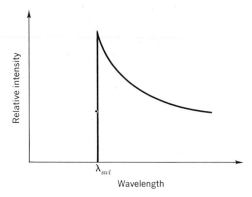

FIG. 8. X-ray spectrum of a thin target. The decline of the intensity is proportional to $1/\lambda^2$.

and so forth. Each such transition must be accompanied by the emission of a quantum of energy equal to the energy difference of the initial and final quantum states of the atom. Since the binding energies of the K, L, M, . . . electrons are different in each kind of atom, the energies, or frequencies, of the emitted photons are different and characteristic of the originating atom. In a classic experiment, Moseley showed, in 1913, that these frequencies are directly proportional to Z^2, and this dependence is called *Moseley's law.*

The origin of the characteristic spectra can be understood most easily by considering the energy-level diagram of an atom. Figure 9 shows such a diagram for silver. The horizontal lines indicate schematically the energy levels of the atom when a K, L, M, or N electron is missing. Note that there is only one K energy level, because both K electrons have the same energy. The three L levels arise from the three different energies of the atom following removal of a $2s$ electron, or a $2p$ electron having $m = 0$ or $m = \pm 1$. Similarly, there are five possible M levels and seven possible N levels. An atom undergoes a transition from its ground state, where no electrons are missing, to one of the higher quantum states, subsequent to the loss of an inner electron. Suppose a K electron is knocked out so that the atom is in its K state. It can then undergo a

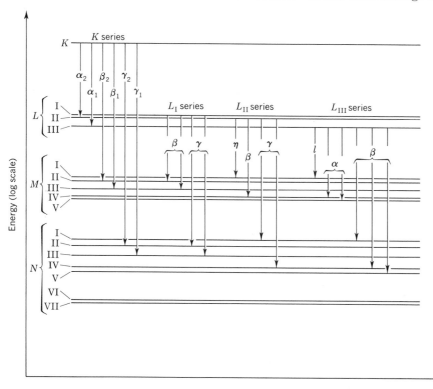

Fig. 9. Energy-level diagram of a silver atom. Note that the energy of the K state of the atom is larger than that of the other states because more energy is required to remove a K electron than any other electron in an atom. The transitions permitted by the selection rules are indicated by vertical arrows.

transition to a lower state by having an outer electron fall into the K vacancy. The x-ray photon emitted will have one of the following possible energies

$$
\begin{aligned}
E_K - E_{L_{II}} &= h\nu_{\alpha_2} \\
E_K - E_{L_{III}} &= h\nu_{\alpha_1} \\
E_K - E_{M_{II}} &= h\nu_{\beta_2} \\
E_K - E_{M_{III}} &= h\nu_{\beta_1} \\
E_K - E_{N_{II}} &= h\nu_{\gamma_2} \\
E_K - E_{N_{III}} &= h\nu_{\gamma_1}
\end{aligned}
\tag{54}
$$

depending on which outer electron falls into the K shell. The experimental observation of all the possible transitions listed in (54) therefore serves to confirm the electronic structure of atoms determined by quantum mechanics.

It is possible to conceive of other transitions in Fig. 9 also. The actual transitions that can occur are determined by *selection rules* derived from quantum mechanics. It turns out that transitions are allowed only when the quantum numbers of the quantum states differ by $\Delta l = \pm 1$ and $\Delta j = \Delta(|m| + m_s) = \pm 1$ or $\Delta j = 0$. Because the initial quantum state of the atom in (54) is the K state, the allowed transitions described in (54) constitute the K series of the characteristic spectrum. As can be seen in Fig. 7, the most probable transition is $E_K \rightarrow E_{L_{III}}$, so that the most intense spectral line is denoted $K\alpha_1$. The relatively large intensity of the $K\alpha$ lines is the reason why they are selected for x-ray diffraction studies requiring monochromatic radiation.

A similar situation arises when an L electron is knocked out of an atom. Since there are three different kinds of L electrons in an atom, there are three possible sets of allowed transitions shown in Fig. 9. The allowed transitions are determined by the selection rules cited above, and their probability of occurrence can be calculated by quantum-mechanical means. In each case, the energy (wavelength) of the emitted x-ray quantum is equal to the energy difference of the initial and final state of the atom (Exercise 14.) It is important to realize that in order for any spectral line to appear, it is first necessary to knock out an inner electron. Once such an electron is knocked out from an atom, the energy of the actual x-ray emitted is determined by the relative probabilities for the allowed transitions. For a target containing a large number of like atoms, this means that either the entire characteristic spectrum is produced in an x-ray tube or none of it.

Periodic table. One of the early triumphs of quantum mechanics was its ability to give a rational explanation for the periodic table of elements. Probably the most important contribution to this end was made by Pauli when he postulated the exclusion principle. As stated in the preceding section, Moseley discovered quite early that the characteristic x-ray spectra of elements regularly shifted to larger frequencies with increasing atomic number. The existence of regular periods in this sequence, however, cannot be understood until the detailed distribution of electrons among the allowed energy levels has been determined. The ground state of hydrogen is $1s$, described by $n = 1$, $m = l = 0$, and $m_s = \pm\frac{1}{2}$. The energy of this quantum state is not affected by the sign of the spin quantum number, so that either can be chosen for the one electron in hydrogen. When an atom contains two or more electrons, their distribution among the available states is no longer trivial. The two electrons in helium both can have $n = 1$, and $l = m = 0$ in the ground state $1s$. In the absence of an external field, their energies are the same regardless of their spins. Since each spinning electron has associated with it a magnetic moment, however, the atom must have a net magnetic moment if the spins are parallel, whereas the spin moments cancel if they are antiparallel. It is possible to distinguish these two cases, therefore, by placing a helium atom in a magnetic field and observing its interaction with the field (Zeeman effect). Repeated measurements have confirmed that a splitting of the ground state of helium does not occur, so that it must be concluded that the ground state contains two electrons with antiparallel spins. Additional measurements have similarly shown that any quantum state containing two electrons with parallel spins is unlikely. In the case of helium, the antiparallel spins of the two $1s$ electrons also account for the fact that helium is diamagnetic as discussed further in Chap. 14. Observations of this type led Pauli to conclude that the occupation of a single quantum state, defined by n, l, and m, by two electrons having parallel spins was statistically unlikely and constituted a forbidden state for an atom.

It is now possible to determine the distribution of electrons among the allowed states in all atoms. The ground state of an atom is defined as the one in which its electrons occupy the allowed states having the lowest energies. The relative energies of the various states are shown in Fig. 4, and the number of states having the same energy is indicated in Table 1. According to the Pauli exclusion principle, therefore, when $n = 1$, the $1s$ state can contain two electrons provided their spins are not parallel. For $n = 2$, the $2s$ state can contain two electrons having opposite spins and three $2p$ states, each of which can contain two electrons with opposite spins, and so on, for larger values of n. It is convenient to represent this diagrammatically by drawing a box for each allowed quantum state and by indicating the pair of electrons occupying the state by vertical arrows pointing in opposite directions. The s state is thus represented by one box, the p state by three boxes, the d state by five boxes, etc. Accordingly, the ground states of hydrogen and helium can be represented

$$\text{Hydrogen} \quad 1s\,\boxed{\uparrow} \quad \text{or} \quad 1s\,\boxed{\downarrow}$$

$$\text{Helium} \quad 1s^2\,\boxed{\uparrow\downarrow} \tag{55}$$

Note that the superscript 2 added to the spectrographic designation of the ground state of helium denotes that two electrons occupy the $1s$ state. (Similarly, the symbol $3p^4$ means that four electrons occupy the six available $3p$ states, and so forth.)

The next atom, lithium, has three electrons. According to Fig. 4, the lowest energy state available to these electrons is the 1s state.   In keeping with the Pauli exclusion principle, however, only two electrons can occupy this orbital.   Consequently, the third electron in lithium must occupy the next-highest energy state, which is 2s.   The diagrammatic representation of the ground state of lithium, therefore, is

$$1s^2 \boxed{\uparrow\downarrow}$$
$$2s^1 \boxed{\uparrow} \qquad 2p^0 \boxed{}\boxed{}\boxed{} \tag{56}$$

Following a similar procedure, the ground states of the next two elements can be represented as follows:

$$\text{Beryllium} \quad 1s^2 \boxed{\uparrow\downarrow}$$
$$2s^2 \boxed{\uparrow\downarrow} \qquad 2p^0 \boxed{}\boxed{}\boxed{}$$

$$\text{Boron} \quad 1s^2 \boxed{\uparrow\downarrow}$$
$$2s^2 \boxed{\uparrow\downarrow} \qquad 2p^1 \boxed{\uparrow}\boxed{}\boxed{} \tag{57}$$

The next element, carbon, has six electrons, of which two occupy the 1s state, two the 2s state, and two the 2p state.   These last two, however, can be distributed among the three boxes in essentially nine different ways:

$$\boxed{\uparrow\downarrow}\boxed{}\boxed{} \qquad \boxed{\uparrow}\boxed{\uparrow}\boxed{} \qquad \boxed{\uparrow}\boxed{}\boxed{\downarrow}$$
$$\boxed{}\boxed{\uparrow\downarrow}\boxed{} \qquad \boxed{\uparrow}\boxed{}\boxed{\uparrow} \qquad \boxed{\uparrow}\boxed{}\boxed{\downarrow} \tag{58}$$
$$\boxed{}\boxed{}\boxed{\uparrow\downarrow} \qquad \boxed{}\boxed{\uparrow}\boxed{\uparrow} \qquad \boxed{}\boxed{\uparrow}\boxed{\downarrow}$$

Note that the two spins are parallel in the central column and antiparallel in the left and right columns.   When the interaction between the electrons is taken into account, the energies of the three basically different arrangements can be shown to be different.   It turns out that the energy is lowest when the electrons in the same energy level have parallel spins.   This is sometimes called *Hund's rule* in quantum mechanics.   Accordingly, the electronic distribution in the ground state of carbon is

$$1s^2 \boxed{\uparrow\downarrow}$$
$$2s^2 \boxed{\uparrow\downarrow} \qquad 2p^2 \boxed{\uparrow}\boxed{\uparrow}\boxed{} \tag{59}$$

The diagrammatic representation of the electronic distribution for the next four atoms, N, O, F, Ne, is similar to (59), except that all three boxes of the p states are occupied by three parallel arrows in nitrogen, one set of paired and two parallel arrows in oxygen, all the way to three paired sets in neon.

The Bohr concept of electronic shells now can be given a new meaning.

The K shell can contain up to two electrons having opposite spins. These two electrons complete, or *close*, the K shell and are both equally tightly bound to the nucleus, so that they have the same energy. The L shell can contain two $2s$ electrons and six $2p$ electrons. Since the $2s$ and $2p$ energy levels differ slightly but not as much as, say, the $1s$ and $2s$ or $2s$ and $3s$ energy levels, it is reasonable to assign the $2s$ and $2p$ states to a common shell. Similarly, the M shell can contain two s electrons, six p electrons, and ten d electrons, for a total of eighteen electrons. In filling this shell, however, it should be noted in Fig. 4 that the $3d$ energy level is slightly higher than that of the $4s$ state. Thus, in the ground state of potassium, having a total of nineteen electrons, the $3p$ states are filled but the nineteenth electron occupies the $4s$ state of the N shell. This can be represented by the formula $1s^22s^22p^63s^23p^64s^1$, or diagrammatically by

$$1s^2\boxed{\uparrow\downarrow}$$
$$2s^2\boxed{\uparrow\downarrow}\quad 2p^6\boxed{\uparrow\downarrow|\uparrow\downarrow|\uparrow\downarrow}$$
$$3s^2\boxed{\uparrow\downarrow}\quad 3p^6\boxed{\uparrow\downarrow|\uparrow\downarrow|\uparrow\downarrow}\quad 3d^0\boxed{\ |\ |\ |\ |\ }\ .$$
$$4s^1\boxed{\uparrow}$$

$$(60)$$

Similarly, the twentieth electron in calcium also occupies the $4s$ state. The twenty-first electron in scandium, however, goes into the $3d$ state because the energy level of the $4p$ state is higher (Fig. 4). The progressive occupation of the upper energy levels in the ground states of potassium $(Z = 19)$ to zinc $(Z = 30)$ is shown in Table 2. The next elements, gallium $(Z = 31)$ to krypton $(Z = 36)$, have an identical electronic structure to zinc except that the $4p$ states progressively contain from one to six electrons.

Table 2
Electronic structure of elements in the first long period

Element	K	Ca	Sc	Ti	V	Cr	Mn	Fe	Co	Ni	Cu	Zn
Number of electrons in $3d$	0	0	1	2	3	5	5	6	7	8	10	10
Number of electrons in $4s$	1	2	2	2	2	1	2	2	2	2	1	2

An interesting feature of Table 2 is that the $3d$ states in scandium, titanium, and vanadium contain one, two, and three electrons, respectively, whereas the next element, chromium has five $3d$ electrons. This can be explained by the exchange interaction between unpaired electrons (Hund's rule), which favors the formation of closed *half-shells* containing electrons with parallel spins. Thus when the $3d$ half-shell contains four electrons, one of the 4*s* electrons is "demoted" to a 3*d* state in order to close the half-shell. The energy required to demote a 4*s* electron decreases from 2.5 eV in calcium to 0.3 eV in vanadium and again from 2.1 eV in manganese to 0.02 eV in nickel. This energy difference reflects the energy separations of the 3*d* and 4*s* energy levels in these atoms. A similar demotion occurs in copper, in which the 3*d* states are completely filled but the 4*s* state is half empty. This suggests that the binding energy should be similar in chromium and copper. This is confirmed by

H 1	He 2

Li 3	Be 4	B 5	C 6	N 7	O 8	F 9	Ne 10
Na 11	Mg 12	Al 13	Si 14	P 15	S 16	Cl 17	A 18

K 19	Ca 20	Sc 21	Ti 22	V 23	Cr 24	Mn 25	Fe 26	Co 27	Ni 28	Cu 29	Zn 30	Ga 31	Ge 32	As 33	Se 34	Br 35	Kr 36
Rb 37	Sr 38	Y 39	Zr 40	Nb 41	Mo 42	Tc 43	Ru 44	Rh 45	Pd 46	Ag 47	Cd 48	In 49	Sn 50	Sb 51	Te 52	I 53	Xe 54
Cs 55	Ba 56	La 57 †	Hf 72	Ta 73	W 74	Re 75	Os 76	Ir 77	Pt 78	Au 79	Hg 80	Tl 81	Pb 82	Bi 83	Po 84	At 85	Ry 86
Fr 87	Ra 88	Ac 89	Th 90	Pa 91	U 92	Np 93	Pu 94	Am 95	Cm 96	Bk 97	Cf 98						

† Rare earth metals	Ce 58	Pr 59	Nd 60	Pm 61	Sm 62	Eu 63	Gd 64	Tb 65	Dy 66	Ho 67	Er 68	Tm 69	Yb 70	Lu 71

FIG. 10

the heats of sublimation for Cr and Cu, which are 80.0 and 80.3 kg-cal/mole, respectively.

The above discussion illustrates how quantum mechanics explains the occurrence of regular periods in the periodic table shown in Fig. 10 (see also Appendix 4). The periods simply reflect the filling of a particular shell. Thus the first period starts with hydrogen having one electron in the *K* shell and ends with helium having a closed *K* shell containing two electrons. The second period starts with lithium having one *L* electron and terminates with neon having a closed *L* shell. Similarly, the third period starts with sodium having one *M* electron. The last atom in this period, argon, "closes" the *M* shell because the next electron added (in potassium) starts the occupation of the *N* shell, and so on. Note that atoms having similar outer electron distributions lie above each other in the periodic tables. This similarity in the elements lying in the same column of the periodic table is reflected in the similarity of their properties. The atoms having such similar outer electronic structures are connected by broken lines in Fig. 10. Thus the inert gases lying in the extreme right-hand column, He, Ne, A, . . . , all have closed shells formed by s^2p^6 states. Similarly, the elements lying in the first column, H, Li, Na, . . . , have one s electron lying in a newly started shell; the elements in the second column, Be, Mg, Ca, . . . , have two s electrons beyond the last closed shell, and so on. Note that Cu, Ag, Au, . . . also have one s electron in a newly started shell, while Zn, Cd, Hg, . . . have two s electrons and are therefore similar in this respect to the elements in the first two columns. These parallelisms express themselves in the properties of the atoms. For example, the normal valence of the alkali metals Li, Na, K, . . . and of Cu, Ag, Au, . . . is $+1$. The binding energies of the s electrons are also similar, as can be seen from their first ionization potentials:

Element	Li	Na	K	Cu	Ag
First ionization potential	5.36	5.13	4.32	7.68	7.54 eV

By comparison, the first ionization potentials of He, Ne, A, . . . are much higher, 24.46, 21.47, 15.68 eV, respectively.

Another important similarity between the atoms in the first long row of the periodic tables should be noted. As shown in Table 2, the elements scandium to nickel in the first row have partly filled $3d$ and $4s$ states. The comparative ease with which a $4s$ electron can be demoted to a $3d$ state is a consequence of the fact that both kinds of electrons are bound to the nucleus with comparable binding forces. This can be verified by comparing the ionization potentials of some of these elements:

Element	Sc	Ti	Mn	Fe	Co
$4s$ ionization potential	6.76	6.76	7.43	7.84	7.92 eV
$3d$ ionization potential	7.29	6.88	9.16	8.11	8.51 eV

This suggests that these *transition elements* exhibit unusual properties. It turns out that they can have more than one stable valence and otherwise behave in unusual ways in different materials, as discussed further in later chapters. Similar transition series, obviously, occur in the second and third long periods in the periodic table also.

Suggestions for supplementary reading

W. Heitler, *Elementary wave mechanics*, 2d ed. (Oxford University Press, London, 1956).

Robert B. Leighton, *Principles of modern physics* (McGraw-Hill Book Company, Inc., New York, 1959).

F. K. Richtmeyer, E. H. Kennard, and T. Lauritsen, *Introduction to modern physics*, 5th ed. (McGraw-Hill Book Company, Inc., New York, 1955).

Suggestions for further reading

David Bohm, *Quantum theory* (Prentice-Hall, Inc., Englewood Cliffs, N.J., 1951).
John C. Slater, *Quantum theory of matter* (McGraw-Hill Book Company, Inc., New York, 1951).
John C. Slater, *Quantum theory of atomic structure*, vol. 1 (McGraw-Hill Book Company, Inc., New York, 1960).

Exercises

1. Equation (9) states that the energy of an electron in the first excited state of a hydrogen atom ($n = 2$) has a magnitude equal to one-fourth the magnitude of the ground-state energy. How is this made compatible with the fact that the energy of an atom in an excited state is greater than its energy in the ground state?

2. The translational kinetic energy of a gas molecule is equal to $\frac{3}{2}kT$, where k is the Boltzmann constant and T is the temperature measured on the absolute scale. What is the de Broglie wavelength of a hydrogen atom at a temperature of 27 deg C? (ANSWER: 1.49 Å.)

3. The usual operating voltage of an electron microscope is 50 kV. What is the de Broglie wavelength of an electron accelerated by such a potential?

4. Suppose a photographic plate consists of sensitive square grains, 10,000 Å on a side, and an exposure of 10^{-8} J/cm² is required to produce noticeable blackening for light whose wavelength is 5,000 Å. How many photons must fall on each grain to produce such blackening?

5. The mass of a hydrogen nucleus (proton) is 1.67×10^{-27} kg. Suppose a certain excited state of a hydrogen atom is known to have a lifetime of 2.5×10^{-14} sec, what is the minimum error with which the energy of the excited state can be measured? If the center of a hydrogen atom can be located with a precision of 0.01 Å, what is the corresponding uncertainty in its velocity? (ANSWER: 4.22×10^{-19} J; 6.39×10^{4} m/sec.)

6. What is the uncertainty in the velocity of an automobile having a mass of 2 tons when its center of mass is located with an uncertainty no greater than 1 mm? (ANSWER: 1.163×10^{-30} m/sec = 2.62×10^{-34} mph.)

7. Starting with the general equation (12) for a vibrating string, derive the time-independent form (13). To do this, assume a solution of the form $U(x,t) = u(x) \sin \omega t$.

8. The Schrödinger equation (32) for a hydrogen atom can be solved by assuming solutions of the form $\psi = xf(r)$. Show that this leads to an equation

$$\frac{\partial^2 f}{\partial r^2} + \frac{4}{r}\frac{\partial f}{\partial r} + \frac{8\pi^2 m}{h^2}\left(E + \frac{e^2}{4\pi\epsilon_0 r}\right)f = 0$$

which can be solved similarly to Eq. (41). The wave function $\psi_x = xf(r)$ is not spherically symmetric, and a similar result can be obtained by replacing x by y or z. The three wave functions ψ_x, ψ_y, and ψ_z represent three possible solutions of (33) having the same energy and illustrate the case of three-fold degeneracy.

9. The frequencies ν of the spectral lines emitted by hydrogen were observed by Balmer to obey the relation

$$\nu = \left(\frac{1}{m^2} - \frac{1}{n^2}\right) \times \text{constant}$$

where m and n are any two integers. Derive this formula from the quantum theory.

10. The wave function $\psi(x,t) = e^{(-2\pi i/\lambda)x}e^{2\pi i\nu t}$ is a possible solution of the time-dependent Schrödinger equation. Assume two solutions having slightly different frequencies and wavelengths; that is, $1/\lambda = 1/\lambda_0 \pm d(1/\lambda)$ and $\nu = \nu_0 \pm d\nu$. Show that the sum of such waves is an "average" wave having an amplitude of 2 when $x = d\nu/d(1/\lambda)t$. The quantity $d\nu/d(1/\lambda) = v_g$ is the group velocity of this average wave; that is, the wave travels along x with this velocity without changing its form. An analogous process can be used to construct the wave packets described in the text.

11. The absorption of x-rays by the atoms in a material can be expressed by an absorption coefficient μ. Assuming that this absorption process consists entirely of knocking out the inner electrons in the atoms and that $\mu \propto \lambda^3$, prepare a schematic plot of the absorption coefficient as a function of wavelength. (When the energy of the incident x-rays is less than that required to knock out a K electron, μ decreases discontinuously and then resumes the λ^3 dependence.)

12. The discontinuity in μ described in Exercise 11 is called the K *absorption edge*. Similar discontinuities occur at the L edge. How many actual discontinuities do you expect to observe in the L edge? The M edge? Why?

13. In order to produce the $K\alpha$ lines of copper ($\lambda_{K\alpha} = 1.54$ Å), it is necessary that the energy used to excite a copper atom exceed a certain value. This energy corresponds to the K absorption edge (Exercise 11) of copper. If the wavelength at this edge is 1.377 Å, what is the minimum potential that can be applied to the x-ray tube in order to observe the $K\alpha$ lines of copper? In order to observe the $K\beta$ line of copper? ($\lambda_{K\beta} = 1.39$ Å for copper.) (ANSWER: 8.998 kV.)

14. Making use of the transitions indicated in Fig. 9, determine the energies and frequencies of the L spectra of silver. Do this by writing relations like those in (54) in the text. The three subsets are usually denoted the L_I, L_II, and L_III spectra of silver, depending on the initial states. Which of these three series lies at longer wavelengths? How does the wavelength of the $L\alpha$ lines compare with that of $K\alpha$ in silver?

15. The discontinuous behavior of the absorption coefficient (Exercise 11) can be utilized to absorb the characteristic spectra of an element having certain wavelengths while transmitting others. Thus a metal foil of one element can be used to absorb the $K\beta$ radiation emanating from the target of another element while transmitting the $K\alpha$ rays with very little attenuation. Making use of Moseley's law, should the metal foil used as such an *x-ray filter* consist of a lighter or heavier element than the target?

16. Prepare a table showing the distribution of electrons in the allowed energy states for the atoms in the second long period in the periodic table (rubidium to xenon). All these atoms have the so-called krypton core, which need not be repeated each time.

4. atomic bonding

There is a fundamental law of nature which states that the total energy of a system, having reached equilibrium with respect to its environment, tends toward a minimum value. This means that it is possible for the constituents in a system to have individual energies that are higher than their respective ground states provided that the assembly as a whole is in its minimum-energy state. In the case of atomic assemblies, this can happen when the interaction energies between atoms in excited states is sufficiently greater (more negative) than that between the same atoms in their ground states. It is not surprising, therefore, that a particular kind of atom can exist in more than one energy state in different materials, depending on the nature of its interaction with its neighboring atoms. The nature of this interaction is said to determine the type of bond formed between the atoms. In the absence of such a bond, obviously, the atoms will not stay together to form a solid material.

There are several possible ways that the bonding between atoms can be classified. The most informative, and therefore most commonly used, classification is based on the detailed interactions between an atom and its nearest neighbors. These are called *primary bonds*, and essentially three different kinds, the *ionic*, *covalent*, and *metallic* bonds, can be distinguished. In addition, there are *secondary bonds* possible, the most important being the *van der Waals* bond. These four bond types and the resulting forces exerted by atoms on their neighbors are considered below. Making use of quantum mechanics, the origins of the primary-bond types can be understood in the following way. As discussed in the preceding chapter, Hund's rule favors closed shells or half-shells in atoms. Thus the atoms lying at the two extreme sides of the periodic table tend to lose or gain valence electrons, so that the resulting *ions* have completed outer shells. The elements in the first or second columns of the periodic table lose electrons, so that they form positive ions called *cations*. The elements in columns 6 and 7 tend to gain electrons, so that they become negatively charged ions called *anions*. Such oppositely charged ions can attract each other, and this electrostatic interaction is called an ionic bond.

Atoms lying in the central columns of the periodic table similarly tend to form closed half-shells. In the case of carbon, for example, this means that the electronic configuration can be $1s^2 2s 2p^3$ instead of the ground-state configuration $1s^2 2s^2 2p^2$. The resulting four unpaired electrons in a carbon atom can then pair up with similarly unpaired electrons in other atoms to form a covalent bond. As discussed further in subsequent sections of this chapter, the bonding in many solids is seldom wholly ionic or covalent in nature, but rather is a mixture of both, with one or the other bond type predominating. A somewhat different "crossing" between these two bond types occurs in metals. The closing of the outer shells in metal atoms, particularly of those lying in the first and second columns of the periodic table, takes place by a sharing of the valence electrons by all the atoms in the metal instead of between nearest neighbors only. Thus the interactions causing a metallic bond to form arise from forces between the positive metal ions and the collectively shared negative electrons. Finally, the atoms lying in the last column of the periodic table already have closed outer shells in their ground state. Consequently, these atoms do not readily bond to each other (or other atoms) and are called inert gases for that reason. Under very high pressures and at extremely low temperatures, however, they can form solids. The interactions between the inert-gas atoms consist of very weak electrostatic attractions between the shifting electric fields produced by the moving electrons in each atom and are said to form a van der Waals bond. Similar interactions are possible between ions in ionic compounds and, in particular, between molecules in organic materials, as discussed further below.

Bond types

Forces between atoms. In addition to the attractive interactions mentioned above, repulsive interactions between atoms also exist. Under equilibrium conditions, obviously, the attractive forces and repulsive forces must balance each other. Before proceeding to a detailed consideration of the bond types that are encountered in actual materials, the general nature of interatomic forces is considered first. The potential, or stored, internal energy of a material is a sum of the individual energies of the atoms plus their interaction energy. Since the zero of potential energy can be chosen arbitrarily, let $V = 0$ when the atoms are all in their ground state and infinitely far apart so that they do not interact with each other. In order to form a solid, these atoms must be brought together, and if the resulting assembly is to be more stable than the initially dispersed one, the potential energy of the assembly V must be less than zero. This means that the total interaction energy must be negative and that it should be inversely proportional to some power of the interatomic separation r. There are essentially two interactions possible when the atoms are brought together. Either they attract each

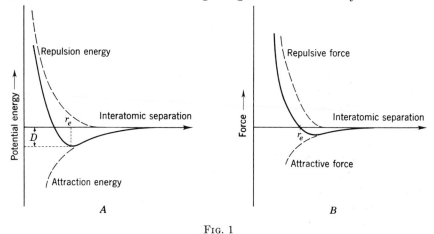

FIG. 1

other or they repel each other. The potential energy of attraction is negative because the atoms do the work of attraction. It is inversely proportional to some power of r, so that $V_{\text{attract.}} = -\alpha/r^n$, where α is the proportionality constant. At large separations, this term is zero, and it decreases gradually, as shown by the lowermost dashed curve in Fig. 1A. Conversely, the potential energy of repulsion is positive because external work must be done to bring the atoms together when they repel each other. Normally, the repulsion between two atoms is not significant until they are so close to each other that the outer electrons of each atom can interact. Its magnitude is also inversely proportional to some power of r (but not necessarily the same power), so that $V_{\text{repuls.}} = \beta/r^m$. It increases more rapidly (upper dashed curve in Fig. 1A) than the attractive potential because the positive charges on the nuclei repel each other very strongly when r becomes small. The net potential energy is the sum of both terms,

$$V = V_{\text{attract.}} + V_{\text{repuls.}}$$

$$= -\frac{\alpha}{r^n} + \frac{\beta}{r^m} \tag{1}$$

and is represented in Fig. 1A by the solid curve.   (The physical basis for the shapes of these curves is discussed in subsequent sections.)

The forces between the atoms can be derived directly from (1) by recalling that the force is the derivative of potential energy.

$$F = -\frac{dV}{dr} = -\frac{n\alpha}{r^{n+1}} + \frac{m\beta}{r^{m+1}}.$$
(2)

The form of (2) is similar to (1), and so is the plot of force as a function of interatomic separation shown in Fig. 1B.

The solid curves in Fig. 1 have the following meaning.   At large separations, the atoms do not interact with each other, so that $V = 0$ and $F = 0$.   As the atoms approach each other, they exert attractive forces on each other, primarily because of the attraction of positive and negative charges in the atoms.   As the interatomic separation decreases to the order of one to two atomic diameters, the repulsive forces come into effect and tend to increase the potential energy of the system and to prevent further decrease of the interatomic separation.   At some separation, called the equilibrium separation r_e, the attractive and repulsive forces just balance, so that F in (2) is equal to zero, and the assembly is at its minimum energy.   Any further decrease in separation causes the potential energy to increase until at some slightly smaller value than r_e it becomes positive and the assembly becomes unstable.   This is the reason why two atoms cannot interpenetrate each other, a reassertion of the old maxim that two objects cannot occupy the same space at the same time.

Ionic bond.   The foregoing discussion is quite general and does not take into account the detailed nature of the charge distributions in the different atoms.   As stated earlier, an atom can become either a positively or a negatively charged ion as a result of the loss or gain of the number of valence electrons required to give it a closed outer shell.   Thus ions have the same electronic structure as the nearest inert-gas atoms, except that they are electrically charged.   This charge is spherically distributed and represents the difference between the charge on the nucleus and the sum of the electronic charges surrounding it.   An ion, therefore, is not unlike a small charged sphere, and the packing of ions follows the rules for closest packings of spheres discussed in Chap. 1.   When oppositely charged ions are brought together, each one tends to neutralize its charge by surrounding itself with ions having an opposite charge.   Because ions having like charges repel each other, the stable packing attained is determined by the relative sizes of the ions and their respective charges. A periodic array results, therefore, in which the environment of all similar atoms is the same and the sum of all positive and negative charges adds up to zero, so that the ionic solid is electrically neutral.

The compounds formed by the elements in the first column (alkali metals) and in the seventh column (halogens) are probably the best examples of a truly ionic bond. These alkali halides crystallize with either the halite structure shown in Fig. 2 or the CsCl structure shown in Fig. 3. The particular structure adopted is determined by the radius ratio of the two ions as discussed in Chap. 1. Note that, as a consequence of the spherical charge distribution in ions, the ionic bond is nondirectional. Thus, for example, a potassium ion can form ionic bonds with

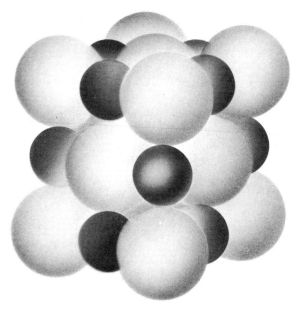

Fig. 2. Halite, NaCl, structure. It consists of a cubic closest packing of the larger anions in which all the octahedral voids are occupied by the smaller cations.

the 6 chlorine ions in KCl (Fig. 2) or with as many as 12 anions in more complicated structures. In each case, the *strength of the electrostatic bond* formed is determined by the *valency* of the ion z divided by the number of ions surrounding it as determined by its coordination number CN. Accordingly,

$$S = \frac{z}{CN} \tag{3}$$

and the electrostatic bond that is formed is Se, where e is the charge of one electron. Similarly, the sum of the bonds reaching an anion $-\zeta e$ must equal the electric charge of the anion. If the anion is surrounded by j cations and the strength of the electrostatic bond joining the ith cation is S_i, then

$$\zeta = \sum_{i=1}^{j} S_i. \tag{4}$$

As an illustration of this principle of overall charge neutrality, consider the compound $MgAl_2O_4$. The valency of each magnesium ion $z = +2$,

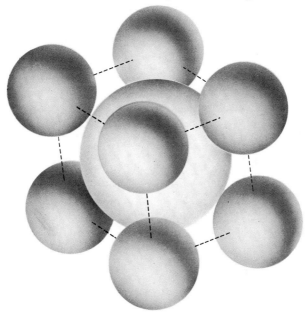

Fig. 3. CsCl structure. Note that this is a primitive cubic structure containing two atoms (Cs and Cl) per lattice point and should not be confused with the body-centered cubic packing of like atoms shown on page 20.

and it occupies tetrahedral voids in a cubic closest packing of O^{2-} ions. According to (3),

$$S_{Mg} = \frac{+2}{4} = +\frac{1}{2}. \tag{5}$$

Each aluminum atom has $z = +3$ and occupies octahedral voids, so that

$$S_{Al} = \frac{+3}{6} = +\frac{1}{2}. \tag{6}$$

Each of the closest-packed oxygen atoms, $z = -2$, receives four bonds like (5) and (6), so that

$$\zeta = \frac{1}{2} + \frac{1}{2} + \frac{1}{2} + \frac{1}{2} = +2 \tag{7}$$

and charge neutrality is preserved.   This analysis is diagrammatically illustrated in Fig. 4.

The ionic bond is fairly strong, as attested to by the amount of work required to dissociate an ionic solid into its component atoms.   The magnitude of the binding energy in NaCl is 7.8 eV, and in LiF it is 10.4 eV.   This strong bonding means that ionic crystals are hard and have relatively high melting points.   The valence electrons are also bound quite tightly to the ionic nuclei, so that electrical conductivity via electrons is not possible and ionic crystals are insulators at room temperature.   At elevated temperatures, however, the ions themselves become more mobile and ionic conductivity becomes possible, as discussed further in Chap. 12.

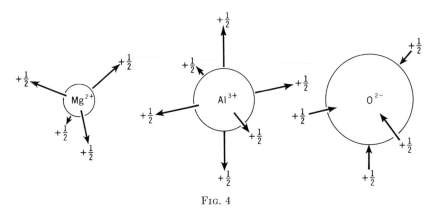

Fig. 4

Although a typical ionic bond is nondirectional, there are three special cases of ionic bonds which have some directional character.   These are as follows:

1. The *hydrogen bond* is formed by a hydrogen ion located between two anions.   Since hydrogen has only one electron, it can lose it to either of the two adjoining ions, with the result that there is an equal probability of finding the electron on either ion.   The positive hydrogen ion tends to draw the two anions more closely together than their normal separation in crystals, so that such a shortening of their interatomic separation serves to indicate the presence of a hydrogen bond.   Recent careful diffraction measurements have shown that the hydrogen ion is not located halfway between the two anions, but rather occupies either of two lower energy positions, each respectively one-third of the way along the interionic separation.

2. The *hydroxyl bond* is a special case of the hydrogen bond in which the H^{1+} ion attaches itself to an oxygen ion to form an $(OH)^{1-}$ ion.   The hydroxyl ion is pictured as containing four charges, each of magnitude

$\frac{1}{2}e$, disposed at the corners of a tetrahedron. Charge neutrality requires that three charges be negative and one positive. Two of the negative charges usually are directed toward a cation, whereas one positive and one negative charge serve to bond adjacent hydroxyl ions in a continuous manner. A typical hydroxide structure consists of hydroxyl layers, in which the hydroxyl ions are bonded to each other, and cations located on either side of such layers. The hydroxyl bond, therefore, is a directed bond between adjacent OH^{1-} ions. It can be distinguished from the hydrogen bond by comparing the resulting O-O separations:

	Hydrogen bond	Hydroxyl bond	
O-O separation	2.54 Å	2.70 Å	(8)

By comparison, in the absence of either type of bond, the interoxygen distance in most oxide crystals is about 3.20 Å.

3. The *ion-dipole bond* is found to occur in certain complex ions, such as $Fe(H_2O)_6^{3+}$, for example. The electrically neutral water molecule actually contains two positive and two negative charges of magnitude $\frac{1}{2}e$ tetrahedrally arrayed as in the hydroxyl ion. (This is the reason why H_2O is a liquid and not a gas at room temperature.) The excess positive charge on the central Fe^{3+} cation attracts the two negative half charges, so that the H_2O molecules become electric dipoles. The resulting $Fe(H_2O)_6^{3+}$ complex thus is simply a means by which the $3+$ charge on the central ion can be distributed over a larger surface formed by the surrounding water molecules. Such structurally bound water is called *water of crystallization* and cannot be driven off the way interstitially absorbed water molecules can.

Covalent bond. The tendency of an atom to have a closed outer shell can be satisfied in another way also. The atoms lying on the extreme right side of the periodic table lack one or more electrons from having a closed outer shell. According to Hund's rule, such unpaired electrons occupy separate orbitals and have parallel spins. Suppose two such like atoms are brought together until the orbitals of one unpaired electron in each of the atoms begin to overlap. If the two electrons have, respectively, antiparallel spins, they attract each other and form an *electron-pair bond*. Since the valence of both atoms forming this bond is the same, it is also called the covalent bond. This is illustrated for two chlorine atoms below.

$$:\overset{..}{\underset{..}{Cl}}\cdot \; + \; \cdot\overset{..}{\underset{..}{Cl}}: \longrightarrow \; :\overset{..}{\underset{..}{Cl}}:\overset{..}{\underset{..}{Cl}}: \tag{9}$$

Note that the two electrons lying between the two chlorine atoms are shared equally by both atoms and serve to complete the outer shells of both atoms. If the two unpaired electrons on the left of (9) have parallel spins, however, such bonding cannot take place without violating the Pauli exclusion principle. The energies for both cases, as a function of interatomic separation, are plotted in Fig. 5. The quantum-mechanical analysis of covalent bonding is deferred to a later section.

Unlike the ionic bond, the covalent bond is strongly directional since the electron-pair bond is formed along a line joining the two atoms whose orbitals overlap. Furthermore, the number of electron-pair bonds that an atom can form is obviously limited by the number of unpaired electrons it has. This number is determined by the 8-N rule, where N is the number of the column in the periodic table containing the atom. Thus fluorine can form one bond, oxygen can form two bonds, nitrogen three, and so forth. The bonds are formed by an overlap of the $2p$ orbitals of the unpaired electrons in these atoms, which are directed along three mutually perpendicular directions as illustrated on page 73. Now according to the 8-N rule, carbon can form four electron-pair bonds, even though there are only two unpaired electrons in the ground state of carbon $1s^2 2s^2 2p^2$. As already noted, this becomes possible when one of the $2s$ electrons is excited to a $2p$ state, giving four unpaired electrons in the sp^3 hybrid state. The higher energy of the carbon atom is then compensated in a crystal by an increase in the negative bonding energy result-

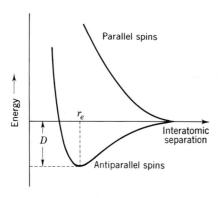

Fig. 5. Relative energies of bonding orbitals (lower curve) and antibonding orbitals (upper curve) when the two unpaired electrons have antiparallel and parallel spins, respectively. The minimum in the bonding energy occurs at the equilibrium interatomic separation r_e, and D represents the dissociation energy or binding energy of the molecule.

ing from the formation of four instead of two bonds. The four orbitals of the sp^3 hybrid are elliptical and are directed toward the four corners of a tetrahedron. Thus each carbon atom in a diamond crystal is tetrahedrally coordinated by four other carbon atoms. It should be noted that other hybrid bonds are also possible, for example, the dsp^2 hybrid formed by copper in tenorite, CuO, in which the four bonds are coplanar and directed toward the corners of a square. Hybrids containing five bonds can be formed, for example, sp^3d and spd^3 hybrids occurring in $MoCl_5$, $NbCl_5$, and $NbBr_5$, in which the five bonds are directed toward the corners of a trigonal bipyramid.

As might be expected, the covalent bond is also a very strong bond. The binding energies of carbon in diamond and of SiC are 7.4 and 12.3 eV, respectively. The cohesive strength of diamond is well known, and the melting points of covalently bonded crystals are correspondingly high. Because the valence electrons are localized in the electron-pair bonds by fairly strong forces, electrical conductivity is normally not possible and such crystals are insulators at ordinary temperatures. Ionic conductivity at elevated temperatures is similarly not possible. As discussed in later chapters, however, the electronic binding forces in some covalent crystals are weaker than in others, so that a limited amount of electronic conductivity is possible in such crystals at temperatures sufficiently high to free some of the valence electrons.

Metallic bond. The elements lying on the left side of the periodic table have one or more valence electrons lying outside a closed shell. The closed shell serves to screen the positive nucleus from these electrons, so that such atoms lose their valence electrons quite easily and become positive ions. A comparison of the observed first ionization potentials of a number of elements in Table 1 verifies this. Note that the ionization potential gradually increases on going from left to right along a row of the periodic table, say, potassium to krypton. (The discontinuity following zinc is a consequence of the filling of $4p$ states in gallium.) The approximate radial distances to the maximum charge density of the valence-electron orbitals are also listed in Table 1. As expected, the radial distances are inversely proportional to the ionization potentials, signifying that the weaker the nuclear attraction becomes, the larger the outer electron orbitals.

When two metal atoms, say, two sodium atoms ($Z = 11$), approach each other, the $3s$ orbitals begin to overlap at comparatively large interatomic separations. As already discussed, if the spins of the two unpaired electrons are antiparallel, such overlap can lead to the formation of an electron-pair bond. Suppose a third sodium atom, also having an unpaired $3s$ electron, approaches such a pair. According to the Pauli exclusion principle, such an atom must be repelled because both $3s$ quantum states are already occupied. It turns out, however, that the energy levels of the $3p$ states in sodium are sufficiently close to the $3s$ levels, so that the third electron can be easily excited into the higher energy state. (As discussed in subsequent chapters, this is not really necessary, because when a large number of atoms are brought together to form a solid, the energy levels of the outer electrons become *bands* containing barely separated energy states whose total number exactly equals the number of such states in the same number of isolated atoms.) Thus the third electron can occupy a different quantum state without violating the Pauli exclusion principle. Consequently, it can also form an electron-pair bond with either of the two other valence electrons. In fact, it turns out that a fairly large number of atoms can thus surround a single

Table 1
Properties of valence electrons

Element	First ionization potential, eV	Radius of electron orbital, Å
H	13.53	0.53
Al	5.96	1.21
Si	8.12	1.06
P	8.75	0.92
S	10.30	0.82
Cl	12.93	0.75
A	15.68	0.67
K	4.32	2.20
Ca	6.09	2.03
Sc	6.70	1.80
Ti	6.81	1.66
V	6.71	1.52
Cr	6.74	1.41
Mn	7.41	1.31
Fe	7.83	1.22
Co	7.81	1.14
Ni	7.61	1.07
Cu	7.68	1.03
Zn	9.36	0.97
Ga	5.97	1.13
Ge	8.09	1.06
As	10.50	1.01
Se	8.70	0.95
Br	11.80	0.90
Kr	13.93	0.86

sodium atom. Since the central atom has only one unpaired electron, this electron must "take turns," forming electron-pair bonds with each of the surrounding atoms. Consequently, it forms less than a whole electron-pair bond with each neighbor, a situation that can be described as an *unsaturated* covalent bond. Solid sodium has a body-centered cubic structure, so that, on the average, each sodium atom forms one-eighth of an electron-pair bond, or one-quarter of an *electron bond,* with each of its neighbors.

The above described sharing of valence electrons by metal atoms leads to the following picture of the metallic bond. The atoms, bereft of their valence electrons, are positive ions. They are surrounded by the negative electrons, which can be thought of as constituting an electron "gas" which permeates the periodic array of positive ions. The entire array is held together by electrostatic attractions between the positive metal ions and the negative electron gas. Such a *free-electron model* for metals was actually proposed by Drude at the beginning of this century, even before the origins of this model were clearly understood. This model is eminently successful in explaining qualitatively typical metallic properties. For example, the ability of metals to form alloys, which violate Dalton's rule of simple integral ratios for the constituents of a compound, is a direct result of their ability to share electrons in virtually any ratio. Most alloys, therefore, are simple solid solutions in which the replacement of an atom having a fixed number of valence electrons by another atom having a different number of valence electrons does not appreciably alter the bonding situation. Detailed variations are observed, of course, and these can be explained by considering all the electrons collectively by statistical methods as discussed further in Chap. 6.

The metallic bond, therefore, is an unsaturated covalent bond and resembles, somewhat, the ionic bond. It is obviously not as strong as either; the binding energies of sodium and aluminum, for example, are 1.13 and 3.23 eV, respectively. In the transition metals, the bonds may involve hybrid orbitals since the unpaired electrons lie in d and s orbitals. Such hybridization has the effect of increasing the bond strengths, so that the binding energies in transition metals are somewhat higher (4.0 eV in iron and 4.6 eV in cobalt). Thus the metallic bond ranges from the free-electron model in alkali metals, having only one valence electron, to nearly covalent bonding in metals having large numbers of valence electrons. The effect that this has on the properties of metals is discussed in Chap. 6.

Van der Waals bond. The van der Waals attraction was satisfactorily explained for electrically neutral gas molecules by Debye, who postulated that the moving electric charges in such molecules create instantaneous electric dipoles which can interact with each other. For example, consider the HF molecule consisting of a hydrogen atom firmly attached to one side of a fluorine atom by an electron-pair bond. Because both positive nuclei are not located at the effective center of the molecule, the distribution of negative and positive charges is no longer homogeneous. Thus the HF molecule has associated with it a dipole field which can interact with similar dipoles in other HF molecules. This interaction force is relatively weak and is inversely proportional to the seventh power of the intermolecular separation. The larger the molecule, the larger the total interaction forces, so that small molecules are gases at room temperature, medium-size molecules are liquids, and large molecules can form solids. Organic solids are of this type and are characterized by their low melting points and softness.

The importance of van der Waals forces in other kinds of solids was not recognized until much later, when it was discovered that the cohesive energies of ionic solids could not be fully explained unless a small van der Waals contribution to the energy was included. It comes about, similarly to the case of molecular solids, because of small variations in the electro-static fields of the ions. As discussed in a later section, the primary electrostatic attraction between ions is inversely proportional to the first power of the interatomic separation, so that the van der Waals contribution is an almost negligible quantity in such solids.

Cohesion in crystals

Quantum-mechanical approach. In order to gain an insight into the quantum-mechanical calculations required to determine the cohesive or binding energies in solids, consider a molecule formed by joining together a hydrogen ion with a hydrogen atom. This results in a hydrogen molecule-ion consisting of two nuclei each having a charge $+e$ and one electron of charge $-e$. Labeling the two nuclei A and B and the respective distances separating them from the electron r_A and r_B, the potential energy of the electron moving in the field of both nuclei is

$$V = -\frac{e^2}{4\pi\epsilon_0 r_A} - \frac{e^2}{4\pi\epsilon_0 r_B}. \tag{10}$$

Substituting this value for the potential energy in the Schrödinger equation, it is possible to solve this problem exactly, similarly to the case of the hydrogen atom discussed in Chap. 3. It will be recalled that the ground-state solution in that case was $\psi(r) = e^{-ar}$, where $a = \pi m e^2/\epsilon_0 h^2$. It follows, therefore, that the wave function correctly describing the motion of the electron very close to nucleus A of the hydrogen molecule-ion has the form $\psi_A = e^{-ar_A}$; and very close to nucleus B, $\psi_B = e^{-ar_B}$. The proper wave function for the electron moving in the presence of both nuclei must reduce to either of these two cases in their vicinity. One way to meet this requirement is to form a linear combination of these two functions, for example,

$$\psi_+ = \psi_A + \psi_B = e^{-ar_A} + e^{-ar_B}$$
or
$$\psi_- = \psi_A - \psi_B = e^{-ar_A} - e^{-ar_B}. \tag{11}$$

It can be shown that wave functions like (11) actually do constitute solutions of the Schrödinger equation for the hydrogen molecule-ion. The energies of the two solutions in (11) are not the same, however, and depend on whether the plus or minus sign is used. Their specific values can be determined with the aid of *perturbation* theory. The two unperturbed solutions ψ_A and ψ_B are first obtained by assuming no internuclear interactions. A charge density is next calculated for the unperturbed wave functions, and the internuclear interaction is introduced as a perturbation potential.† This leads to a quadratic equation for the allowed energy levels whose two roots differ by plus or minus a term called the *exchange energy*. A detailed discussion of perturbation theory is outside the scope of this book, so that qualitative arguments will be used below to arrive at the same result.

The two wave functions in (11) are shown in Fig. 6 for not too large a value of r_{AB}. Note that ψ_+ is symmetric about the midpoint

$$r_A = r_B = \tfrac{1}{2} r_{AB},$$

whereas ψ_- passes through zero at that point and is antisymmetric. (At very large internuclear separations, ψ_+ also falls to zero at the midpoint.)

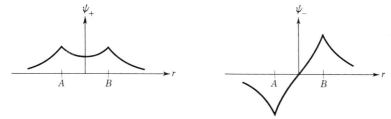

FIG. 6. Two possible wave-function solutions of Schrödinger's equation for one electron moving in the potential field of two positive hydrogen nuclei, along a line passing through the nuclei.

Also note that the symmetric wave function is approximately twice as high at the midpoint as either of the two functions e^{-ar} alone would be. The charge density of the symmetric wave function $|\psi_+|^2$, therefore, is four times larger at the midpoint. By comparison, the charge density of the antisymmetric wave function is zero at the midpoint. The significance of this result can be deduced by considering the potential energy of an electron in the field of two positive nuclei shown in Fig. 7. The potential-energy variation along a line joining the two nuclei can be expressed by an equation like

$$V = -\frac{e^2}{4\pi\epsilon_0|x + r_{AB}/2|} - \frac{e^2}{4\pi\epsilon_0|x - r_{AB}/2|} \tag{12}$$

with $x = 0$ chosen halfway between the two nuclei. Each of the terms in (12) expresses the potential energy of the electron in the field of one nucleus similarly to (10). Suppose only one nucleus, say, at B, is

† It should be noted that this is a fairly general procedure used in many quantum-mechanical calculations. The unperturbed wave functions represent a first approximation. The perturbation potential, which may be an interatomic interaction or an external potential, is then used to arrive at a more exact solution. It frequently leads to a splitting of the degenerate energy levels obtained in the absence of the perturbation.

present and the other is missing. The potential energy of the electron
when $r_B = r_{AB}/2$, that is, at $x = 0$, then is equal to $-2e^2/4\pi\epsilon_0 r_{AB}$. In
the presence of both nuclei, however, its potential energy is $-4e^2/4\pi\epsilon_0 r_{AB}$
at the midpoint, as can be seen by substituting $x = 0$ in (12). This
means that the excess charge density given by the symmetric wave function lies at a lower potential energy between two nuclei when they are
fairly close together than when they are very far apart. The symmetric
wave function therefore represents a state of the hydrogen molecule-ion
favoring a close approach of the two nuclei, and the energy of this state
as a function of internuclear separation can be expected to follow a curve
similar to the lower curve in Fig. 5.

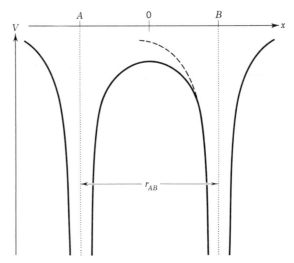

FIG. 7. Potential energy of an electron in the field of two hydrogen nuclei at A and B,
respectively. The potential energy due to one nucleus at B only is shown by the
dashed curve.

The above discussion also explains the shape of the lower curve. At
large values of r_{AB}, the total energy of the molecule-ion tends to zero by
definition, since the nuclei cannot interact. As r_{AB} is decreased, the
symmetric case leads to a lower energy. The total energy of the molecule-ion must include a repulsive potential $+e^2/4\pi\epsilon_0 r_{AB}$ due to a Coulomb
repulsion of the nuclei, and this term starts to dominate when r_{AB}
becomes very small, so that the total energy increases. Thus there must
be a minimum point in this curve corresponding to the equilibrium internuclear separation of the hydrogen molecule-ion.

By comparison, the charge density of the antisymmetric wave function
equals zero midway between the two nuclei and has maxima at the nuclear
sites. As the two nuclei approach closer, the two charge densities should

repel each other, so that the total energy of this state follows the upper curve in Fig. 5. The symmetric wave function therefore represents a bonding state of the molecule-ion, whereas the antisymmetric one represents an antibonding state. The cohesive, or binding, energy of the hydrogen molecule-ion, obviously, is the energy D in Fig. 5 required to dissociate the stable molecule. The above analysis also can be extended to include the possible excited states. As long as only one electron is present, the problem can be solved exactly and leads to similar results for the p and d states to those already described.

Covalent crystals. When a molecule contains two or more electrons, an exact solution is no longer possible. Methods for handling the many atom problems in molecules and crystals have been developed, and some of their features are described in later chapters. In order to learn the basic principles involved, consider a simple diatomic molecule, specifically, H_2. The hydrogen molecule differs from the hydrogen molecule-ion in that it consists of two hydrogen nuclei and two electrons. As a first approximation, assume that each electron moves in the field of two positive nuclei which are equally screened from it by the second electron. The potential energy of the electron is different from what it was in the hydrogen molecule-ion, but its symmetry should be the same since it seems reasonable that the screening effect on both nuclei is the same. The wave functions also should be similar to those in (11). Regardless of whether a symmetric or antisymmetric wave function is chosen, therefore, the charge density of the screening electron about the positions of both nuclei is the same. Thus the field obtained from these wave functions is self-consistent with the initial assumption of equal screening of both nuclei.

This symmetry of the hydrogen molecule means that the wave-function solutions of Schrödinger's equation are the same for both electrons. It also follows that both electrons can have symmetric and antisymmetric wave functions similar to those shown in Fig. 6. If both electrons are to have symmetric wave functions, then both will give rise to a charge density midway between the two nuclei. This can happen only if the two electrons have opposite spins, so that the Pauli exclusion principle is not violated. When this is the case, the two electrons can be said to be paired up, so that this case represents the formation of an electron-pair bond. The antisymmetric wave functions, similarly to the hydrogen molecule-ion case, represent an antibonding state. The energy dependence on interatomic separation can be expected to follow two curves like those shown in Fig. 5, except that each curve now represents the energies of both electrons in the molecule. Thus the two electrons can occupy a total of four possible quantum states (two energy levels) in the H_2 molecule, one pair corresponding to covalent bonding. This is an important result because it shows that *the total number of quantum states in a molecule is the same as the sum of allowed states in the individual atoms.*

The above analysis consisted of an attempt to find wave-function solutions for the molecule by considering the motion of the electrons in the presence of both nuclei directly. It is usually called the *method of molecular orbitals* for that reason. There is another way to approach this problem, first proposed by Heitler and London. Instead of starting with either a symmetric or an antisymmetric molecular orbital, the *Heitler-London method* starts with the unperturbed one-electron-one-nucleus wave functions. Denoting the two hydrogen atoms by A and B, respectively, and the electron belonging to atom A by a and that belonging to atom B by b makes it possible to describe the state of the two electrons by the following wave functions:

ψ_{Aa} if a electron is in an orbital about nucleus of A
ψ_{Ab} if b electron is in an orbital about nucleus of A
ψ_{Bb} if b electron is in an orbital about nucleus of B
ψ_{Ba} if a electron is in an orbital about nucleus of B.

The probability that the a electron is in an orbital around the nucleus of A is $|\psi_{Aa}|^2$; that b is around B, $|\psi_{Bb}|^2$; and so forth. The probability that two events occur simultaneously is given by the product of their individual probabilities. This is called the joint probability of two events. The joint probability that each nucleus is surrounded by one electron thus is equal to $|\psi_{Aa}|^2 \times |\psi_{Bb}|^2$ or $|\psi_{Ab}|^2|\psi_{Ba}|^2$. (The probability that both electrons are in orbitals about the same nucleus is finite but very small and is ignored in the following discussion.) The wave functions for the hydrogen molecule in which one electron surrounds each nucleus must have corresponding forms, namely, $\psi_{Aa}\psi_{Bb}$ or $\psi_{Ab}\psi_{Ba}$. Either one of these two forms alone is not sufficient, however, because it is not possible in practice to distinguish the two electrons as being either a or b. The indistinguishability of the two electrons requires, therefore, that the wave function of a hydrogen molecule must be a linear combination of these two cases. Two possible wave functions are thus obtained:

$$\psi_+ = \psi_{Aa}\psi_{Bb} + \psi_{Ab}\psi_{Ba}$$
and
$$\psi_- = \psi_{Aa}\psi_{Bb} - \psi_{Ab}\psi_{Ba}. \tag{13}$$

Note that interchanging a and b in ψ_+ does not change this wave function in any way, whereas interchanging a and b in the second wave function in (13) makes ψ_- negative. Thus ψ_+ is a symmetric wave function and ψ_- is antisymmetric. As in the method of molecular orbitals, the necessity of forming linear combinations like (13) is a consequence of perturbation theory.

The joint probabilities corresponding to the wave functions in (13) are

$$|\psi_+|^2 = (\psi_{Aa}\psi_{Bb})^2 + (\psi_{Ab}\psi_{Ba})^2 + 2\psi_{Aa}\psi_{Bb}\psi_{Ab}\psi_{Ba}$$
$$|\psi_-|^2 = (\psi_{Aa}\psi_{Bb})^2 + (\psi_{Ab}\psi_{Ba})^2 - 2\psi_{Aa}\psi_{Bb}\psi_{Ab}\psi_{Ba}.$$
(14)

Similar to the case of the hydrogen molecule-ion, the four possible wave functions describing one electron around one nucleus have essentially the same magnitude at the midpoint between the two nuclei. The last term in both wave functions represents an interaction between the two atoms due to the possibility of an exchange of their respective electrons. This is a purely quantum-mechanical result, and its contribution to the energy of the molecule is usually called the *exchange energy*. Since the four wave functions in the first two terms in (14) have equal magnitudes midway between the nuclei, it becomes clear that $|\psi_+|^2$ has a large value halfway between the two nuclei whereas $|\psi_-|^2$ is equal to zero. Physically, this means that ψ_+ leads to a charge density lying between the two nuclei, so that this is the quantum-mechanical description of the electron-pair bond when the two electrons have opposite spins. (Note that this bond is localized between the two atoms.) Similarly, ψ_- describes a repulsive state for the molecule in which the electrons do not bond. This is the case when their spins are parallel. The energies of these two states differ only by the amount of the exchange energy, that is, whether it is positive or negative, and depend on the interatomic separation as shown in Fig. 5.

It is readily seen from the above discussion that the method of molecular orbitals and the Heitler-London method lead to the same result for the hydrogen molecule. Actually, each has its own shortcomings. The method of molecular orbitals assumed that one-half electron screens each nucleus. When the two nuclei are very far apart, this cannot be correct because they should then revert to being two isolated hydrogen atoms. The Heitler-London method, on the other hand, breaks down when the two nuclei merge, that is, when $r_{AB} = 0$, because then it should result in the helium atom, for which the wave functions in (14) are not appropriate. Both of these shortcomings can be overcome by making use of perturbation theory, so that the differences between the two methods are mostly procedural and do not affect the final results. As shown in subsequent sections, both methods find applications in solids. The particular method used depends on whether it is more convenient, in a given case, to start with individual atomic wave functions or to formulate molecular functions directly.

A similar procedure can be used to describe the bonding between unlike atoms. As long as the orbitals of two unpaired electrons having opposite spins overlap, they can form electron-pair bonds. The number of unpaired electrons that an atom has determines the number of bonds it can form. These bonds can be formed with several atoms or with the same atom. For example, oxygen forms two electron-pair bonds with two hydrogen atoms in H_2O or with one other oxygen atom in the O_2 molecule. The undirected nature of the ionic and metallic bonds prevents them from forming finite molecules in the solid state. Conversely, saturated covalent bonds can exist between groups of atoms in crystals, forming complex ions or molecules; however, these groups must then be joined in infinite three-dimensional arrays by other kinds of bonds. On the other hand, atoms can be joined by covalent bonds without forming discrete molecules, so that the entire crystal is one big molecule. In principle it is possible to calculate the energies of different kinds of crystal structures for each type of binding. If the wave function describing the most stable configuration for one kind of bond is ψ_1 and that for another kind is ψ_2, then the linear combination of the two wave functions can be written

$$\psi = a\psi_1 + b\psi_2. \tag{15}$$

The ratio of the coefficients a/b, which gives the wave function (15) corresponding to the lowest energy for a particular structure, then determines the stable configuration of the aggregate. If a is much larger or much smaller than b, one or the other type of bond predominates. On the other hand, if the coefficients have comparable magnitudes, then the structure is said to resonate between the two possible configurations. This postulate of resonance is, of course, a direct consequence of the statistical nature of the results of quantum-mechanical analysis. It should be understood that resonance does not imply that the aggregate literally divides itself into a state described by ψ_1 for $a^2/(a^2 + b^2)$ part of the time and into the state described by ψ_2 for the rest of the time, but rather that, on the average, the assembly behaves as if it were rapidly alternating between the two states, with the relative predominance of each determined by the relative magnitudes of a^2 and b^2.

Ionic crystals. The bonding in ionic crystals is due to the electrostatic attraction of oppositely charged ions for each other. Thus it is not necessary to consider the detailed orbitals of the electrons since it can be assumed that the excess positive or negative charge is spherically distributed. In fact, the spherical symmetry of the ions is the reason why ionic structures are based on the different kinds of closest packings described in Chap 1. A calculation of the cohesive energy of ionic structures, therefore, can be based on the electrostatic forces between point charges of positive and negative electricity arrayed at the atomic centers in such packings. The Coulomb attraction between two point charges z_1e and z_2e, separated by r, is $z_1ez_2e/4\pi\epsilon_0r$. In a crystal, however, there are present repulsive forces which also are assumed to be inversely proportional to some power of the separation r. The potential energy of the crystal can be written, therefore,

$$V = -A \frac{z_1 z_2 e^2}{4\pi\epsilon_0 r} + B \frac{1}{r^n} \tag{16}$$

where A and B are constants to be determined. For the simplest case of the univalent alkali halides, (16) reduces to

$$V = -\frac{Ae^2}{4\pi\epsilon_0 r} + \frac{B}{r^n}. \tag{17}$$

At the equilibrium separation r_e, the potential energy is at a minimum and the first derivative of (17) must vanish.

$$\left(\frac{dV}{dr}\right)_{r=r_e} = 0 = +\frac{Ae^2}{4\pi\epsilon_0 r_e^2} - \frac{nB}{r_e^{n+1}}. \tag{18}$$

From this it follows that

$$\begin{aligned} B &= \frac{Ae^2}{4\pi\epsilon_0 r_e^2} \frac{r_e^{n+1}}{n} \\ &= A \frac{e^2 r_e^{n-1}}{4\pi\epsilon_0 n}. \end{aligned} \tag{19}$$

Substituting (19) in (17),

$$\begin{aligned} V_{r=r_e} &= -\frac{Ae^2}{4\pi\epsilon_0 r_e} + \frac{Ae^2 r_e^{n-1}}{4\pi\epsilon_0 n r_e^n} \\ &= -\frac{Ae^2}{4\pi\epsilon_0 r_e}\left(1 - \frac{1}{n}\right). \end{aligned} \tag{20}$$

This type of analysis was originally applied to ionic crystals by Born and Landé. The constant A was first evaluated by Madelung and bears his name. It depends on the exact crystal structure and is very difficult to calculate exactly; however, it can be approximated quite easily for most simple structures (Exercises 10 and 11). The value of n can be determined empirically from the compressibility of the crystal. It can be shown that the compressibility is given by the relation

$$\kappa = \frac{72\pi\epsilon_0 r_e^4}{Ae^2(n-1)} \tag{21}$$

in which κ is measured experimentally and all the other factors except n are known. It should be noted that the potential energy in (20) is not greatly affected by small errors in n.

One of the requirements of a good theory of cohesion is that it be able to predict the structural arrangements that a group of atoms will assume. This, the most stable of several possible crystal structures, is the structure

having the lowest value for the potential energy in (20).   It turns out that
the form of (20) is not adequate for this purpose because the repulsive
term is incorrectly formulated and the small but nevertheless finite
van der Waals forces must be taken into account.   These alterations of
the theory were worked out by Born and Mayer, giving the following
relation for the potential energy:

$$V = -\frac{Ae^2}{4\pi\epsilon_0 r} + Be^{-r/n} - \frac{C}{r^6} + \delta \tag{22}$$

in which A, B, C are constants
$\qquad\qquad$ n is determined from compressibility data as before
$\qquad\qquad$ δ is a small contribution expressing energy of structure at
$\qquad\qquad\quad$ absolute zero of temperature.

Notice that the attraction term is unaltered, the repulsion term is
proportional to an exponential in r/n, and the third term expresses the
van der Waals forces.   Actual computations based on (22) give energy
values not too different from those obtained using (20); however, the
agreement with experimentally determined values is improved.   The
different terms in (22) are shown plotted in Fig. 8.   The resultant energy
curve has the by now familiar form of a stable structure whose energy
corresponds to the minimum in energy at r_e.

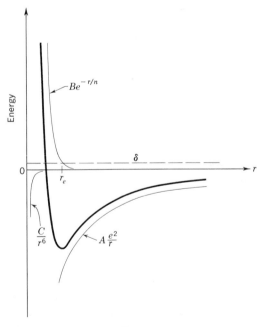

Fig. 8

Quantum mechanics was first applied to calculating cohesive forces in ionic crystals by Hylleraas for lithium hydride and by Landshoff for sodium chloride. More recently, the cohesive energies of NaCl and KCl were calculated by Löwdin. Such calculations make use of the Heitler-London method because it is more convenient in this case to start with the wave functions of the isolated ions. Landshoff calculated the energy of the structure of NaCl with respect to the energy of the ions in a free state. This makes it difficult to estimate the accuracy of his solution on a theoretically absolute basis; however, his value agrees very closely with that determined experimentally. Löwdin calculated the cohesive energies of NaCl and KCl directly from the wave functions of the free ions, and the results of his calculations are compared with experimental values below.

	NaCl	KCl	
Theoretical cohesive energy	7.948	7.241 eV	(23)
Experimental cohesive energy	7.934	7.134 eV	

Metallic crystals. According to the free-electron model proposed for metals by Drude and Lorentz, the cohesive energy of a metal should be based on the electrostatic attraction between the positive metal ions and the negative free-electron gas. A semiempirical procedure, based on Born's theory for ionic crystals, was developed by Grüneisen, who combined such a formulation with the condition that his expression for energy give the observed values of atomic volume, cohesive energy, and compressibility of the solid at the absolute zero of temperature. For example, Eq. (21) is one such relation, and Grüneisen applied it to the calculation of the elastic properties of several monatomic metals with fair success. These methods have been extended by others, and they are quite valuable in the calculation of elastic properties such as thermal expansion and compressibility of metals. The cohesive energy predicted by Grüneisen's method does not agree too well with experimental values, however, even for the simple alkali metals.

A semiquantitative quantum-mechanical explanation of the cohesive energy in metals was proposed by Pauling. It will be recalled from an earlier section that the quantum-mechanical model of the metallic bond can be likened to an unsaturated electron-pair bond. The bonding electrons occupy one of several closely spaced energy levels and can form partial electron-pair bonds with each other. This leads to a resonance of electrons among the various levels, and the larger the coordination number of an atom, the larger this resonance becomes, resulting in a lower energy for the structure. This is in agreement with the observed tendency of metals to form closest-packed structures (CN = 12), but it fails to explain why certain metals crystallize in a body-centered cubic

packing (CN = 8).   Pauling's model for, say, the metals of the first long
period postulates the formation of hybridized 3*d*, 4*s*, and 4*p* orbitals for
the valence electrons.   As discussed below, it is useful for explaining
qualitatively some of their properties, including cohesion.   If all the
valence electrons form covalent bonds, however, it is not possible to
explain the high electrical conductivity of metals.   Pauling gets around
this by postulating a nonuniform distribution of electrons in which some
of the atoms become temporarily ionized.   Although the ionization of an
atom requires an increase in its energy, it turns out that the overall
energy of the crystal is lowest for this type of distribution.

It is possible to correlate qualitatively the cohesive properties of metal
structures to the metallic valences obtained from Pauling's theory.

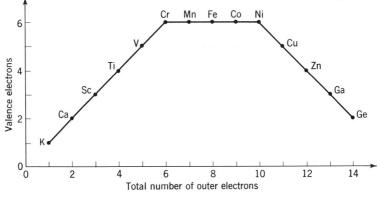

Consider the elements of the first long period from potassium to germanium.   In each of these metal atoms there are five 3*d* orbitals, one 4*s*
orbital, and three 4*p* orbitals available for the outer electrons.   According to Pauling, the electrons can occupy these hybridized nine available
orbitals as follows: One orbital, which is set aside so that the electrons can
resonate into it, is called the metallic orbital.   The other orbitals are
successively occupied by unpaired electrons, with the further restriction
that the number of valence electrons can never exceed six.   The resulting
valences for these metal atoms are plotted in Fig. 9. The number of
valence electrons increases gradually from one in potassium to six in
chromium, then levels off at six until nickel, after which it decreases
again.   The reason for this decrease is seen by considering the copper
atom whose 11 outer electrons are schematically distributed as follows:

$$3d \boxed{\uparrow\downarrow\,\uparrow\downarrow\,\uparrow\downarrow\,\uparrow\downarrow\,\uparrow} \qquad 4s\boxed{\uparrow} \qquad 4p\boxed{\uparrow\,\uparrow\,} \tag{24}$$

where the unoccupied 4*p* state is reserved for the metallic orbital.   The
elastic properties of the elements show a marked similarity to the graph

in Fig. 9; that is, the elastic properties, as evidenced by, say, critical-stress values, increase from potassium to chromium, are similar for the remaining transition metals, and decrease for Cu, Zn, Ga, and Ge, in that order. Quite similarly, Pauling explains the decreasing interatomic separation from 3.21 Å in scandium to 2.63 Å in vanadium by assuming that the increasing number of electrons present makes possible the formation of increasing numbers of hybridized bonds. This also accounts for the stronger binding, say, of titanium as compared with scandium. When other metallic properties such as ferromagnetism are considered, however, it becomes necessary to postulate nonintegral numbers of electrons in the various orbitals. Because of the somewhat artificial nature of these postulates, Pauling's model has been criticized as being physically untenable.

A quantitative calculation of the cohesive energy based on the free-electron model has been carried out by Wigner and Seitz. Since the atoms in a metal crystal are periodically repeated, it is possible to divide the crystal volume into cells of equal volume and to consider only the binding taking place within each cell. Wigner and Seitz formed these cells by bisecting the distances separating an atom from its nearest

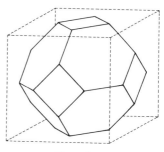

Fig. 10

neighbors by planes which are terminated at their mutual intersections. The *atomic polyhedra* thus formed completely occupy all the available space in a crystal; a typical atomic polyhedron for an atom in a body-centered cubic packing is shown in Fig.10. The Wigner-Seitz approximation consists of the assumption that the valence electron of the atom inside such a polyhedron is influenced only by a spherically symmetric potential field due, solely, to the positively charged ion at its center; that is, the fields of neighboring polyhedra do not extend past their boundaries. This makes it possible to replace the more complicated boundaries of the polyhedron by a sphere without introducing appreciable errors in the calculation. For alkali metals, the wave functions of the free electrons then have constant values throughout 90 per cent of the atomic volume and exhibit pronounced variations from this value only in the immediate vicinity of the nucleus. Thus the valence electrons in the cellular model of Wigner and Seitz are free to move without change in energy throughout most of this volume, not unlike the assumption of the early Drude-Lorentz theory. In the case of other monovalent metals such as copper, silver, and gold, the wave functions are constant throughout very small portions of the atomic volumes, indicating that the free-electron model is not applicable to them in the same degree.

It is possible to give this a simple physical interpretation. If an electron is really influenced only by the field of the positive metal ion, this field falls off as the electron moves farther away from the nucleus. Accordingly, when the ratio between the ionic radius and one-half the separation between atoms in the metal structure is small, then the electron is relatively more free than when this ratio is large. Some values of this ratio are as follows:

$$\begin{array}{ccccccc} & \text{Li} & \text{Na} & \text{K} & \text{Cu} & \text{Ag} & \text{Au} \\ \text{Ratio} & 0.39 & 0.51 & 0.58 & 0.78 & 0.88 & 0.95 \end{array} \quad (25)$$

It is clearly evident from this comparison that the naive picture of a metal consisting of positive ions imbedded in a "sea" of negative electrons is not too bad an approximation for the alkali metals but cannot be carried over in detail to the other metals.

Thus, in general, the potential energy of a metal crystal is the sum of the interaction energies of the charge within each atomic polyhedron plus the energy of interaction of the polyhedra with each other. In applying the cellular method to alkali metals, it is assumed that the potential field due to the cation is limited to the volume of one polyhedron which contains one valence electron and is therefore electrically neutral. The potential energy of the crystal is determined directly by the kinetic energy of each electron, E_0, and the potential energy of each electron, V_0, in the field of the positive ion. The cohesive energy E_c is then given by the expression

$$E_c = -(V_0 + \tfrac{3}{5}E_0 + V_1) \quad (26)$$

where V_1 is the first ionization potential of the atoms. Actually, this equation is an oversimplification, since it does not include electron-electron interactions nor van der Waals interactions. Nevertheless, the approximate treatment is in fairly good agreement with experimental values, as shown by a comparison between calculated and observed values:

$$\begin{array}{cccc} & \text{Li} & \text{Na} & \text{K} \\ \text{Calculated } E_c & 1.51 & 1.06 & 0.71 \text{ eV} \\ \text{Observed } E_c & 1.69 & 1.13 & 1.00 \text{ eV} \end{array} \quad (27)$$

Fuchs has attempted to extend the Wigner-Seitz approximation to the noble metals. It turns out in these cases that the outermost filled shells extend beyond the boundaries of an atomic polyhedron along certain directions and hence contribute to the binding almost as much as the valence electrons. The exact treatment of this situation becomes very complex; however, by means of several approximations, Fuchs was able to calculate the cohesive energy of copper and obtained a value of 1.43 eV per atom. When this value is compared with the experimentally determined energy of 3.51 eV, it becomes clear that the cellular method is not well suited for such metals.

An alternative approach to cohesion in metals is provided by the Heitler-London method previously described. It will be recalled that this approximation starts out with the wave functions of free atoms and is sometimes called the method of linear combination of atomic orbitals, normally abbreviated LCAO. The valence electrons are considered to be bound to their respective nuclei, so that a slight modification of this approach is also called the *tight-binding approximation* in order to distinguish it from the nearly free electron model of the cellular method. An important result of applying this method to crystals (very large molecules) is that it leads to a splitting of the energy levels of the valence or bonding electrons. Consider a one-dimensional "crystal" comprised of a linear array of six hydrogen atoms. The energy dependence of the 1s level on interatomic separation is shown in Fig. 11. At the equilibrium interatomic separation in such a fictitious crystal, the 1s level splits up into six distinct levels each of which can accommodate two electrons with opposite spins. The energy separation of these levels is extremely small, and, as described in a later chapter, the levels jointly form an *energy band* containing as many discrete levels in the crystal as there were in the isolated atoms. The wave functions of this "crystal," corresponding to the lowermost and uppermost allowed energy levels, are shown in Fig. 12. Note their similarity to the symmetric and antisymmetric wave functions of the hydrogen molecule.

The tight-binding approximation has been used with limited success in metals. It is quite useful for calculations involving inner electrons, but does not seem to be applicable to the conduction or free electrons. Thus it can be used for determining the energies of the d electrons in transition metals in which the 4s electrons are the conduction electrons. Recently, Stern has combined this approach with the cellular method in calculating the cohesive energy of body-centered cubic iron at the absolute zero of temperature. In essence, he assumes that the potential field acting on an electron is that of the ion core and the other valence electrons present in the atomic polyhedron and ignores possible contributions from adjacent cells. This is not too bad an approximation for iron because the 3d orbitals do not extend as far outward as they do in copper. Stern then formulates suitable electronic wave functions and uses the tight-binding approximation to calculate the energies for such atomic orbitals. The cohesive energy of iron obtained by this procedure is 5.85 eV per atom as compared with the measured value of 4.35 eV. It is interesting to note that the Wigner-Seitz model, as applied to copper by Fuchs, underestimates the cohesive energy, whereas the tight-binding approximation tends to overestimate it. In concluding this discussion, it should be noted that the energy difference between two equally probable structures,

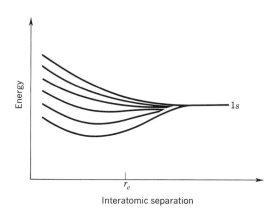

Interatomic separation

Fɪɢ. 11.   Energy levels for a row of six hydrogen atoms.   At large interatomic separations, the $1s$ energy level is the same in all six atoms.   As they are brought closely together, the levels split up forming a $1s$ "band" of energies at the equilibrium interatomic separation r_e.

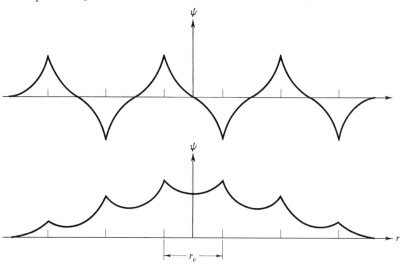

Fɪɢ. 12.   Wave functions for a linear array of six hydrogen atoms along the row joining their centers.   The upper wave function corresponds to the uppermost energy level in Fig. 11, whereas the bottom one corresponds to the lowermost energy level in Fig. 11.

say, a body-centered cubic packing and a face-centered cubic closest packing, is very much smaller than the cohesive energies of either.   In the absence of a quantitatively more exact theory, therefore, it is not yet possible to account for the different structures adopted by metals from cohesive-energy calculations.   Despite this shortcoming, due in part to the very complex nature of such calculations, the success of the above-described methods in accounting for the cohesive energy of known structures, notably the alkali metals (27), has greatly aided the development of modern solid-state concepts.

Suggestions for supplementary reading

R. C. Evans, *An introduction to crystal chemistry* (Cambridge University Press, London, 1948), especially pp. 9–88.

Linus Pauling, *The nature of the chemical bond*, 2d ed. (Cornell University Press, Ithaca, N.Y., 1948).

John C. Slater, *Introduction to chemical physics* (McGraw-Hill Book Company, Inc., New York, 1939), especially pp. 352–376.

Robert L. Sproull, *Modern physics* (John Wiley & Sons, Inc., New York, 1956), especially pp. 190–233.

Suggestions for further reading

J. Callaway, Electron energy bands in solids, in *Solid-state physics*, vol. 7 (Academic Press Inc., New York, 1958), pp. 99–212.

Per-Olov Löwdin, Quantum theory of cohesive properties of solids, in *Advances in physics*, vol. 5 (Taylor and Francis, Ltd., London, 1956), pp. 1–171.

John C. Slater, Electronic structure of metals, in *Encyclopedia of physics*, vol. 19 (Springer-Verlag OHG, Berlin, 1956), pp. 1–48.

Exercises

1. Using Eqs. (1) and (2), derive an expression for the potential energy at the equilibrium separation involving only the constants α, n, and m. Show that the derivative of this potential energy with respect to r is equal to zero; that is, justify that the force is zero when $r = r_e$.

2. Draw a schematic representation of the bonding in the structure of $NaHF_2$ (like Fig. 4), assuming that sodium has octahedral coordination; that is, six fluorine ions surround each sodium ion.

3. Draw a schematic representation of $Al(OH)_3$, assuming octahedral coordination for Al^{3+} and that the aluminum octahedra form sheets which are joined to adjacent sheets by hydroxyl bonds.

4. In the crystal structure of tenorite, CuO, each copper atom has four coplanar neighbors with the Cu-O distance equal to 1.95 Å and two oxygen neighbors at Cu-O equal to 2.32 Å. Why are the bond distances in the resulting distorted octahedron unequal? What kind of bonds do you think Cu forms with its six oxygen neighbors? In Cu_2O, copper is tetrahedrally coordinated by oxygen with equal Cu-O distances. What kind of bonding explains this structure?

5. Describe briefly what happens when lithium atoms approach each other to form a body-centered cubic crystal structure.

6. Making use of the relative-energy diagram of an atom (Fig. 4 in Chap. 3), show why it is not possible for more than two hydrogen atoms to combine in a single molecule.

7. Calculate the potential energy in electron volts of an electron in the hydrogen molecule-ion located midway between the two nuclei when they are at their equilibrium separation of 1.06 Å. Compare this value with the potential energy of a single electron 0.53 Å away from a hydrogen nucleus. What is the potential energy of the electron in H_2^+, 0.53 Å along x away from an end hydrogen nucleus? (ANSWER: -57.63 eV; -28.82 eV; -38.42 eV.)

8. Assume that the cohesive energy of the hydrogen molecule-ion is the sum of an attractive term due to electron sharing by the two nuclei and a Coulomb-type repulsive term due to the nuclei alone. If the cohesive energy of the one-electron bond formed is -2.65 eV, what is the actual value of the attractive term at an internuclear separation of 1.06 Å?

9. Consider the formation of a KCl molecule. Assuming $V = 0$ when the two atoms are infinitely far apart:

(*a*) How much energy is required to ionize both atoms if the electron affinity of chlorine is 3.82 eV?

(*b*) What is the ionic attractive energy between the two ions when their interatomic separation is 2.79 Å?

(*c*) The measured dissociation energy of a KCl molecule is -4.40 eV. If the equilibrium interionic distance is 2.79 Å, what is the magnitude of the repulsive energy, ignoring van der Waals forces? HINT: The cohesive energy must be calculated relative to the energy of the neutral atoms infinitely far apart.

(*d*) From the magnitude of the repulsive energy determined above, do you think it is primarily due to nuclear or electronic repulsion? Why should the electrons repel each other instead of forming electron-pair bonds?

10. Bragg has shown that the Madelung constant for NaCl can be calculated as follows: Starting with any atom, it is surrounded by 6 nearest neighbors of opposite sign at a distance r, 12 next-nearest neighbors of like sign at a distance of $2r$, 8 next-next-nearest neighbors of unlike sign at $3r$, etc. The value of A is given by adding up such terms in a series of the form

$$A = \frac{6}{\sqrt{1}} - \frac{12}{\sqrt{2}} + \frac{8}{\sqrt{3}} - \frac{6}{\sqrt{4}} + \frac{24}{\sqrt{5}} - \cdots .$$

Ignoring van der Waals forces and zero-point energy, what is the potential energy of sodium chloride, at equilibrium, if the interionic separation is 2.76 Å and the value of n from compressibility data is 9.1? What is the magnitude of the neglected energy terms?

11. What is the potential energy of cesium chloride, at equilibrium, if the interionic distance is 3.56 Å and $n = 11.5$? Compute the Madelung constant for this structure following the procedure outlined in Exercise 10 for NaCl. HINT: Remember that the structures of NaCl and CsCl are different. (ANSWER: -6.50 eV.)

12. The potential energy of a valence electron within an atomic polyhedron in lithium can be approximated by $V_0 = \alpha e^2/4\pi\epsilon_0 r$, where $\alpha \simeq 2$ and has the units of kg-m^3/(C-sec)^2. Using the approximate equation (26), calculate the cohesive energy for lithium metal if $r = 3.21$ Å, $E_0 = 1.896$ eV, $V_1 = 5.365$ eV. How does this value compare with that given in (27) in the text?

13. Pauling has shown that such physical properties as hardness and melting point, and others, change in proportion to the interatomic separations in the structures of the elements. Specifically, this can be illustrated by computing a quantity Pauling calls the *ideal density*. The ideal density for the elements from potassium to germanium is equal to 50/(gram-atomic volume) and is the density that these elements would have if they all had an atomic weight of 50 and crystallized in a closest packing. Look up the necessary metal radii in Appendix 3, calculate the ideal densities of K to Ge, and plot the calculated values against the number of outer electrons in each atom. Compare this plot with Fig. 9 in the text.

14. The cohesive energy of crystals is frequently expressed in kilocalories per kilogram molecular weight. If 1 K-cal = 4.18×10^3 J, how many kilocalories per kilogram molecular weight equal 1 eV/molecule? (ANSWER: 23.06×10^3.)

5. statistical mechanics

The use of Newtonian, or classical, mechanics to describe the motion of a single large particle is well known. When dealing with a collection of many small particles such as atoms or molecules, however, it is no longer meaningful to attempt describing the motion of each and every individual particle. Instead, the assembly of particles is considered as a whole. Suppose a monatomic gas is comprised of N total atoms. The energy of the ith atom in the gas can be designated E_i. The total energy of this gas E then is given by a sum of the individual energies of the N atoms:

$$E = \sum_{i=1}^{N} E_i. \tag{1}$$

Since it is not possible to keep track of individual particles in an assembly, it is convenient to select a representative particle having an average energy. In order to determine such an average energy, or the total energy in (1), it is first necessary to know what the energy distribution of the particles is, that is, how many particles have each possible energy value. This can be done most easily by making use of a branch of mathematical physics called *statistical mechanics*. The energy distribution depends on whether the particles are completely free to move independently of their environment or whether they are acted on by an external force. When their motion is unrestrained, their energy values are predicted by the *Maxwell distribution law*. If their motion is affected by an external force, then their energy values follow the *Boltzmann distribution.* Actually, the Maxwell distribution is a special case of the Boltzmann distribution, so that they can be considered together. Both distributions are derived on the assumption that the particles do not interact with each other directly except for occasional chance collisions.

Whether neutral particles in an assembly will interact with each other can be determined by calculating their de Broglie wavelengths. If this wavelength is shorter than the average particle separation, then the particles will not interact. Conversely, when the interparticle distance is smaller than the de Broglie wavelength, the particles do interact and classical statistical mechanics cannot be used. This is the case, for example, for electrons in a solid (Exercise 1). It is necessary, therefore, to utilize the quantum statistical mechanics developed by Fermi and by Dirac. The different kinds of statistical distributions are described in the first part of this chapter. Use is made of statistical mechanics to explain certain properties of crystalline materials in the second half of this chapter and also in subsequent chapters in this book.

Elements of theory

Maxwell-Boltzmann distribution. It can be shown by thermodynamic arguments that the probability that a particle has energy E_i is proportional to $e^{-E_i/kT}$, where k is the Boltzmann constant and T is the temperature measured on the Kelvin, or absolute, scale. The inverse exponential dependence predicts that, at a fixed temperature, this probability decreases as the energy E_i increases. This means that it is more likely that the particles have a low energy rather than a high energy. Moreover, the tendency to have low energies, or to occupy low-energy states, increases as the temperature decreases. The fraction of particles that have the energy E_i, averaged over the entire assembly, is then given by

$$\frac{e^{-E_i/kT}}{\sum_i e^{-E_i/kT}} \tag{2}$$

and is called the *Maxwell-Boltzmann distribution law.* Relation (2) is particularly useful because it tells what the probability is that the ith energy state is occupied by a particle in the assembly at any temperature T.

The Maxwell-Boltzmann distribution law can be used successfully for calculations involving gases because the assumption of no interparticle interactions is valid in this case. For example, consider a monatomic gas containing N total atoms at some temperature T. The kinetic energy of an atom can be expressed in terms of its momentum since

KE $= p^2/2m$. Momentum is a function of the direction in which an atom travels so that it can be resolved into three components.

$$p_x = mv_x, \qquad p_y = mv_y, \qquad p_z = mv_z$$

and
$$p^2 = p_x^2 + p_y^2 + p_z^2 \tag{3}$$

where m is the atom's mass, and v_x is the velocity in the x direction, v_y in the y direction, and v_z in the z direction.

The form of (3) suggests that different values of the momentum p can be represented graphically by plotting its three components, p_x, p_y, and

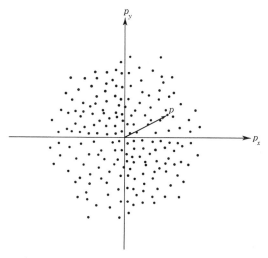

FIG. 1.   Two-dimensional momentum space.   Each point represents a possible value of the momentum p for a particular combination of p_x and p_y.

p_z, along three mutually perpendicular axes.   For simplicity in drawing, a two-dimensional plot in *momentum space* is shown in Fig. 1.   Each point in this momentum space represents a possible value of

$$p = (p_x^2 + p_y^2)^{\frac{1}{2}}$$

that an atom can have.   According to classical mechanics, there are no restrictions on the possible values of p^2 that an atom can have, except for the probability expressed by (2), so that the distribution is a continuous one.   It is possible, therefore, to replace the summation in (2) by an integration of all infinitesimal volumes $dp_x\,dp_y\,dp_z$.   The fraction of atoms dN/N having momenta that lie within each such volume element in momentum space is then given by

$$\frac{dN}{N} = \frac{e^{-p^2/2mkT}dp_x\,dp_y\,dp_z}{\displaystyle\int\!\!\!\int\!\!\!\int_{-\infty}^{+\infty} e^{-p^2/2mkT}\,dp_x\,dp_y\,dp_z}. \tag{4}$$

The integrals in the denominator of (4) can be separated by substituting (3) for p so that

$$\iiint_{-\infty}^{+\infty} e^{-p^2/2mkT} \, dp_x \, dp_y \, dp_z = \int_{-\infty}^{+\infty} e^{-p_x^2/2mkT} \, dp_x \int_{-\infty}^{+\infty} e^{-p_y^2/2mkT} \, dp_y$$
$$\int_{-\infty}^{+\infty} e^{-p_z^2/2mkT} \, dp_z. \quad (5)$$

Consider one of the integrals on the right side of (5), and let $x = p_x$ and $a = 1/2mkT$, so that

$$\int_{-\infty}^{+\infty} e^{-p_x^2/2mkT} \, dp_x = \int_{-\infty}^{+\infty} e^{-ax^2} \, dx = \sqrt{\frac{\pi}{a}} \quad (6)$$

as can be checked by consulting a table of definite integrals. Since the three integrals in (5) are equivalent, the product in (5) is equal to $(2\pi mkT)^{\frac{3}{2}}$. This means that the fraction of atoms whose momentum lies in the range $dp_x \, dp_y \, dp_z$ is

$$\frac{dN}{N} = (2\pi mkT)^{-\frac{3}{2}} e^{-p^2/2mkT} \, dp_x \, dp_y \, dp_z \quad (7)$$

which expresses the Maxwell distribution of atomic momenta in a monatomic gas.

Alternatively, it may be desirable to know what fraction of the atoms has velocities lying in some range between v and $v + dv$. Since

$$v = (p_x^2 + p_y^2 + p_z^2)^{\frac{1}{2}}/m,$$

this can be determined by considering the distribution of states in momentum space. The volume in momentum space which corresponds to velocities in the range v to $v + dv$ lies, according to (3), between two spherical shells whose radii are mv and $mv + m \, dv$, that is, $4\pi (mv)^2 \, mdv$. Substituting this volume for $dp_x \, dp_y \, dp_z$ in (7),

$$\frac{dN}{N} = (2\pi mkT)^{-\frac{3}{2}} e^{-p^2/2mkT} (4\pi m^3 v^2) \, dv$$

and rearranging terms,

$$\frac{dN}{N} = 4\pi \left(\frac{m}{2\pi kT}\right)^{\frac{3}{2}} v^2 e^{-p^2/2mkT} \, dv. \quad (8)$$

The form of the Maxwell distribution of velocities in (8) can also be used to express the fraction of atoms having energy values lying in the range between E and $E + dE$. To do this, remember that the kinetic energy $E = p^2/2m = \frac{1}{2}mv^2$ and $dE = d(\frac{1}{2}mv^2) = mv\,dv$. Substituting for dv in (8),

$$\frac{dN}{N} = 4\pi \left(\frac{m^3}{8\pi^3 k^3 T^3}\right)^{\frac{1}{2}} v^2 e^{-E/kT} \frac{dE}{mv}$$

$$= \frac{4\pi}{2\pi} \left[\frac{\frac{1}{2}mv^2}{\pi(kT)^3}\right]^{\frac{1}{2}} e^{-E/kT}\,dE$$

or
$$\frac{dN}{dE} = 2N \left[\frac{E}{\pi(kT)^3}\right]^{\frac{1}{2}} e^{-E/kT}. \tag{9}$$

To help understand the meaning of this more familiar form of the Maxwell distribution (9), it is plotted in Fig. 2. The shape of the distribution is controlled by the two terms on the right in (9), which are both function of energy. The distribution goes to zero very rapidly as $E \to 0$ because it is directly proportional to $E^{\frac{1}{2}}$, while the exponential approaches unity as its exponent goes to zero. When the energy increases, however,

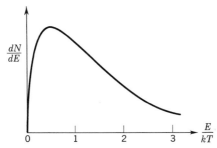

FIG. 2.   Maxwell distribution of kinetic energies.

the exponential term becomes dominant. As $E \to \infty$, the exponential goes to zero much faster than the corresponding rate of increase for $E^{\frac{1}{2}}$. Thus the distribution in (9) tails off for higher energies. The energy at which the curve reaches its maximum value can be determined by differentiating the right side of (9) and setting it equal to zero (Exercise 2). This is the most probable value for the energy of an atom in the gas; that is, more atoms have $E = \frac{1}{2}kT$ than any other energy. Note that, as the temperature increases, the maximum shifts to higher energy values.

   The discussion so far has assumed that the kinetic energy of a particle is also its total energy. This assumption is valid provided that the particle is not acted upon by an external force so that its potential energy is constant and can be set equal to zero for convenience. Next consider what happens when an external force acts on the particles. Electrons emitted from a hot metal surface into an evacuated space, as discussed in Chap. 11, can be considered as a gas of electrically charged particles. If an electric field ε perpendicular to the surface is present in the space outside the metal, the potential energy of the electrons at a distance x

above the surface is

$$V = e\mathcal{E}x \tag{10}$$

where e is the charge on the electron. Consider a cylindrical column of gas bounded by two planes at x and $x + dx$ whose cross section is A as shown in Fig. 3. The force pushing up on the bottom of the cylinder is F, and the force pushing up on the top is $F + dF$. The increment of force dF is due to the action of the electric field on the electrons in the cylinder and equals the force exerted by each electron, which, according to (10), is $-dV/dx = -e\mathcal{E}$ times the number of electrons in the cylinder.

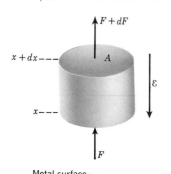

If the density of electrons in the cylinder is n, the total number of electrons in the volume is $nA\,dx$, so that the differential force is

$$dF = -e\mathcal{E}nA\,dx. \tag{11}$$

The force can also be determined from the pressure as given by the gas law

$$P\mathcal{V} = N_0 kT \tag{12}$$

where P is pressure of gas in a volume $\mathcal{V}$ at temperature T

N_0 is Avogadro's number

k is Boltzmann's constant.

Fig. 3

Metal surface

Although the gas law strictly applies only to a perfect gas in which the particles do not interact directly except through occasional chance collisions, it is permissible to use it for an electron gas in which the number of particles per unit volume is relatively small, as in the present case. Since the density of electrons is simply $n = N_0/\mathcal{V}$, the differential pressure is found by differentiating both sides of (12),

$$dP = kT\,dn$$

$$\frac{dF}{A} = kT\,dn. \tag{12a}$$

Using (11), this becomes

$$\frac{dF}{A} = -e\mathcal{E}n\,dx = kT\,dn$$

or

$$\frac{dn}{n} = -\frac{e\mathcal{E}}{kT}\,dx \tag{12b}$$

which can be integrated, assuming that the density of electrons immediately outside the metal surface at $x = 0$ is n_0,

$$\int_{n_0}^{n} \frac{dn}{n} = -\frac{e\mathcal{E}}{kT} \int_{0}^{x} dx.$$

The result is

$$\ln n - \ln n_0 = -\frac{e\mathcal{E}}{kT} x$$

or
$$n = n_0 e^{-e\mathcal{E}x/kT}. \tag{13}$$

Substituting the potential energy from (10) in the exponent of (13),

$$n = n_0 e^{-V/kT} \tag{14}$$

which shows that the electron density decreases exponentially as the potential energy increases in agreement with the assumption made at the beginning of this section. Equation (14) is the general form of the Boltzmann distribution.

The discussion so far has ignored the availability of states for an assembly and has concentrated on their occupation in compliance with the Maxwell-Boltzmann distribution law. The actual distribution of the available states in momentum space, as distinct from the distribution of occupied states, is determined by the nature of the assembly considered. Some examples of the limitations imposed by specific assemblies are given in later sections of this chapter and in subsequent chapters in this book.

Fermi-Dirac statistics. When quantum-mechanical assemblies are considered, each available state is specified by the quantum numbers and its occupation is controlled by the Pauli exclusion principle. The observation of the actual occupation of such states is governed, of course, by the Heisenberg uncertainty principle. Although this means that it is not meaningful to prescribe specific distributions for each instant of time, it is nevertheless convenient to consider the occupation probabilities similarly to the above discussion for classical assemblies. The evolution from classical to quantum statistical mechanics can be understood by considering just how quantum-mechanical restrictions affect the assembly. Starting with the Maxwell-Boltzmann distribution law (2) at thermodynamic equilibrium, it is easily seen that the probability that the ith energy state is occupied, $f(E_i)$, has an inverse exponential dependence on the energy of that state, E_i.

$$f(E_i) \propto e^{-E_i/kT}. \tag{15}$$

It now will be demonstrated that this is indeed the case. Consider two similar particles having energies E_1 and E_2, and suppose that the two particles collide. One particle gains a small amount of energy δ, while the other one loses the same amount of energy since the law of conservation of energy must be satisfied. Accordingly,

$$\begin{aligned}
E_1 + E_2 &= (E_1 + \delta) + (E_2 - \delta) \\
&= E_3 + E_4
\end{aligned} \tag{16}$$

where E_3 and E_4 are the respective energies after the collision. Next, suppose that F such collisions occur per second. The probability of this happening is proportional to the joint probability that one particle has the energy E_1, namely, $f(E_1)$, and that the other particle has the energy E_2, namely, $f(E_2)$. Therefore

$$F = Cf(E_1)f(E_2) \tag{17}$$

where C is a proportionality constant. Under equilibrium conditions, the number of similar particles having energies E_1, E_2, E_3, and E_4 must remain unchanged by collisions, so that for each collision $(1,2) \rightarrow (3,4)$ there must be a collision $(3,4) \rightarrow (1,2)$. The number of such collisions occurring per second clearly is

$$F = Cf(E_3)f(E_4) \tag{18}$$

and, at equilibrium, the number of collisions of opposite kinds must equal, so that

$$\begin{aligned} f(E_1)f(E_2) &= f(E_3)f(E_4) \\ &= f(E_1 + \delta)f(E_2 - \delta) \end{aligned} \tag{19}$$

according to (16).

If the probabilities in (19) are correctly expressed by (15), then

$$\begin{aligned} e^{-E_1/kT}e^{-E_2/kT} &= e^{-(E_1+\delta)/kT}e^{-(E_2-\delta)/kT} \\ e^{-(E_1+E_2)/kT} &= e^{-(E_1+E_2)/kT}e^{-(\delta-\delta)/kT} \end{aligned} \tag{20}$$

and the equality in (20) is self-evident. In fact, it can be shown that, except for a constant, the probability function in (15) is the only one that satisfies the condition expressed in (19).

According to classical mechanics, the number of particles in an assembly having energies E_i is unrestricted and is governed solely by (2). According to quantum mechanics, however, the energies of the particles in an assembly are determined by quantum numbers which must satisfy the Pauli exclusion principle. This means that two particles having energies E_1 and E_2 can undergo transitions to energy states E_3 and E_4 only if the latter two states are not already occupied, and a collision between two similar particles, expressed by (16), can occur only if both energy states E_3 and E_4 are empty. The probability that each of these states is occupied is, respectively, $f(E_3)$ and $f(E_4)$, so that the probability that each is empty is $[1 - f(E_3)]$ and $[1 - f(E_4)]$, respectively. The joint probability that the initial two states are occupied and the final two states are empty is given by the product of their individual probabilities. This means that for F collisions per second to occur,

$$F = Cf(E_1)f(E_2)[1 - f(E_3)][1 - f(E_4)]. \tag{21}$$

Under equilibrium conditions, the reverse transitions must occur with the same frequency, so that

$$F = Cf(E_3)f(E_4)[1 - f(E_1)][1 - f(E_2)]. \tag{22}$$

Equating the right sides of (21) and (22) and dividing through by $Cf(E_1)f(E_2)f(E_3)f(E_4)$,

$$\left[\frac{1 - f(E_3)}{f(E_3)}\right]\left[\frac{1 - f(E_4)}{f(E_4)}\right] = \left[\frac{1 - f(E_1)}{f(E_1)}\right]\left[\frac{1 - f(E_2)}{f(E_2)}\right]. \tag{23}$$

Finally, substituting $E_3 = E_1 + \delta$ and $E_4 = E_2 - \delta$ as required by the conservation law (16),

$$\left[\frac{1 - f(E_1 + \delta)}{f(E_1 + \delta)}\right]\left[\frac{1 - f(E_2 - \delta)}{f(E_2 - \delta)}\right] = \left[\frac{1 - f(E_1)}{f(E_1)}\right]\left[\frac{1 - f(E_2)}{f(E_2)}\right]. \tag{24}$$

The equality in (24) can be satisfied if

$$\left[\frac{1 - f(E_i)}{f(E_i)}\right] = Ce^{\beta E_i} \tag{25}$$

as can be verified by substituting the appropriate form of (25) for each of the functions in brackets in (24).

The two constants C and β can have any value and still satisfy the mathematical condition in relation (24). In order to deduce their physical equivalence, the terms in (25) are rearranged so as to solve for the probability that the ith state is occupied:

$$f(E_i) = \frac{1}{Ce^{\beta E_i} + 1} \tag{26}$$

and $f(E_i) \leq 1.0$, according to the usual definition of probability. For large values of the energy E_i, $Ce^{\beta E_i} \gg 1.0$ and the unity in the denominator can be neglected, so that

$$f(E_i) = \frac{1}{C} e^{-\beta E_i}. \tag{27}$$

By comparison with the Maxwell-Boltzmann law (15), it can be seen that $\beta = 1/kT$. Note that, according to (26) or (27), the probability that a particle has a very large energy is small, in agreement with the general principle that the particles in an assembly tend to occupy the lowest energy states first. Finally, let $C = e^{-E_0/kT}$, where E_0 is a constant called the *Fermi energy*. The probability that the ith quantum state is occupied is then given by the *Fermi-Dirac distribution function*

$$f(E_i) = \frac{1}{e^{(E_i - E_0)/kT} + 1}. \tag{28}$$

The physical meaning of the constant E_0 and its magnitude are discussed in subsequent chapters where the Fermi-Dirac statistics are applied to the valence electrons in crystals. Relation (28) is quite general, however, and can be used to describe the energy distribution for most assemblies obeying quantum mechanics rather than classical mechanics.

The meaning of the distribution in (28) can be understood best by plotting the Fermi-Dirac distribution (Fig. 4) as a function of $(E_i - E_0)$ for two different temperatures. First consider what happens as $T \to 0$ deg K. For $E_i < E_0$, the exponent in the exponential term is negative and the exponential tends to zero, since $e^{-\infty} = 0$. Thus, provided that $(E_i - E_0)$ is negative, $f(E_i) = 1.0$ at absolute zero. This means that all states whose energy is less than E_0 are completely filled. When $E_i > E_0$,

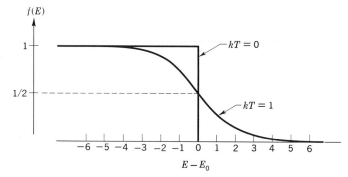

FIG. 4. Fermi distribution function at two temperatures. (*After Slater.*)

however, the exponential becomes infinitely large, since $e^{+\infty} = \infty$ and $f(E_i) = 0$ regardless of the actual magnitude of E_i. Consequently, all states having energies larger than E_0 are empty at absolute zero of temperature. At any other temperature $T > 0$ deg K, some of the particles may have energies greater than E_0 and the distribution changes more smoothly. This is shown by the second curve in Fig. 4, where the temperature was chosen such that $kT = 1.0$ for convenience in plotting.

The fraction of unoccupied states to the left of E_0 can be calculated with the aid of (28).

$$1 - f(E_i) = 1 - \frac{1}{e^{(E_i-E_0)/kT} + 1}$$
$$= \frac{e^{(E_i-E_0)/kT} + 1 - 1}{e^{(E_i-E_0)/kT} + 1}$$
$$= \frac{1}{e^{-(E_i-E_0)/kT} + 1}. \tag{29}$$

It can be seen from (29) that the distribution function is symmetrical about the point $E_i = E_0$ except for a change of sign in the exponent.

Note that the inflection in the distribution curve occurs at $f(E_0) = \frac{1}{2}$ regardless of the temperature.

Applications of statistical mechanics

Kinetic energy of a free particle. The statistics of Fermi and Dirac are very useful in the electron theory of crystals. For this reason, further discussion of this topic is postponed until the next chapter. The application of Maxwell-Boltzmann statistics to several problems in solid materials not involving electronic processes is considered in the remainder of the chapter. Returning to the Maxwell-Boltzmann distribution shown in Fig. 2, it can be seen that it is asymmetric about its maximum. This means that the average energy of a particle is not the same as the most probable energy corresponding to the position of the maximum. To determine the mean energy of a free particle such as an atom in a gas, multiply the kinetic energy of an atom $p^2/2m$ by the corresponding fraction of atoms dN/N given by (7) and integrate over all momentum space (Fig. 1). This gives a suitably weighted average energy

$$\bar{E} = \int\!\!\!\int\!\!\!\int_{-\infty}^{+\infty} \frac{p^2}{2m} (2\pi mkT)^{-\frac{3}{2}} e^{-p^2/2mkT} \, dp_x \, dp_y \, dp_z. \tag{30}$$

In order to carry out the integration, let $a = 1/2mkT$ and

$$p^2 = p_x^2 + p_y^2 + p_z^2,$$

so that (30) becomes

$$\bar{E} = \frac{1}{2m} \left(\frac{\pi}{a}\right)^{-\frac{3}{2}} \int\!\!\!\int\!\!\!\int_{-\infty}^{+\infty} (p_x^2 + p_y^2 + p_z^2) e^{-a(p_x^2 + p_y^2 + p_z^2)} \, dp_x \, dp_y \, dp_z. \tag{31}$$

The triple integral in (31) is a sum of three equivalent terms like

$$\int_{-\infty}^{+\infty}\!\int_{-\infty}^{+\infty}\!\int_{-\infty}^{+\infty} p_x^2 e^{-ap_x^2} e^{-ap_y^2} e^{-ap_z^2} \, dp_x \, dp_y \, dp_z. \tag{32}$$

The two integrations over the y and z components of momentum have the form

$$\int_{-\infty}^{+\infty} e^{-ap^2} \, dp = \sqrt{\frac{\pi}{a}}$$

so that (32) reduces to

$$\frac{\pi}{a} \int_{-\infty}^{+\infty} p_x^2 e^{-ap_x^2} \, dp_x = \frac{\pi}{a} \times \frac{1}{2a} \sqrt{\frac{\pi}{a}} = \frac{1}{2a} \left(\frac{\pi}{a}\right)^{\frac{3}{2}}$$

and (31) becomes

$$
\begin{aligned}
\bar{E} &= \frac{1}{2m}\left(\frac{\pi}{a}\right)^{-\frac{3}{2}} \times 3 \left(\frac{1}{2a}\right)\left(\frac{\pi}{a}\right)^{\frac{3}{2}} = \frac{3}{4ma} \\
&= \frac{3(2mkT)}{4m} \\
&= \tfrac{3}{2}kT.
\end{aligned}
\tag{33}
$$

The above analysis can be carried out alternatively by considering each of the three components of momentum separately. In that case, the average energy associated with each momentum (or space) coordinate has the form of (32) and is equal to $\frac{1}{2}kT$. The total average energy is then equal to the sum of the average energies associated with each degree of freedom, $\frac{1}{2}kT + \frac{1}{2}kT + \frac{1}{2}kT = \frac{3}{2}kT$. This illustrates a very important principle of classical statistical thermodynamics called the *equipartition of energy*. According to this principle, the total kinetic energy of a particle is equally distributed between its three degrees of freedom, on the average, so that each has the same average energy. It is sufficient, therefore, to calculate the average energy for one degree of freedom in order to determine the total energy. This not only simplifies such calculations, but also points out certain relations between the behavior of individual particles and an "average" particle in the assembly. A particle moving in a particular direction may have two components of momentum equal to zero, but the third component has some finite value, so that it is unlikely that any particle has zero energy (Fig. 2). When the average energy for the particles in an assembly is evaluated in (33), all particles are considered jointly and their momenta, rather than their directions of motion, are important. This means that the distribution of the momenta (velocities) of the particles is independent of their coordinate (position) distribution; that is, the average energy does not depend on the choice of the reference system.

Specific-heat theories. It is possible to characterize the thermal properties of a material by its ability to absorb heat. The *specific heat*, or heat capacity, of a material is defined as the amount of thermal energy absorbed by a unit mass when its temperature is raised by one degree. In measuring this quantity, it is necessary to keep either the volume or the pressure constant. The specific heat at constant volume is defined by

$$
C_V = \frac{dE}{dT}
\tag{34}
$$

where dE is the increment of internal energy produced by an increase in the material's temperature of amount dT. The atoms in a monatomic gas have three degrees of freedom, so that, according to the equipartition of energy (33), they have an average energy of $\frac{3}{2}kT$ per atom, or

$$N_0 \times \tfrac{3}{2}kT = \tfrac{3}{2}RT \tag{35}$$

per kilogram atomic weight, where N_0 is Avogadro's number and $R = N_0 k$ is called the *gas constant*. The specific heat for an ideal gas, therefore, is

$$C_V = \frac{d(\frac{3}{2}RT)}{dT}$$
$$= \tfrac{3}{2}R. \tag{36}$$

The atoms in a solid material are not free to move about like atoms in a gas, so that they absorb heat by vibrating about their mean positions in the solid. It is convenient, therefore, to liken their vibration to that of harmonic oscillators. Each atom still has three degrees of freedom, which can be expressed by three harmonic oscillators vibrating along three mutually perpendicular directions. The average kinetic energy of a classic harmonic oscillator is $\frac{1}{2}kT$. Its potential energy also is $\frac{1}{2}kT$, so that the total average energy that can be absorbed by one atom is $1kT$ per degree of freedom. Making use of the equipartition law, the average energy absorbed by a kilogram of the material is $3N_0 kT$, so that its specific heat is

$$C_V = \frac{d(3N_0 kT)}{dT}$$
$$= 3R. \tag{37}$$

This is known as the *rule of Dulong and Petit*, who had observed that the specific heat of all solids tended to the same constant value at elevated temperatures.

At lower temperatures, particularly as $T \to 0$ deg K, the specific heat of crystalline materials tends toward zero, in clear violation of (37), which is independent of temperature. This is another example of the limitation of the applicability of classical mechanics to solids at high temperatures or large energies; note, for example, that the Fermi-Dirac distribution has an exponential, or Boltzmann-like, tail at large energies and temperatures in Fig. 4. At lower temperatures and energies, however, it is necessary to make use of quantum mechanics. The energy states that a quantum harmonic oscillator can have are given by

$$E_n = (n + \tfrac{1}{2})h\nu \qquad \text{where } n = 0, 1, 2, 3, \ldots \tag{38}$$

The lowest energy state corresponding to $n = 0$ has an energy of $\frac{1}{2}h\nu$. This is called the *zero-point* energy, because it is independent of temperature and persists down to absolute zero. Unlike classical mechanics, therefore, quantum mechanics predicts that atoms have a small but finite energy at absolute zero. [Note that the Maxwell-Boltzmann distribution law (15) predicts that $f(E_i) = 0$ when $T = 0$ deg K, unless $E_i = 0$; that is, only zero energy states can be occupied.]

Since the energy states of a quantum oscillator are quantized, a transition from one energy state described by n_1 to the next higher state n_2 can be written

$$E_{n_2} - E_{n_1} = (n_2 + \tfrac{1}{2})h\nu - (n_1 + \tfrac{1}{2})h\nu$$
$$= (n_2 - n_1)h\nu. \qquad (39)$$

The integers n_1 and n_2 differ by unity, so that the above transition requires the absorption of one quantum of energy of amount $h\nu$. If the transition proceeds in the reverse order, that is, from a higher energy state to one of lower energy, a quantum of energy is emitted. By analogy to similar transitions in which quanta of electromagnetic energy called photons are absorbed or emitted, the term *phonon* has been adopted for the description of quanta of thermal energy. The use of such a representation has the great advantage that the results of the studies of interactions of electromagnetic radiations with solids can be transposed directly to yield results concerning the elastic vibrations in solids. Although a phonon is actually an elastic wave propagating through a solid (just as the photon is also an electromagnetic wave), it is convenient to think of the associated quantum of energy as a particle present in the solid that is capable of interacting with other particles in the solid.

The first to apply quantum mechanics to the specific-heat theory of solids, Einstein assumed that each atom can vibrate independently of all other atoms. Although admittedly this is an oversimplification, since the vibration of an atom is affected by the vibrations of its neighboring atoms, it is mathematically very convenient. Thus a solid containing N atoms can be represented by $3N$ harmonic oscillators all vibrating with the same frequency ν. In place of (38), Einstein used the original Planck postulate that $E_n = nh\nu$ for an oscillator and calculated an average vibrational energy for one kg/mole of material,

$$\bar{E} = 3N_0 \frac{h\nu}{e^{h\nu/kT} - 1}. \qquad (40)$$

The specific heat for any temperature, then, is obtained by differentiating (40) with respect to temperature:

$$C_V = 3R \left(\frac{h\nu}{kT}\right)^2 \frac{e^{h\nu/kT}}{(e^{h\nu/kT} - 1)^2}. \qquad (41)$$

At very high temperatures, $KT \gg h\nu$, so that the denominator in (40) can be expanded to give

$$e^{h\nu/kT} - 1 = 1 + \frac{h\nu}{kT} + \cdots - 1 \simeq \frac{h\nu}{kT}$$

and on substituting back in (40),

$$\bar{E} = 3N_0 \frac{h\nu}{h\nu/kT} = 3N_0 kT \tag{42}$$

which leads to the Dulong-Petit rule (37) for the specific heat at elevated temperatures. At very low temperatures, $kT \ll h\nu$ and $e^{h\nu/kT} \gg 1$, so that (40) becomes

$$\bar{E} = 3N_0 h\nu e^{-h\nu/kT} \tag{43}$$

from which the specific heat at constant volume

$$C_V = \frac{d\bar{E}}{dT} = 3N_0 k \left(\frac{h\nu}{kT}\right)^2 e^{-h\nu/kT}. \tag{44}$$

According to (44), the specific heat should decrease exponentially with decreasing temperature, and suitable values of the frequency ν can be found to give a fair agreement with measured values. These values of frequency, called the *Einstein characteristic frequency* ν_E, are commonly used to define the *Einstein characteristic temperature* of a solid $\Theta_E = h\nu_E/k$. Careful experiments have shown, however, that the decline in the specific heat is more gradual than that predicted by Einstein and is more nearly proportional to T^3.

A more sophisticated model was adopted by Debye to represent a solid. Debye assumed that the atomic vibrations were not independent of each other and considered the three normal modes of vibration of a continuous solid made up of individual atoms. It can be shown that the frequencies of atomic vibrations in such a solid are quantized according to

$$\nu = \frac{v}{2} \sqrt{\left(\frac{n_x}{X}\right)^2 + \left(\frac{n_y}{Y}\right)^2 + \left(\frac{n_z}{Z}\right)^2} \tag{45}$$

where v is the velocity of propagation, and n_x, n_y, n_z are integers in the directions of the three orthogonal axes X, Y, Z. An alternative description of this is to say that only those phonons whose energies or frequencies are given by (45) can be absorbed by the solid.

It is possible to think of the quantity under the square-root sign in (45) as representing a vector in reciprocal space. This is so because $1/X$, $1/Y$, $1/Z$ can be thought of as three reciprocal-lattice vectors defining a reciprocal lattice whose lattice points represent the allowed frequency values determined by the three integers n_x, n_y, n_z. The magnitude of a vector $\mathbf{R}$ from the origin of this lattice to any lattice point, then, is given by

$$R^2 = \left(\frac{n_x}{X}\right)^2 + \left(\frac{n_y}{Y}\right)^2 + \left(\frac{n_z}{Z}\right)^2 \tag{46}$$

and (45) can be written

$$\nu = \frac{v}{2} R. \tag{47}$$

It follows from this that the lattice points in the reciprocal lattice are the allowed frequencies and that the frequency increases as the length of **R** in reciprocal space increases.

Consider a rectangular-shaped crystal having one atom per lattice point of a primitive cubic lattice. The atoms are spaced a unit-cell edge a apart, so that the dimensions of the crystal are $X = N_x a$, $Y = N_y a$, $Z = N_z a$, where N_x is the number of atoms along the X edge, N_y is the number of atoms along Y, and N_z along Z. The total number of atoms and lattice points in the crystal is thus

$$N_x N_y N_z = N. \tag{48}$$

It turns out that the total number of allowed frequencies given by (45) just equals the total number of atoms present in the crystal. Consequently, the maximum value of n_x is $N_x = X/a$, and the maximum values of n_y and n_z are $N_y = Y/a$ and $N_z = Z/a$, respectively. The total number of allowed frequencies

$$N = \frac{XYZ}{a^3} = \frac{\mathcal{v}}{a^3} \tag{49}$$

according to (48), and $\mathcal{v} = XYZ$ is the volume of the crystal. Although a primitive cubic lattice was chosen for simplicity, similar results are obtained for more complicated structures.

Returning to the reciprocal-lattice intepretation of (47), the volume of a unit cell in reciprocal space is $1/a^3$, so that the number of allowed frequencies per lattice point $n_x n_y n_z$ is $\mathcal{v}$ according to (49). The number of allowed frequencies dN lying in a shell between R and $R + dR$ is contained within its volume, which is $4\pi R^2 \, dR$. All unique values of the frequency in (45) or (47) are obtained by considering positive integers, so that only the part of the shell lying in the positive octant need be considered. The volume of this part is $\frac{1}{2}\pi R^2 \, dR$, and it contains $4\pi \nu^2 \, d\nu / v^3$ lattice points according to (47). Since each lattice point has $\mathcal{v}$ allowed frequencies associated with it, the total number of allowed frequencies for the range ν to $\nu + d\nu$ is

$$dN = \frac{4\pi \nu^2 \, d\nu}{v^3} \mathcal{v}. \tag{50}$$

Note that the frequency density of allowed states $dN/d\nu$ is directly proportional to the crystal volume $\mathcal{v}$. Thus doubling the crystal volume

doubles the number of allowed frequencies but does not otherwise change
the distribution function in (50).

Only one possible mode of vibration has been considered so far. Actually, three independent modes of vibration must be considered, one
longitudinal and two transverse modes, so that a crystal containing N
atoms can have $3N$ frequencies, in agreement with the Einstein theory
discussed above. In general, the propagation velocities will be different
for these vibrations, and (50) becomes

$$dN = 4\pi\nu^2 \, d\nu \left(\frac{1}{v_l^3} + \frac{2}{v_t^3} \right) \upsilon \tag{51}$$

where the subscripts l and t refer to longitudinal and transverse vibrations, respectively. The distribution of frequencies in reciprocal space

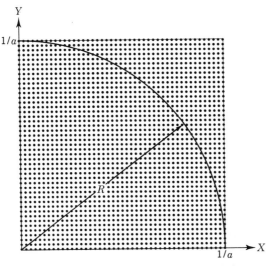

Fig. 5

can be determined directly by examining the reciprocal-lattice construction. Figure 5 shows the two-dimensional reciprocal lattice of a fictitious
crystal having a primitive square lattice. The number of allowed frequencies increases as R or ν increases until $R = 1/a$. Further increase in
R causes the number of allowed frequencies contained in the area $1/a^2$ to
decrease, reaching zero when $R \geq \sqrt{2}/a$. The distribution of allowed
frequencies for the analogous three-dimensional case is shown in Fig. 6,
in which it was assumed that $v_l = 2v_t$ in calculating $dN/d\nu$ according to
(51).

The distribution of allowed frequencies in Fig. 6 can be used next to
calculate the average frequency or average energy of vibration and, from

it, the specific heat. In order to simplify the calculations, Debye assumed that this frequency distribution can be approximated by the dashed curve in Fig. 6. The position of the cutoff frequency ν_D, called the *Debye characteristic frequency*, is chosen so that the total number of frequencies enclosed by the broken curve is exactly equal to the total number of allowed frequencies enclosed by the correct curve. This leads to an expression for the specific heat of the form

$$C_V = 9Nk \frac{1}{x_0^3} \int_0^{x_0} \frac{x^4 e^x}{e^x - 1} \, dx \tag{52}$$

where $x = h\nu/kT$, and $x_0 = h\nu_D/kT$. Finally, defining the *Debye temperature* $\Theta_D = x_0 T$, it can be seen that (52) expresses the specific heat in terms of a ratio of the actual temperature to the Debye temperature.

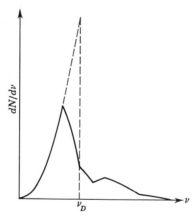

FIG. 6. Allowed frequency distribution for a cubic crystal having one atom per lattice point of its primitive lattice (for $\nu_l = 2\nu_t$).

Both the characteristic cutoff frequency and the Debye temperature are constant for any material. Typical values of the Debye temperature for some elements are as follows:

	C(diamond)	Na	Al	K	Cu	Ag	Au
Θ_D	1840	159	398	99	315	215	180 deg K

It is possible to determine the specific heat by integrating relation (52) numerically. For temperatures considerably smaller than the Debye temperature, $x_0 \gg 1$, and the integration can be carried out from zero to infinity. This definite integral has been evaluated by Debye and is equal to $(\frac{4}{15})\pi^4$. At very low temperatures, therefore, the specific heat becomes

$$C_V = \tfrac{12}{5}\pi^4 Nk \frac{T^3}{\Theta_D^3}. \tag{53}$$

This is the well-known *third-power law* of Debye and represents more correctly the change in the specific heat at extremely low temperatures. The specific-heat curves calculated by Einstein and Debye are compared in Fig. 7. As can be seen, they are quite similar, the chief differences arising at temperatures well below the characteristic temperature.

Imperfections in crystals. As a final example of the application of statistical mechanics in this chapter, the formation of imperfections in an otherwise perfect lattice array of atoms is considered. The presence of imperfections in a crystalline material affects its properties in different

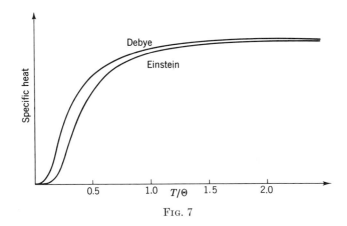

FIG. 7

ways, some of which are considered in later chapters. There are essentially three kinds of imperfections that can occur in crystals:

1. *Point defects*, such as interstitial atoms and missing atoms, also called vacancies

2. *Line defects*, which extend along some direction in a crystal and are commonly called *dislocations* in the otherwise perfect lattice array

3. *Plane defects*, which are two-dimensional arrays of the above two imperfection types, for example, grain boundaries in a polycrystalline material.

The electronic properties of most materials are affected by all three imperfection types; however, it is sufficient for present purposes to consider only the point defects. This is done in order to illustrate the application of statistical mechanics to the calculation of the number of imperfections present in a crystal under equilibrium conditions and in order to demonstrate its temperature dependence. As an example, a calculation is carried out below for atomic vacancies. The details of this calculation are similar to those carried out whenever statistical thermodynamics can be used to solve solid-state problems.

According to thermodynamics, the *Helmholtz free energy* of a crystal F is determined by

$$F = E - TS \qquad (54)$$

where E is the internal energy of the crystal

T is absolute temperature

S is its entropy.

It is left to the reader (Exercise 7) to demonstrate that the entropy can be correctly expressed in terms of the number of different ways of arranging atoms in a crystal W by the Boltzmann relation

$$S = k \ln W. \qquad (55)$$

Consider a perfect crystal composed of equal numbers of positively and negatively charged ions. In order for a cation vacancy to occur, a positive ion must somehow migrate out of its proper position in the structure to the crystal's exterior. If only positive ions migrate out of the crystal and collect on its surface, the surface will become positively charged. This positive surface charge opposes the migration of additional positive ions out of the crystal's interior. Simultaneously, the excess negative charge created inside the crystal is conducive to the formation of negative vacancies. In the absence of external forces, therefore, the number of oppositely charged vacancies inside a crystal tends to be equal.

Suppose that the crystal contains a total of N atoms and that n vacancy pairs are produced by removing n cations and n anions from the crystal's interior. The number of different ways in which each kind of ion can be removed is given by

$$\frac{N(N-1)(N-2) \cdots (N-n+1)}{n!} = \frac{N!}{(N-n)!\,n!}. \qquad (56)$$

The different ways in which n vacancy pairs can be formed is then obtained by squaring the expression in (56), since the numbers of cation and anion vacancies are equal. The creation of n vacancy pairs thus increases the crystal's entropy, according to the Boltzmann relation (55), by an amount

$$\Delta S = k \ln \left[\frac{N!}{(N-n)!\,n!} \right]^2. \qquad (57)$$

This in turn produces a change in the Helmholtz free energy:

$$\Delta F = \Delta E - T\,\Delta S$$

$$= nE_p - kT \ln \left[\frac{N!}{(N-n)!\,n!} \right]^2 \qquad (58)$$

where E_p is the energy required to remove a pair of atoms from the crystal's interior, so that nE_p represents the total change in its internal energy.

The logarithmic term in (58) containing factorials can be simplified by using Stirling's approximation, $\ln x! \simeq x \ln x - x$, so that

$$\ln \left[\frac{N!}{(N-n)!n!} \right]^2 \simeq 2[\ln N! - \ln (N-n)! - \ln n!]$$

$$\simeq 2[N \ln N - N - (N-n) \ln (N-n)$$
$$+ (N-n) - n \ln n + n]$$
$$\simeq 2[N \ln N - (N-n) \ln (N-n) - n \ln n]. \tag{59}$$

When equilibrium is attained at a given temperature T, the Helmholtz free energy is at a minimum and its first derivative, therefore, is equal to zero.

$$\frac{d(\Delta F)}{dn} = 0 = E_p - 2kT \left[\ln (N-n) - \ln n\right]$$

$$= E_p - 2kT \ln \left(\frac{N-n}{n} \right) \tag{60}$$

where the differentiation is with respect to n since the total number of atomic positions in the crystal N is not altered.

Rearranging the terms in (60),

$$E_p = 2kT \ln \left(\frac{N-n}{n} \right) \tag{61}$$

and

$$\frac{N-n}{n} = e^{E_p/2kT}. \tag{62}$$

The number of vacancies in a crystal is much smaller than the number of atoms; that is, $n \ll N$ and $N - n \simeq N$. It is possible, therefore, to use Eq. (62) to determine the approximate number of defects present at any temperature:

$$n \simeq Ne^{-E_p/2kT}. \tag{63}$$

For example, the energy required to remove a pair of oppositely charged ions from a sodium chloride crystal is approximately 2 eV, so that the number of vacancy pairs present in NaCl at room temperature is approximately 10^{12} defects/m³. The number of Na^{1+} and Cl^{1-} ions in 1 m³ of salt is approximately 10^{28}, so that, on an average, there is one defect present for each 10^{16} ions. Thus the neglect of n as compared with N in going from Eq. (62) to (63) is fully justified.

As can be seen from the above analysis, the probability that an energy state is occupied, or the fraction of occupied states, is given by expressions quite similar to the Maxwell-Boltzmann distribution law. It also follows from (63) that the number of defects present in a crystal increases as its temperature increases. Note that a certain amount of defects is present at all temperatures above absolute zero, which is the reason why imperfections are so important when the properties of materials are considered. The point defects in crystals normally do not interact with each other, except possibly in pairs, so that Maxwell-Boltzmann statistics are applicable. As already stated, when the interaction between particles becomes appreciable, it is necessary to resort to Fermi-Dirac statistics. This is the case for electronic interactions in crystals, so that Fermi-Dirac statistics are used further in the subsequent chapters of this book.

Suggestions for supplementary reading

Ronald W. Gurney, *Introduction to statistical mechanics* (McGraw-Hill Book Company, New York, 1949).

E. H. Kennard, *Kinetic theory of gases* (McGraw-Hill Book Company, Inc., New York, 1938).

Robert L. Sproull, *Modern physics* (John Wiley & Sons, Inc., New York, 1956), especially pp. 24–44.

Suggestions for further reading

Adrianus J. Dekker, *Solid state physics* (Prentice-Hall, Inc., Englewood Cliffs, N.J., 1957).

Charles Kittel,. *Introduction to solid state physics*, 2d ed. (John Wiley & Sons, Inc., New York, 1956).

J. C. Slater, *Introduction to chemical physics* (McGraw-Hill Book Company, Inc., New York, 1939).

Exercises

1. There are approximately 4.53×10^{24} valence electrons in one kilogram of cesium. The average kinetic energy of these valence electrons at room temperature is 1.55 eV. Calculate the average velocity of a valence electron. If the density of cesium is 1.87×10^3 kg/m^3, show that the de Broglie wavelength of these electrons is of the same order as the interelectron separation, assuming that they are uniformly distributed in the metal. (ANSWER: $\lambda_{\text{de Broglie}} = 9.8$ Å; $e - e = 5.0$ Å.)

2. Show that the Maxwell distribution in Fig. 2 reaches its maximum value when $E = \frac{1}{2}kT$.

3. Plot Maxwell's distribution of velocities (8) as a function of $v(m/kT)^{\frac{1}{2}}$. At what value of the velocity does the curve reach a maximum? What is the average velocity of a particle at some temperature T?

4. Making use of the equipartition of energy, the Maxwell distribution law for the x component of velocity has the form

$$\frac{dN}{N} = v_x{}^2 \left(\frac{m}{2\pi kT}\right)^{\frac{1}{2}} e^{-mv_x{}^2/2kT} \, dv_x.$$

Plot the distribution dN/dv_x as a function of v_x in units of $2(kT/m)^{\frac{1}{2}}$. At what value of the velocity does this distribution have a maximum? How can you make this compatible with the energy distribution in Fig. 2?

5. Suppose that the probability of one kind of particle having the energy E_1 is $\alpha(E_1)$ and of another kind of particle, $\beta(E_2)$. Show that, under equilibrium conditions, $\alpha(E_1) = Ae^{-E_1/kT}$ and $\beta(E_2) = Be^{-E_1/kT}$, verifying the universality of the Maxwell-Boltzmann law.

6. The probability that an assembly of particles is composed of W individual complexions can be expressed by the Maxwell-Boltzmann distribution law $W = Ce^{-F/kT}$, where F is the Helmholtz free energy given by (54). As the temperature increases, it is more probable that the internal energy of the system E increases, so that the proportionality constant C can be set equal to $e^{E/kT}$. Show that this leads to the Boltzmann relation (55) in the text.

7. The density of allowed energy states in a metal $dN/dE = CE^{\frac{1}{2}}$, where C is a proportionality constant, and it must satisfy the condition that $\int_0^\infty (dN/dE)\, dE = N$, the total number of allowed states. Using the Fermi distribution function for $T = 0$ deg K, show that the average energy of an electron in the metal is $\frac{3}{5}E_0$. (HINT: First evaluate E_0 from the above condition that the total number of states for $E < E_0$ must be N at absolute zero of temperature.)

8. Experimental and theoretical investigations have shown that the formation of vacancies in metal crystals is more likely than the formation of interstitial atoms. If the energy required to remove a copper atom from the interior of a crystal to its surface is 4.61 eV, how many vacancies are present in copper at room temperature ($\simeq 300$ deg K)? Estimate the energy required to form an interstitial defect in copper if the number of interstitial atoms equals one-tenth of one per cent of vacancies present.

9. Another common point-defect type consists of an atom that has been displaced from its correct site to an interstitial one. If there are N atoms in the crystal and N_i interstitial sites in its structure, there are

$$\frac{N!}{(N-n)!n!}\frac{N_i!}{(N_i-n)!n!}$$

ways in which such *Frenkel defects* can be formed. By analogy to the analysis of vacancy pairs, also called *Schottky defects*, calculate the number of Frenkel defects formed in a crystal.

10. The energy required to form a Frenkel defect in silver chloride is 1.4 eV. Using the results of Exercise 9, how many Frenkel defects are there present in AgCl at room temperature? Which type of ion do you think goes into the interstitial site, Ag^{1+} or Cl^{1-}? Why?

6. free-electron theory

Early theories

Drude-Lorentz theory. In 1900, Drude made the then very bold proposal that metals are composed of positive metal ions whose valence electrons are free to roam among the ionic array, with the only restriction that they are confined to remain within the boundaries of the metal crystal. As already discussed in Chap. 4, the metal ions are bonded to the electrons by an electrostatic attraction between their positive charges and the negatively charged electron "gas" (Fig. 1). The valence electrons are not merely dissociated from their parent atoms, but are assumed to be as free to move about inside the metal as the atoms or molecules of a perfect gas. In general, their motion is random, but in an electric field, the negatively charged electrons stream in the positive-field direction and produce an electric current within the metal. In order to prevent the electrons from accelerating indefinitely, it was assumed in this model that they collide elastically with the metal ions. This leads to a steady-state current which is proportional to the applied voltage and explains the origin of Ohm's law. Although the details of this model have required modification, it should not be overlooked that it was proposed almost fifteen years before the periodic nature of atomic arrays in metals was firmly established by x-ray diffraction. Its lasting success is entirely due to Drude's perspicacity in realizing that the valence electrons in a metal are free to move through the metal as a whole as distinct from discrete atom-to-atom jumps.

In 1909, H. A. Lorentz carried this model to its logical conclusion by applying Maxwell-Boltzmann statistics to the electron gas. The mutual repulsion between the negatively charged electrons was neglected, and the potential field due to the positive ions was assumed to be everywhere constant. Despite these limiting assumptions of the *classical free-electron theory* of Drude and Lorentz, it was able to explain many properties of metals in a quantitatively satisfactory way. It had certain notable failures, however. Although it correctly predicted the magnitude of electrical resistivity of most metals at room temperature, the predicted resistivity dependence on temperature was proportional to $T^{\frac{1}{2}}$ instead of the observed linear dependence. It also yielded incorrect magnitudes for the specific heat and the paramagnetic susceptibility of

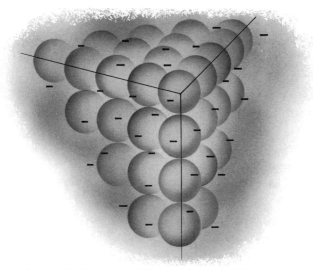

FIG. 1.   Drude model of a metal.   The positive metal ions are assumed to be imbedded in a negatively charged electron gas.

metals.   These quantitative errors were removed by Sommerfeld, in 1928, who substituted the quantum statistics of Fermi-Dirac in place of the classical statistics used by Lorentz.   Finally, it should be noted that even this *quantum free-electron theory* fails to explain two important aspects of materials.   It is incapable of explaining why some crystals have metallic properties and others do not and why the atomic arrays in crystals, including metals, should prefer certain structures and not others. These properties of crystals can be explained by the more general *zone theory* discussed in the next chapter.

Applications of classical theory.   The postulate that a metal contains *free* electrons is sufficient to explain most metallic properties.   The ease with which the free electrons can move under the influence of an applied field explains one of the most outstanding properties of a metal, namely, its high electrical and thermal conductivity. Since the theory does not depend on structure, it further indicates that the ratio of electrical to thermal conductivity should be constant for all metals at the same temperature. This relation is called the *Wiedemann-Franz law* and is borne out by experimental measurements.

Another characteristic property of metals is their high luster and complete opacity. Suppose an electromagnetic field such as a light beam falls on a metal crystal; the free electrons are set into forced oscillations having the same frequency as the incident light. Thus all the incident energy is absorbed by the free electrons, and the metal appears to be opaque. When an electron thus excited returns to its initial state, it does so by emitting a photon of the same energy as it absorbed initially. The light given off by the electron goes off equally in all directions, but only the light rays directed toward the metal's surface can get through. Thus the metal appears to reflect virtually all the light that is incident on it, giving it the characteristic metallic luster.

The specific heat and paramagnetism of metals also can be understood on the basis of interactions between the free electrons and the external energy source, whether thermal or magnetic in nature. The difficulty in the classical theory arises because the Maxwell-Boltzmann statistics permit all the free electrons to gain energy, leading to much larger predicted quantities than are actually observed. When quantum statistics are used, however, it turns out that only about one per cent of the free electrons can thus absorb energy. The resulting specific-heat and paramagnetic-susceptibility values are in much better agreement with experimental values. The introduction of Fermi-Dirac statistics into these calculations, therefore, is the principal accomplishment of the Sommerfeld theory, discussed next.

Sommerfeld theory

Quantum mechanics. Before considering the detailed modifications of the classical free-electron theory proposed by Sommerfeld, it is necessary to determine the restrictions imposed by quantum mechanics on the energies that a free electron can have inside a metal crystal. As shown in Chap. 3, the allowed energy levels of an electron bound to a single atom are quantized. In this section, the permissible energy levels are determined for an electron that is free to roam within the crystal but is prevented from leaving it by very high energy barriers at its surfaces. The barrier actually extends over a few atomic layers near the surface, and this complicates the analysis, so that a rigorous solution of this problem is not attempted here. Instead, an analogous model is used to obtain an approximate solution which turns out to give quantitatively acceptable results.

For the sake of mathematical simplicity, consider first an electron limited to remain within a one-dimensional "crystal" of length L. Next, assume that the potential energy everywhere within this crystal is constant and equal to zero. At the two ends of the crystal the electron is prevented from leaving the crystal by a very high potential energy

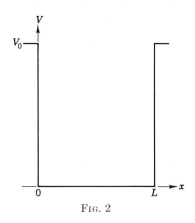

Fig. 2

barrier represented by V_0 in Fig. 2. Since the potential energy inside the crystal is zero, the Schrödinger equation has the form

$$\frac{d^2\psi}{dx^2} + \frac{8\pi^2 mE}{h^2}\psi = 0. \tag{1}$$

The general solution of this equation, by analogy to Chap. 3, is

$$\psi = C_1 \sin 2\pi \frac{\sqrt{2mE}}{h}x + C_2 \cos 2\pi \frac{\sqrt{2mE}}{h}x. \tag{2}$$

The values of the two coefficients can be determined according to the boundary conditions existing at $x = 0$ and $x = L$.

Before investigating the boundary conditions, consider the form of Schrödinger's equation on either side of the crystal, that is, in the regions where the potential energy is very large. Including the potential-energy term V_0, the Schrödinger equation can be written

$$\frac{d^2\psi}{dx^2} - \frac{8\pi^2 m(V_0 - E)}{h^2}\psi = 0. \tag{3}$$

The general solution of this equation for $V_0 > E$ is

$$\psi = C_3 e^{2\pi \frac{\sqrt{2m(V_0-E)}}{h}x} + C_4 e^{-2\pi \frac{\sqrt{2m(V_0-E)}}{h}x} \tag{4}$$

An examination of the solution in (4) shows that, as x increases in either a positive or negative sense, the value of the wave function also increases. This obviously cannot be, since this means that the probability of finding the electron outside the crystal increases with increasing distance from the crystal, becoming infinite in the limit as $\pm x$ approaches infinity. Consequently, $C_3 = 0$ when $x \geq L$ and $C_4 = 0$ when $x \leq 0$. This gives two different solutions, one for each side of the crystal.

When $x \geq L$:
$$\psi = C_4 e^{-2\pi\frac{\sqrt{2m(V_0-E)}}{h}x}. \tag{5}$$

When $x \leq 0$:
$$\psi = C_3 e^{2\pi\frac{\sqrt{2m(V_0-E)}}{h}x}. \tag{6}$$

It is now possible to consider the boundary conditions that solutions (2), (5), and (6) must satisfy. Since there is a finite probability of finding the electron anywhere in space, the wave function must be continuous everywhere. Similarly, it can be shown that $d\psi/dx$ must be continuous. Consider first the requirement that the slope be continuous at $x = L$. Differentiating Eqs. (2) and (5),

$$\left(\frac{d\psi}{dx}\right)_{x=L} = 2\pi\frac{\sqrt{2mE}}{h}C_1 \cos 2\pi\frac{\sqrt{2mE}}{h}L$$
$$- 2\pi\frac{\sqrt{2mE}}{h}C_2 \sin 2\pi\frac{\sqrt{2mE}}{h}L \tag{7}$$

and
$$\left(\frac{d\psi}{dx}\right)_{x=L} = -2\pi\frac{\sqrt{2m(V_0 - E)}}{h}C_4 e^{-2\pi\frac{\sqrt{2m(V_0-E)}}{h}L}$$
$$= -2\pi\frac{\sqrt{2m(V_0 - E)}}{h}\psi_{x=L}. \tag{8}$$

As the potential energy at the crystal surface increases, that is, as $V_0 \rightarrow \infty$, Eq. (8) becomes infinite unless simultaneously $\psi_{x=L} \rightarrow 0$. In the limiting case, $V_0 = \infty$ and $\psi_{x=L} = 0$. A similar argument can be used to show that, for an infinitely high potential barrier at $x = 0$, the wave function must be zero at that boundary also. Although infinitely high barriers are not encountered in crystals, the evaluation of the coefficients is greatly simplified by making this assumption. Considering the solution in (2) at $x = 0$,

$$\psi = 0 = C_1 \sin (0) + C_2 \cos (0)$$
$$= C_2. \tag{9}$$

At the other boundary, therefore,
$$\psi = 0 = C_1 \sin 2\pi\frac{\sqrt{2mE}}{h}L. \tag{10}$$

Excluding the trivial solution that C_1 also is zero, the only way (10) can be satisfied is for the sine to be zero. This occurs whenever

$$2\pi \frac{\sqrt{2mE}}{h} L = n\pi \qquad \text{where } n = 1, 2, 3, \ldots . \qquad (11)$$

From this it follows that the energy can have only the discrete values given by

$$E_n = \frac{h^2}{8mL^2} n^2. \qquad (12)$$

Note that, if L is large, the energy levels are spaced very closely together. For example, if $L = 1$ cm,

$$E_{n+1} - E_n \simeq 3.5 \times 10^{-19} \text{ eV}.$$

The potential barriers confining an electron to the interior of an actual crystal are not infinitely high and are determined in a complex way by the surface energies of the crystal. If this potential barrier at the surface of a crystal is high but not infinite, the wave function has the form shown in Fig. 3 for the case $n = 2$. (Compare this with Fig. 2 in Chap. 3.) Note that the wave function is sinusoidal in the region $0 \leq x \leq L$ and exponential outside this region in accordance with solutions (2), (5), and (6). Although the exact evaluation of the coefficients in these equations is not attempted here, it seems reasonable to expect that the extension of the wave function beyond the potential barrier is inversely related to the height of the barrier. Furthermore, if the barrier is very narrow, it is more likely that the wave function can extend beyond it. In this case there is a small but finite probability ($\propto |\psi|^2$) of finding the electron on the other side of the barrier. This ability of the electron to penetrate a potential barrier is called the *tunnel effect* and is a direct consequence of the application of quantum mechanics to this problem.

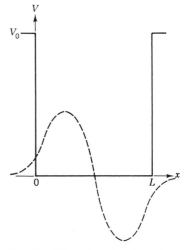

Fig. 3. Wave function for a free electron in a metal for $n = 2$ when V_0 is finite.

In three dimensions, the crystal can be approximated by a cube of edge L inside which the potential energy is zero. It can be shown that the wave function inside such a cube is given by

$$\psi = C \sin \frac{n_x\pi}{L} x \sin \frac{n_y\pi}{L} y \sin \frac{n_z\pi}{L} z \qquad (13)$$

where L is the length of a cube edge, and n_x, n_y, n_z are any three integers greater than zero. The corresponding form of the energy is

$$E_{n_x n_y n_z} = \frac{h^2}{8mL^2} (n_x^2 + n_y^2 + n_z^2). \tag{14}$$

The three integers in (14) are the first three quantum numbers of an electron already encountered in Chap. 3. The fourth, or spin quantum number, must be added to complete the description of the state of an electron, but it does not affect the energy of a free electron in the absence of an external field. Since the forms of (14) and (12) are identical except for the number of integers n, a qualitatively accurate picture of the metallic state can be obtained by considering the one-dimensional case or by letting $n^2 = n_x^2 + n_y^2 + n_z^2$. Note that the same energy value or level in (14) obtains for various combinations of the same three integers n_x, n_y, n_z, for example, 211, 121, and 112. Each combination of integers represents a different wave function having the same energy. Such an energy level is said to be three-fold *degenerate*.

The parabolic variation of energy as a function of n is shown in Fig. 4. Although the energy variation is drawn as a continuous curve, it actually consists of discrete points corresponding to the discrete values of E_n. Because adjacent energy values differ by less than 10^{-18} eV, it is not possible to show actual breaks in this *quasi-continuous* energy distribution drawn in Fig. 4.

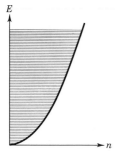

FIG. 4. Allowed energy values for free electrons in a metal.

Momentum space. The kinetic energy of an electron can be related to its momentum by $E = p^2/2m$, and the potential energy is assumed to be equal to zero everywhere inside the metal. Consequently, the total energy of the electron equals its kinetic energy, and Eq. (14) can be written

$$E = \frac{p^2}{2m} = \frac{h^2}{8mL^2} (n_x^2 + n_y^2 + n_z^2). \tag{15}$$

Rearranging the terms in (15),

$$\left(\frac{2L}{h}\right)^2 p^2 = (n_x^2 + n_y^2 + n_z^2) \tag{16}$$

from which it follows that the momentum of an electron similarly can be expressed in terms of its three quantum numbers.

The form of (16) immediately suggests that all possible values of the momentum can be represented in momentum space by a lattice of points corresponding to the three quantum-number values. Such a *momentum lattice* can be constructed by selecting unit vectors along the x, y, and z directions to represent equal increments of momentum. It is then possible to define a vector $\mathbf{R}$ in the momentum lattice, from the origin to any lattice point. The length of this vector is given by

$$\mathbf{R}^2 = \mathbf{n}_x^2 + \mathbf{n}_y^2 + \mathbf{n}_z^2. \tag{17}$$

It follows from (17) and (15) that as the length of $\mathbf{R}$ increases, so does the momentum, and hence the energy of the state represented by a lattice point. In the free-electron model of a metal containing some 10^{23} electrons, the lattice points are so closely spaced that it is not possible to show all of them in a drawing. The presence of discrete points rather than a continuous distribution is in complete agreement with the Pauli exclusion principle. The momentum-lattice construction is particularly useful when the density of states having the same momentum (or energy) values is to be calculated. An example of such a calculation has been given in the preceding chapter.

Fermi-Dirac distribution. The free-electron model of metals has survived to the present time because it is a fairly close approximation to the actual situation in metals, particularly the monovalent elements such as the alkali metals. Quantum mechanics requires that each valence or free electron be indistinguishable but that the state of each electron be specified by the three quantum numbers n_x, n_y, n_z, together with the spin quantum number, which can have either of the two values $\pm\frac{1}{2}$. Moreover, the Pauli exclusion principle does not permit more than one electron to have the same four quantum numbers, so that many of the occupied states in a metal containing some 10^{23} free electrons must be described by fairly large quantum numbers. All this suggests that the most convenient way to discuss the metallic state is to make use of statistical mechanics. As shown in Chap. 5, the probability that a particular quantum state having an energy E is occupied is given by the Fermi-Dirac function

$$f(E) = \frac{1}{e^{(E-E_0)/kT} + 1}. \tag{18}$$

The energy of the state E must be one of the allowed values as determined by (14), and E_0 is usually called the Fermi energy of the metal. For $E \ll E_0$, the exponential term in (18) is very small, so that $f(E)$ is essentially equal to unity. This means that all states having energies much smaller than the Fermi energy are completely occupied. For $E \gg E_0$, the exponential is very large and $f(E)$ rapidly tends to zero as the energy increases. The actual distribution is shown in Fig. 5, and it is evident that the higher energy states are virtually empty.

The important consequence of applying quantum statistics to the free electrons in a metal is that it is now possible to predict correctly how many electrons can gain energy from an external source. The amount of energy that an electron can gain from either a thermal source or an electric or magnetic field is of the order of kT (at room temperature $1kT \simeq 0.03$ eV). Only the electrons whose energy is already very close to the Fermi level can actually gain more energy because electrons occupying lower energy levels would have to undergo transitions to already occupied states, in clear violation of the Pauli exclusion principle. The fraction of electrons that can undergo such transitions can be deduced by considering the shaded region in Fig. 5. Since the height of the curve in Fig. 5 is equal to unity, the ratio of the shaded area to the total area under the curve is approximately

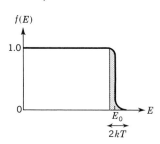

$$\frac{kT}{E_0} \simeq \frac{0.03}{3.00} = \frac{1}{100}$$

selecting 3.0 eV as a representative value of the Fermi energy in a metal.

At the absolute zero of temperature, all states having energies less than E_0 are completely filled and those having greater energies are completely empty. Thus the maximum energy that an electron can have at absolute zero is the Fermi energy E_0. The actual value of this energy can be determined directly by making use of the lattice in momentum space discussed in the previous section. According to this concept, the lattice points representing the allowed states at absolute zero are occupied out to some maximum value of R. Since each state can be occupied by two electrons having opposite spins, there are a total of $N/2$ occupied states in a crystal containing N free electrons. The lattice points representing these states lie in the positive octant of the lattice (because n_x, n_y, and n_z can have positive values only) and are enclosed by a sphere whose radius R_{max} is given by (15) and (17):

$$R_{max} = \left(\frac{8mL^2E_0}{h^2}\right)^{\frac{1}{2}}. \tag{19}$$

The volume of the octant, $\frac{1}{6}\pi R_{max}^3$, contains the $N/2$ occupied states, or $N/2$ lattice points. Because each unit cell in this lattice containing one lattice point has a volume equal to unity, the volume of this octant is

$$\frac{N}{2} = \tfrac{1}{6}\pi \left(\frac{8mL^2E_0}{h^2}\right)^{\frac{3}{2}}$$

so that

$$E_0 = \frac{h^2}{8m}\left(\frac{3N}{\pi L^3}\right)^{\frac{2}{3}}. \tag{20}$$

Note that E_0 is a function only of N/L^3, that is, the number of free electrons per unit volume, since all the other factors on the right side of (20) are constants. Consequently, the Fermi energy does not change when two identical metals are joined together; that is, the value of the Fermi energy is independent of the size of the metal. The Fermi energies for a number of monovalent metals calculated from Eq. (20) are as follows:

Metal	Li	Na	K	Rb	Cs	Cu	Ag
E_0	4.72	3.12	2.14	1.82	1.53	7.04	5.51 eV

The above calculation of the Fermi energy in a metal is strictly correct only at the absolute zero of temperature. Actually, the Fermi level is not constant when electron-energy distributions in metal crystals are considered, but changes slightly with increasing temperature. Denoting the value of the Fermi energy at absolute zero by $E_0(0)$, its level at any other temperature is given by

$$E_0 = E_0(0)\left[1 - \frac{\pi^2}{12}\left(\frac{kT}{E_0(0)}\right)^2\right]. \tag{21}$$

At room temperature, the change in the Fermi level is only about two-tenths of a per cent, so that this effect is not important. At elevated temperatures, however, the decrease in E_0 becomes more significant.

It is often of interest to know the wavelength of an electron at the Fermi level. Since the Fermi energy is equal to the kinetic energy of such an electron, use can be made of the de Broglie relation to determine the minimum wavelength of a free electron.

$$\lambda_{\min} = 2\left(\frac{\pi}{3}\frac{\mathcal{V}}{N}\right)^{\frac{1}{3}} \tag{22}$$

as can be shown by a simple derivation left to Exercise 7. Note that $\pi/3 \simeq 1.0$ and $\mathcal{V}/N = a^3$ in a primitive cubic lattice having one atom per lattice point. Thus the magnitude of $\lambda_{\min}$ is of the order of $2a$.

The allowed states for the free electrons in a metal are given by the parabolic curve shown in Fig. 4. The actual distribution of electrons among the available sites at any temperature is determined by the Fermi-

Dirac function (18). Accordingly, it is possible to define the number of electrons per unit volume having energies between E and $E + dE$:

$$N(E)\, dE = f(E)S(E)\, dE$$
$$= \frac{S(E)\, dE}{e^{(E-E_0)/kT} + 1}. \qquad (23)$$

where $S(E)$ is the number of allowed quantum states in this energy range. Note that this distribution function embodies the Pauli exclusion principle. Since the denominator of (23) can never be less than unity, the density of electrons $N(E)$ having a particular energy E can never exceed the density of available states $S(E)$ having the same energy. The actual energy distribution of free electrons in a metal is illustrated in Fig. 6.

Applications of the free-electron theory

Electrical conductivity in metals. According to the quantum free-electron theory, the valence electrons in a metal occupy the allowed states as determined by the Fermi-Dirac distribution function. Each electron can be assigned a set of the allowed quantum numbers, and hence a specific kinetic energy and a corresponding velocity. In the absence of an external field, the electrons move in all possible directions, so that, on the average, there are as many electrons moving in one direction as there are moving in the exactly opposite direction.

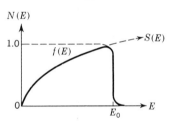

FIG. 6. The density of occupied states $N(E)$ in a metal.

This is illustrated for one direction, say, the x direction, in Fig. 7A. The density distribution for electrons moving in the $+x$ direction with velocities v_x is shown separately from the density distribution for electrons moving in the $-x$ direction. Since the number of electrons moving in each direction is the same, no net current is produced in the metal. When an external field of strength ε is applied along x, as indicated in Fig. 7B, the electrons are accelerated toward the positive side of the field. Only the electrons having energies close to E_0 can actually increase their velocity and gain energy, however, so that the distribution shifts, as shown schematically in Fig. 7B. The excess of electrons having velocity components in the $+x$ direction is then responsible for the electric current produced in the metal.

From the discussion so far, it appears that the electrons can move successively into higher and higher energy states as long as the field is applied. This is obviously contrary to the observed steady-state current in metals. To circumvent this difficulty, it was proposed that the electrons are prevented from attaining such high energies as a consequence of collisions with the metal ions that occupy most of the space in a metal. Such collisions are elastic, and the electrons, because of their much smaller mass, transfer most of their newly gained energy and momentum to the atoms following each collision. Thus the electrons encounter a resistance to current flow in the metal which is inversely related to the average distance that they can travel between collisions, called the

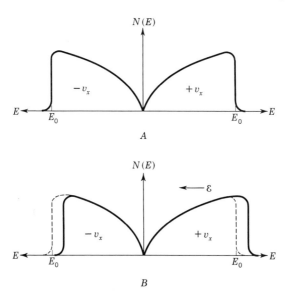

FIG. 7. The change produced in the equilibrium occupied state density (shown in A) by an external electric field ε is shown in B. Note that the net transitions are from states on one side of the Fermi energy to equivalent states on the other side, so that the average energy of the electrons actually contributing to the observed conductivity is E_0.

mean free path l. The electron's acceleration due to the field is proportional to its charge e and inversely proportional to its mass m, so that the *conductivity* σ is proportional to e/m. The *resistivity* of a metal ρ, which is reciprocal to the conductivity, then can be expressed by

$$\rho = \frac{1}{\sigma} = \frac{mv_0}{n_c e^2 l} \tag{24}$$

where v_0 is the average velocity of the electrons actually gaining energy from the field (determined by their average energy E_0), and n_c is the number of *conduction* electrons per unit volume.

Equation (24) can be used to determine the mean free path l once the resistivity of a metal has been measured, since e and m are universal constants and n_c and v_0 can be readily calculated. The resistivity of a metal, however, varies with the absolute temperature as shown in Fig. 8. Drude and Lorentz incorrectly postulated that this temperature dependence was due to the variation of the average energy or velocity of the electrons and that l was of the same order of magnitude as the interatomic spacings. According to quantum statistics, however, the average energy of the conduction electrons E_0 changes very little with temperature. It is necessary to assume, therefore, that the observed variation of resistivity is caused by a change in the mean free path. To understand this variation properly it is necessary to consider the wave nature of an electron and to treat the "collisions" between electrons and atoms by a

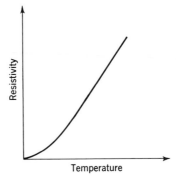

Fig. 8. Resistivity variation with absolute temperature in an ideal metal. Above some characteristic temperature, ρ is linearly proportional to T. At lower temperatures, $\rho \propto T^5$ and goes to zero as $T \to 0$ deg K.

scattering process similar to the diffraction of x-rays, discussed in Chap. 2. When the atoms in a crystal are essentially at rest at absolute zero of temperature, the electron waves are coherently scattered by the periodic array and proceed in a forward direction without disturbance. As shown in the next chapter, this is true whenever the wavelength of the electrons is greater than twice the periodic spacing in the direction of propagation (see also Exercise 13). As the temperature increases, however, the periodicity of the atomic array is disturbed by the atomic displacements resulting from thermal vibrations. These displacements cause an incoherent scattering of the electron waves, since they are not in phase with each other. The mean free path is inversely proportional to the square of the atomic displacements (amplitudes of vibration), which, however, are directly proportional to temperature. Thus $l \propto 1/T$ and $\rho \propto T$ according to (24). Actually, this is strictly correct only above a critical temperature, below which the temperature dependence is nonlinear. The actual variation of l with absolute temperature is illustrated in Fig. 9.

Not only thermal vibrations, but any discontinuity in the periodicity of an atomic array, such as point defects or dislocations, produces scattering and increases the resistivity. The effect of some of these discontinuities is further discussed in the next section. Assuming a perfectly periodic array except for thermal vibrations, quantitative calculations of the mean free path have been made for certain metals using quantum mechanics. These calculations lead to conductivities that are in good agreement with experimental values, particularly for alkali metals.

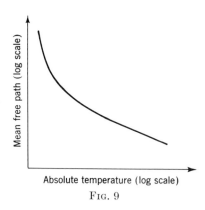

Absolute temperature (log scale)

FIG. 9

The results of some of these calculations are compared with measured values below. (Calculated and measured values are for $T = 20$ deg C.)

	Fe	Na	Al	Cu
Observed σ	1.0	2.2	3.5	5.9×10^7 ohm^{-1}-m^{-1}
Calculated σ	...	2.2	...	16.1×10^7 ohm^{-1}-m^{-1}

The rather large value calculated for copper results from an incorrect estimate of the density of conduction electrons in Cu based on the free-electron model for metals. A more accurate determination can be made when the *zone theory*, discussed in the next chapter, is employed in such calculations. The zone theory similarly provides an explanation for the relatively low conductivity of transition metals such as Fe, which intuitively should have an even higher conductivity than Cu according to the free-electron model since they have a larger number of valence electrons per atom.

The conductivity in crystals belonging to the cubic system is independent of direction; that is, such crystals are electrically isotropic. On the other hand, the conductivity does depend on direction in crystals of lower symmetry. For example, uniaxial crystals belonging to the hexagonal or tetragonal system offer a different resistivity for current flow parallel to the unique c axis than to that normal to it. Denoting the resistivity parallel to c by $\rho_\parallel$ and normal to it by $\rho_\perp$, the resistivity along any direction in a single crystal is given by

$$\rho = \rho_\perp + (\rho_\parallel - \rho_\perp)\cos^2\phi \tag{25}$$

where ϕ is the angle between the direction of current flow and the c axis. Since $a_1 = a_2$ in these crystals, the resistivity has cylindrical symmetry about the c axis. As an indication of the order of magnitude of the differences existing between these two resistivities, their values measured at 20 deg C for three metals having hexagonal closest-packed structures are listed below.

	$\rho_\parallel$	$\rho_\perp$
Cadmium	8.36	$6.87 \times 10^{-8}\,\Omega\text{-m}$
Magnesium	3.85	$4.55 \times 10^{-8}\,\Omega\text{-m}$
Zinc	6.06	$5.83 \times 10^{-8}\,\Omega\text{-m}$

Electrical conductivity in alloys. In addition to thermal scattering which can be treated mathematically either as a scattering of electron

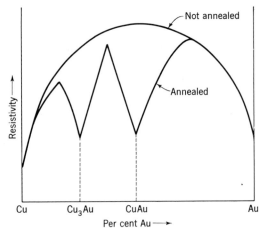

Fig. 10. Resistivity variation with composition in annealed and unannealed copper-gold alloys. (*After Seitz.*)

waves by aperiodicities in the ionic field or by interactions between electrons and phonons (see section on specific heat in Chap. 5), the presence of two or more kinds of atoms in an alloy causes scattering to occur at the point defects. In an ideal solid-solution alloy, these point defects are substitutional atoms. In general, the unlike atoms are more or less randomly distributed, so that the number of scattering centers increases in proportion to the number of substitutional atoms introduced. This is illustrated by the upper curve in Fig. 10, which shows the resistivity variation in copper-gold alloys as a function of composition. In the unannealed or random solid solution, the resistivity reaches a maximum at the 50:50 composition, since this corresponds to the maximum possible distortion of the structure and drops rapidly at both ends as the pure metals are approached. When the alloy is annealed for prolonged periods at an appropriate temperature, the atoms may tend to array themselves among preferred sites corresponding to an ordered, lower-energy structure. The restoration of a periodic array serves to decrease the number of scattering centers and decreases the resistivity of the alloy thereby. This is illustrated by the lower curve in Fig. 10, which shows the resistivity of copper-gold alloys following a suitable anneal. At the compositions of the ordered alloys Cu_3Au and $CuAu$, the curve exhibits definite minima, corresponding to the increased mean free path of the electrons following the ordering process. In practice, annealing produces another effect which serves to decrease the resistivity; namely, it serves to remove other imperfections present which may also act as scattering centers for conduction electrons.

In a random solid solution, the increased resistivity due to the introduction of solute atoms does not disappear at the absolute zero. The resistance that remains is usually called the *residual resistance*. The residual resistance is generally independent of temperature in dilute alloys but is markedly affected by changes in composition of the alloy, as indicated by the upper curve in Fig. 10. According to *Matthiessen's rule*, the resistivity of an ideal alloy is simply the sum of the residual resistance and a temperature-dependent resistivity that is quite similar to the resistivity encountered in pure metals. Note that this term is not a linear function of temperature at very low temperatures, as can be seen in Fig. 8. Deviations from Matthiessen's rule are sometimes observed, particularly in the case of transition metal alloys. These deviations are due to changes in the density of unoccupied states resulting from the alteration of the average electron-to-atom ratio with composition in an alloy.

The introduction of additional scattering centers by any imperfections present in a metal suggests that conductivity measurements can be used to evaluate the number of imperfections present. This is relatively difficult to do in metals because the conductivity is normally so high that even large deformations produce changes in the resistivity only of the order of one-tenth of one per cent. Conversely, the conductivity is quite small in nonmetals, so that the effect of increasing the density of imperfections present in, say, a semiconductor markedly decreases its conductivity. Although resistivity measurements are actually used for measuring imperfection densities in such crystals, the interpretation of measured resistivity values in terms of specific imperfections present is quite complicated because various kinds of imperfections can act as scattering centers.

Thermal conductivity.   Thermal conductivity in metals can proceed either via the atoms in the structure or via the free electrons.   The relative importance of these two mechanisms in metals, therefore, is considered next.   It is convenient to think of each mode of vibration of the atoms as a transient type of imperfection in the periodicity of the crystal, called a phonon, as discussed in the preceding chapter.   The phonons have characteristic energies $h\nu$, and as the temperature of a solid is raised, the increased atomic vibrations can be represented by increasing numbers of phonons flowing through the crystal.   The thermal conductivity of any material is proportional to its specific heat and to the mean free path of the moving particles.   The mean free path of phonons is of the order of 10 to 100 Å, implying that they undergo many collisions.   This is the reason why most materials are poor thermal conductors.   By comparison, the mean free path of conduction electrons in metals is considerably larger (Exercise 12), so that thermal conductivity in metals proceeds quite similarly to electrical conductivity.   In fact, it has been shown that the ratio of thermal to electrical conductivity in metals is equal to a constant times the absolute temperature.   This is called the Wiedemann-Franz law, and it predicts that the proportionality constant is the same for all pure metals and is theoretically equal to 2.45×10^{-8} (V/deg)^2.   Actually, small deviations from this value are observed in most metals, as is shown for several metals at 291 deg K:

	Fe	Cu	Zn	Ag	Pb
Wiedemann-Franz constant	2.88	2.88	2.31	2.36	2.45×10^{-8} (V/deg)^2

The thermal resistivity of a fairly pure metal can be shown to be the sum of two terms, one of which expresses the electron-phonon scattering and the other the electron scattering by impurities in the crystal.   As the concentration of impurities is increased in a metal, say, by alloying, the second term increases until it becomes dominant.   This means that the mean free path of the electrons is reduced to the same order of magnitude as that of the phonons, so that many alloys are relatively poor conductors of heat.   The thermal conductivities of several metals at room temperature are compared with that of stainless steel as follows:

	Na	Al	Cu	Ag	Steel
Thermal conductivity	139	227	307	419	35 W/m-deg K

Electronic specific heat.   According to the classical free-electron theory, all the valence electrons in a metal can absorb thermal energy. This leads to an electronic contribution to the specific heat of a metal of $\frac{3}{2}R$ per kilogram atomic weight.   Accordingly, the total specific heat of a metal can be expressed

$$C_V = C_V \text{ (atomic)} + C_V \text{ (electronic)}$$
$$= 3R + \tfrac{3}{2}R = \tfrac{9}{2}R. \tag{26}$$

As already stated in Chap. 5, Dulong and Petit observed that the specific heat above the characteristic temperature is the same for all solids and equal to $3R$ per kilogram atomic weight. Since the larger specific heat predicted by (26) is not observed in metals, this constitutes an important flaw in the classical free-electron theory. With the advent of quantum mechanics, it became apparent that this shortcoming is due to the incorrect assessment of the number of electrons that can absorb thermal energy.

It is possible to estimate the electronic specific heat in metals by making use of the Fermi-Dirac distribution function in Fig. 5. If it is assumed that only the electrons whose energy differs from E_0 by $\pm kT$ can absorb heat, so that there are approximately $2kT/E_0$ such electrons, then the electronic specific heat is given by

$$C_V \text{ (electronic)} = \tfrac{3}{2}R \times \frac{2kT}{E_0}$$
$$= 3R\,\frac{kT}{E_0}. \tag{27}$$

Since at room temperature kT/E_0 is of the order of 0.01 in most metals, it follows that the electronic contribution to the specific heat in (26) actually is of the order of one per cent. It also follows from (27) that the electronic specific heat is linearly dependent on the absolute temperature. It can be shown by a more exact calculation that for a monovalent metal

$$C_V \text{ (electronic)} = \frac{\pi^2}{2}\,R\,\frac{kT}{E_0} \tag{28}$$

which is of the same order as the qualitatively deduced expression (27).

At very low temperatures, the atomic specific heat is proportional to T^3, according to the Debye theory. The total specific heat of a metal, by analogy to (26), can be written, therefore,

$$C_V = \alpha T^3 + \gamma T \tag{29}$$

where α includes the coefficient of T^3 in the Debye expression derived in the previous chapter, and γ is the coefficient of T in Eq. (28). It is immediately apparent in (29) that the electronic contribution becomes more pronounced as T decreases. Dividing both sides of (29) by the absolute temperature,

$$\frac{C_V}{T} = \alpha T^2 + \gamma \tag{30}$$

so that a plot of the left side of (30) against T^2 should be a straight line whose intercept is γ and slope is α. Such a plot can be used, therefore, to evaluate the electronic specific heat of a metal and to determine the Debye characteristic temperature from α. Although the temperature dependence of the electronic specific heat actually is linear for most metals, the value of the coefficient γ frequently differs from that calculated by (28). This is so because the true density of conduction electrons is not correctly given by the free-electron model, as already noted, and requires the utilization of the zone theory, considered in the next chapter.

Suggestions for supplementary reading

A. H. Cottrell, *Theoretical structural metallurgy*, 2d ed. (St. Martin's Press, Inc., New York, 1957), especially pp. 44–52.

William Hume-Rothery, *Atomic theory for students of metallurgy* (Institute of Metals, London, 1946).

Frederick Seitz, *The physics of metals* (McGraw-Hill Book Company, Inc., New York, 1943).

Robert L. Sproull, *Modern physics* (John Wiley & Sons, Inc., New York, 1956).

Suggestions for further reading

Adrianus J. Dekker, *Solid state physics* (Prentice-Hall, Inc., Englewood Cliffs, N.J., 1957).

Charles Kittel, *Introduction to solid state physics*, 2d ed. (John Wiley & Sons, Inc., New York, 1956).

N. F. Mott and H. Jones, *The theory of the properties of metals and alloys* (Clarendon Press, Oxford, 1936; reprinted by Dover Publications, Inc., New York, 1958).

John C. Slater, *Quantum theory of matter* (McGraw-Hill Book Company, Inc., New York, 1951).

Exercises

1. In order to see at what values of E the Fermi-Dirac function differs appreciably from unity, substitute $E_0 + \varepsilon$ for E in Eq. (18). Solve for ε/kT when $f(E) = 0.2, 0.5, 0.8, 0.9,$ and 0.95.

2. Show that, when $E - E_0 \geq 2kT$, $f(E) \simeq e^{-(E-E_0/kT)}$. [Do this by actually calculating $f(E)$ for $E - E_0 = 2kT, 3kT,$ and $4kT$.]

3. The Fermi energy for Cs is $1.55\,\text{eV}$. By reversing the procedure used in deriving Eq. (20) in the text, determine the number of free electrons in $1\,\text{m}^3$ of cesium. How does this number compare with that obtained from the kilogram atomic weight of Cs, assuming one valence electron per atom? HINT: The density of Cs is $1.87 \times 10^3\,\text{kg/m}^3$; atomic weight of Cs is 132.9 AMU. (ANSWER: $8.26 \times 10^{27}\,\text{m}^{-3}$.)

4. The linear coefficient of thermal expansion of silver is $17 \times 10^{-6}/\text{deg}$ C. Assuming that thermal expansion is the only reason why E_0 changes with temperature, what are the values of the Fermi energy at 100 and 200 deg K as compared with 300 deg K? (Assume one conduction electron per atom in silver.)

5. What is the value of E_0 for sodium at 100, 200, and 300 deg K? Make the same assumptions suggested in Exercise 4 and use 6.2×10^{-5}/deg C as the linear-expansion coefficient.

6. Calculate the Fermi level in copper and silver at 727 deg C. What is the per cent error introduced by neglecting the temperature dependence of the Fermi level?

7. Derive expression (22) in the text for the minimum wavelength of a free electron occupying a state at the Fermi level.

8. Calculate λ_{min} for Na, K, Cu, and Ag. Compare the magnitude of this wavelength with the nearest-neighbor distances in these metals. (Na and K have body-centered cubic structures, while Cu and Ag have face-centered cubic structures.) The unit-cell constants of these metals are 4.24, 5.33, 3.61, and 4.08 Å, respectively. (ANSWER: $\lambda_{Na} = 6.83$ Å, $\lambda_{Cu} = 4.62$ Å, Na $-$ Na $= 3.72$ Å, Cu $-$ Cu $= 2.56$ Å.)

9. Assuming that the average kinetic energy of the conduction electrons according to the free-electron model is given by E_0, calculate their average velocity v_0, for Cs, Na, and Ag.

10. It is sometimes convenient to speak in terms of a *Fermi temperature* $T_0 = E_0/k$. Calculate the ratio of T/T_0 for K, Rb, and Cu at room temperature. (Assume $T = 27$ deg C.) What does this ratio tell you about the density of conduction electrons in a metal? (ANSWER FOR K: 83.)

11. Assuming that the mean free path l is independent of temperature and that the free-electron gas in a metal obeys classical mechanics, as Lorentz did, show that $\rho \propto T^{\frac{1}{2}}$. HINT: Consider the velocity distribution in an ideal gas.

12. In the Drude-Lorentz theory it was assumed that $l \simeq 2a$. For the case of Na, calculate v_0 from E_0 (given in the text) and then calculate l from Eq. (24). For Na, $n = 2.5 \times 10^{28}$ m^{-3} and $\rho = 4.5 \times 10^{-8}$ Ω-m.

13. Assuming that the scattering of electrons is similar to the diffraction of x-rays, for what wavelengths will the electrons be diffracted by a linear periodic array of atoms spaced d apart? How does this compare with the wavelengths that conduction electrons can have in a metal?

14. Estimate the electronic specific heat of aluminum ($E_0 = 11.7$ eV) and of copper ($E_0 = 7.1$ eV) at 1,300, and 1000 deg K.

15. Using the Debye equation in Chap. 5, calculate the specific heat of aluminum and copper at 1 deg K. How does this compare in magnitude with the electronic specific heat determined in Exercise 14?

7. zone theory

Elements of theory

Kronig-Penney model. It has been shown in the preceding chapter that solutions of the Schrödinger equation for a free electron

$$\frac{d^2\psi}{dx^2} + \frac{8\pi^2 m}{h^2}(E - V)\psi = 0 \tag{1}$$

are of the form

$$\psi = Ce^{\pm 2\pi i \frac{\sqrt{2m(E-V)}}{h}x} \tag{2}$$

where the plus or minus sign in the exponent corresponds to electron motion in a positive or negative x direction.

In the free-electron theory, it is assumed that the potential affecting a single valence electron is constant. Yet it is known from x-ray diffraction studies (Chap. 2) that the atomic array is periodic in all crystals, so that it appears that the potential also should vary in a periodic manner. The valence, or free electrons, in a metal are shared by the entire crystal, however, so that they tend to smooth out the periodic variations in the potential due to the metal ions. Thus the assumption of a constant potential is actually not a bad approximation for metals and explains why the free-electron theory is eminently successful in correctly explaining most metallic properties. Since the valence electrons in ionic and covalently bonded crystals are localized near their parent atoms, the assumption of a constant potential in such crystals is not valid. It is not surprising, therefore, that the free-electron model does not work in such crystals. The next step, logically, is to replace V in (1) by a periodically varying potential $V(x)$ and to see what effect this has on the electron wave functions and energies. As discussed below, the effect is quite pronounced and the resulting solutions are very helpful in explaining most of the properties of materials that were not satisfactorily explained by the free-electron model.

It is convenient at this point to define a *wave number* k which is equal to the number of complete wavelengths contained in one full period; that is,

$$k = \frac{2\pi}{\lambda}. \tag{3}$$

The kinetic energy of a free electron, $E - V = p^2/2m$, can be combined with the de Broglie relation

$$p = \frac{h}{\lambda} = \frac{hk}{2\pi} \tag{4}$$

to give

$$E - V = \frac{h^2k^2}{8\pi^2m}. \tag{5}$$

Substituting (5) in the Schrödinger equation (1), the general solution for a free electron moving in a constant potential field becomes

$$\psi(x) = Ce^{\pm ikx}. \tag{6}$$

In addition to simplifying the writing of wave functions like (6), the introduction of the wave number k is very useful when different directions in a crystal, each having its own unique period, are considered.

The Schrödinger equation including a one-dimensional periodic potential $V(x)$ can be written

$$\frac{d^2\psi}{dx^2} + \frac{8\pi^2m}{h^2}[E - V(x)]\psi = 0. \tag{7}$$

The solutions of this equation have been shown by Bloch to have the form

$$\psi(x) = u_k(x)e^{\pm ikx} \tag{8}$$

where $u_k(x)$ is a periodic function having the periodicity of the lattice in the x direction. This means that

$$u_k(x + a) = u_k(x) \tag{9}$$

where a is the period along x.

A comparison of (8) and (6) shows that the two wave functions differ only by the periodic function (9). This has the effect that the wave function in the new solution (8), commonly called a *Bloch function*, is modulated by the periodicity of the lattice, as can be seen by comparing its value at x and one lattice spacing a removed from x.

$$\psi(x + a) = u_k(x + a)e^{\pm ik(x+a)}$$
$$= \psi(x)e^{\pm ika} \tag{10}$$

according to (8) and (9).

The actual form of the periodic function $u_k(x)$ depends on the nature of the periodic field assumed in (7) and is different for various directions and wave numbers k. A variety of differently shaped periodic potentials can be assumed to exist in crystals. Computations based on several likely models have actually been performed, and the results obtained have shown a remarkably close agreement. Consider the one-dimensional potential field shown in Fig. 1, where it is assumed that the potential energy of an electron is zero near the nucleus and equals V_0 halfway between two adjacent nuclei which are periodically spaced a apart.

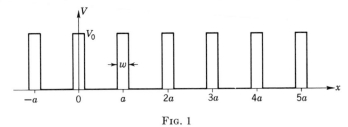

FIG. 1

This model was first postulated by Kronig and Penney, who used it to obtain solutions of the Schrödinger equation in the form of Bloch functions like (8). Similar to the case of the free-electron solutions described in Chap. 6, it is necessary that both ψ and $d\psi/dx$ be continuous throughout the crystal. The attendant computations were further simplified by Kronig and Penney by assuming that the product V_0w is constant; that is, if the height of the barrier V_0 increases, its width w correspondingly decreases. Even so, the mathematics is too complex to be considered in detail here, and only the results are presented. It turns out that solutions are possible only for certain electron energies, which can be determined from the relation

$$\cos ka = P\frac{\sin \alpha a}{\alpha a} + \cos \alpha a \tag{11}$$

where
$$P = \frac{4\pi^2 ma}{h^2}V_0w \tag{12}$$

and
$$\alpha = \frac{2\pi}{h}\sqrt{2mE}. \tag{13}$$

The meaning of the relation in (11) can be understood best by representing it graphically as shown in Fig. 2. The right side of (11) is a

function of the energy, so that it is shown plotted against αa by the solid
oscillating curve in Fig. 2.   The left side of (11) sets a limit, however, on
the values that this function can have.   Since cos ka only can assume
values between $+1$ and -1 unless k is imaginary, the physically meaningful solutions must lie between the two dashed horizontal lines in Fig. 2.
This limitation has a very important consequence; namely, certain
values of α or E do not lead to physically meaningful solutions.   Putting
this another way, an electron moving in the presence of a periodically
varying potential can have only energies lying in certain *allowed zones*.
The allowed energies in these zones are indicated by heavy lines along the
αa axis in Fig. 2.   Other possible energy values, leading to values of the

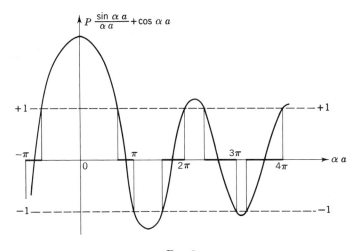

Fig. 2

function lying outside the dashed lines in Fig. 2, constitute *forbidden
zones* of energy for such an electron.   Note also (from 12) that, as the
size of the barrier $V_0 w$ in Fig. 1 increases, P in (11) increases.   This has
the consequence that the first term on the right in (11) becomes dominant
and the allowed energy zones in Fig. 2 become narrower.   On the other
hand, as $V_0 w$ decreases, the allowed zones become broader, until in the
limit, when $V_0 = 0$, the solution degenerates to the free-electron case, as
can be seen by substituting (5) for the energy in (11).   The physical
meaning of these two extremes can be seen by considering the different
kinds of electrons present in a crystal.   The inner electrons in an atom
are tightly bound to the nucleus.   This means that they are restrained
from leaving its vicinity, a situation that can be represented in a model by
two very high potential barriers.   The valence electrons, by comparison,
are bound more loosely, so that the potential barriers are correspondingly

reduced. Finally, the barriers disappear completely when an electron is free to move throughout a crystal, leading to the constant-potential approximation of the free-electron theory.

Allowed-energy zones. It is equally fruitful to explore the meaning of (11) by considering the left side of this relation. It is immediately evident that $\cos ka$ takes on specific values for each allowed energy value E; that is, for each E there is a corresponding k. Note, however, that $\cos ka$ is an even periodic function of ka, so that it has the same value

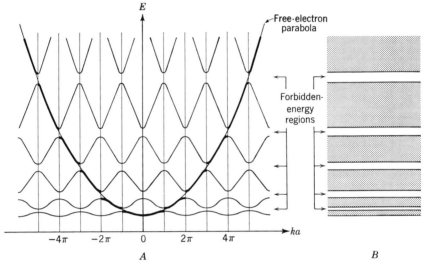

FIG. 3. Allowed- and forbidden-energy regions for an electron in a periodic field. (*A*) The allowed-energy values are indicated by the heavy lines along the free-electron parabola. Note that they depart slightly from those of a free electron near $\pm n\pi$. This means that the electron behaves like a free electron for most values of k except when $k \to n\pi$. (*B*) The shaded regions represent the allowed-energy values as they extend throughout the crystal. Note that both the allowed and forbidden regions become wider as the energy increases.

regardless of whether ka is positive or negative and whether integer multiples of 2π are added to ka. Consequently, the total energy E of the electron is an even periodic function of k with a period equal to $2\pi/a$. This periodic repetition of the allowed energies in several zones is shown in Fig. 3A. For comparison purposes, the parabolic energy dependence of a free electron given by (5) is also shown in this figure. As can be seen therein, the assumption of a periodic potential leads to discontinuities in this parabola. These energy gaps correspond, of course, to imaginary k values in (11) and represent forbidden-energy values. As discussed later in this book, it is sometimes convenient to represent the allowed and forbidden zones of energy diagrammatically, as shown in Fig. 3B.

When the free-electron parabola in Fig. 3A is compared with the energy curves for an electron in a periodic field, it can be seen that discontinuities in the parabola occur when k has the values

$$k = \frac{n\pi}{a} \qquad \text{where } n = \pm 1, \pm 2, \pm 3, \ldots \qquad (14)$$

Substituting (3) for k in (14),

$$\frac{2\pi}{\lambda} = \frac{n\pi}{a}$$

or

$$n\lambda = 2a. \qquad (15)$$

This equation can be compared with the Bragg law in Chap. 2 for the case of diffraction at a Bragg angle of 90 deg. In other words, if a is the interplanar spacing between planes that are normal to the propagation direction of the electron, then (15) defines the wavelength (or energy) for which total reflection backward occurs. Use is made of this equivalence in the next section.

Brillouin zones. The parallelism between Eq. (15) and the Bragg law is a direct consequence of the wave nature of electrons. In order to understand this relationship more clearly, rewrite (14) for the case $n = 1$, using crystallographic notation. The direction of movement of the electrons is then denoted by the subscript $[uvw]$, so that

$$k_{[uvw]} = \frac{\pi}{d_{(hkl)}} \qquad (16)$$

where $d_{(hkl)}$ is the interplanar spacing of the planes which are normal to $[uvw]$ (see Chap. 1 for a discussion of these relationships). In two dimensions, directions are denoted $[uv]$ and the "planes" become lines (hk).

Consider the two-dimensional "crystal" having a square lattice, shown in Fig. 4. Next, consider electrons moving in the direction [21] which is normal to the dotted lines (21). According to relation (16), total reflection of the electrons occurs when the wave number satisfies the condition

$$k_{[21]} = \frac{\pi}{d_{(21)}}. \qquad (17)$$

Alternatively, consider the solid lines (01) forming the angle θ with [21]. An enlarged view of the triangle relating a_2, the interlinear spacing of (01) to $d_{(21)}$, which is measured in a direction parallel to [21], is shown in Fig. 5. It is clear from this that $d_{(21)} = a \sin \theta$. Substituting this value for $d_{(21)}$ in Eq. (17),

$$k_{[21]} = \frac{2\pi}{\lambda} = \frac{\pi}{a \sin \theta}$$

which is Bragg's law for reflection by lines having an interlinear spacing
a. It is clear from the above that condition (16), which determines the
discontinuities in the allowed-energy ranges of the electrons in a crystal,
can be interpreted to mean either total reflection of the electrons by
planes that are normal to their direction of propagation or as Bragg

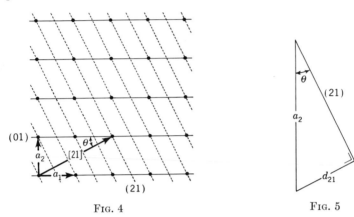

Fig. 4   Fig. 5

reflection by other crystallographic planes. Obviously, the planes that
are chosen for the latter interpretation are the planes that most strongly
reflect the electrons. For a crystal structure consisting of one atom
placed at the lattice points of a primitive cubic lattice, these planes are
the cube faces belonging to the form ((100)). In a simple face-centered

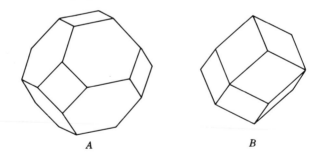

A   B

Fig. 6. Brillouin zones in crystals. (*A*) First zone for a face-centered cubic crystal;
(*B*) first zone for a body-centered cubic crystal.

cubic crystal, the planes are ((111)) and ((200)), and in a simple body-
centered cubic crystal, ((110)). The polyhedra formed by these planes
are called the first *Brillouin zones* of these crystals. Figure 6*A* and *B*,
respectively, shows the first Brillouin zones of a face-centered cubic
crystal and a body-centered cubic crystal.

It should be realized that other planes in the crystal can reflect electrons also. In fact, for larger energies, and consequently larger wave numbers, it is clearly seen in (16) that smaller d values are required to maintain the equality. Thus another set of planes (more closely spaced) is necessary to reflect the electrons having energies lying in the second allowed zone, and so on. The polyhedra formed by these planes are called the second, third, etc., Brillouin zones of the crystal. Use is made of these polyhedra in discussing the energies of electrons in crystals.

k **space.** In discussing the Brillouin zones of cubic crystals, it is conventional to select three axes, X, Y, Z, parallel to the three equivalent crystallographic axes a, and to denote directions parallel to these axes by the subscripts x, y, z. Accordingly, condition (14) for these three directions can be written

$$k_x = \pm \frac{\pi}{a}$$
$$k_y = \pm \frac{\pi}{a} \tag{18}$$
$$k_z = \pm \frac{\pi}{a}$$

where the plus or minus sign corresponds to positive or negative values of x, y, and z. It is more convenient to use the wave numbers k_x, k_y, k_z as coordinate axes for the construction of the Brillouin zones. The space defined by these axes is usually called k *space*, and since the distances in this space are reciprocal to distances in the crystal, according to (16), it is also called *reciprocal space*. (Note that for a free electron this is proportional to the momentum space discussed in Chap. 6.)

The two-dimensional k space defined by k_x and k_y is shown in Fig. 7. The first Brillouin zone for a simple cubic crystal is a cube and is shown in cross section in this figure. The second Brillouin zone for a simple cubic crystal is a dodecahedron bounded by ((110)). Its cross section in the XY plane is also indicated in Fig. 7. Note that the second zone lies outside the first zone in k space because it is bounded by ((110)) planes which have a smaller interplanar spacing than the ((100)) planes. This, of course, is because of the reciprocal relationships between distances in the crystal and distances in k space.

Fermi surfaces. A very useful property of k space is that it can be used to show the distribution of the allowed-energy values for the valence electrons. This is so because energy is related to k by (11), as shown also in Fig. 3A. Now a plot of E as a function of k, like Fig. 3A, can be prepared for various directions in a crystal. The resulting curves have similar shapes, except that the maximum and minimum energy values within a zone will differ with direction because the periodicity of the potential field and the magnitude of the barriers in Fig. 1 generally change with direction. After determining the energy variation with direction, it can be plotted in k space by noting the energy values at different points. When all the points having the same energy values are joined by lines, a set of *energy contours* is obtained. (In three dimensions, it is possible to construct equienergy surfaces by a parallel procedure.) A typical set of such contours for a nearly free electron in a simple cubic crystal is shown in Fig. 8. The inner contours are circles (spheres in three dimensions) because the electrons having these energies have wave numbers that are far removed from the critical values of (18). These electrons, therefore, have the same energy regardless of the direction in which they move within the crystal, not unlike the free-electron case

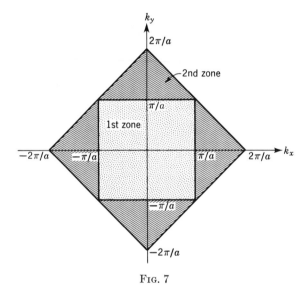

Fig. 7

discussed in the preceding chapter. When the energy contours near the zone boundaries are considered, however, this is no longer true. As the wave numbers k approach the critical values determined by (18), the corresponding energy values increase very slowly (Fig. 3A) and the contours in Fig. 8 begin to bulge toward the zone boundaries. Finally, the energy contours in the corners of the zone terminate on the zone boundaries because they correspond to k values that are larger than the wave numbers elsewhere in the zone.

Figure 8 also shows the first two energy contours in the second Brillouin zone. These contours do not join with any of the contours in the first zone because of the energy discontinuity that takes place at a zone boundary. The first contour of the second zone may correspond to an energy that is either greater or less than the outermost contour of the first zone. To see how this can happen, consider the energy curves for the first and second zones plotted in Fig. 9 as a function of k for two crystallographic directions. Figure 9A shows the energy curve for movement along the a_1 axis. As expected, this curve has a discontinuity at $k_x = \pi/a$, indicated by the top energy value for the first zone E_1 and the bottom energy of the second zone E_2. Figure 9B shows a similar plot for electron movement along [12]. It is clear from Fig. 8 that the highest energy in the first zone is greater in this direction than in the [10] direction; consequently, Fig. 9B shows two possible cases designated by E_1' and

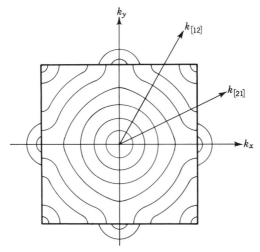

FIG. 8. Contours of equal energy in a simple cubic crystal. The contours lying within the first Brillouin zone are enclosed by the square. The first pair of contours lying in the second zone is also indicated.

E_1''. The lowest energies in the second zone are also higher for this direction, as indicated by E_2' and E_2'' for the two cases, respectively.

Essentially, two possibilities exist regarding the relative magnitudes of the highest energy level of the first zone and the lowest energy level of the second zone; namely, either the first is greater or the reverse is true. Both possibilities are illustrated in Fig. 9. Consider the discontinuity in Fig. 9B represented by E_2' and E_2''. Figure 9C shows schematically the case when the highest energy in the first zone, represented by E_1', lies below the bottom of that in the second-zone region. The width of the resulting gap is determined by the lowest energy in that zone, E_2. Figure 9D shows the other case, where the top energy in the first energy zone, as determined by E_1'', lies above the bottom of the second zone, that is, $E_1'' > E_2$. As can be seen in Fig. 9D, the two zones appear to overlap and there is no forbidden-energy region separating the two zones. The significance of *overlapping zones* becomes apparent when the occupation of the available energy levels by electrons is considered.

According to the usual boundary conditions restricting an electron to the interior of a crystal, the wavelengths that a free electron can have are limited to integral submultiples of the length of the crystal in the direction of propagation. Thus $k_x = 2\pi/\lambda = 2\pi n_1/L_1$; similarly, $k_y = 2\pi n_2/L_2$ and $k_z = 2\pi n_3/L_3$, and the volume in k space corresponding to a single energy level is $(2\pi)^3/L_1 L_2 L_3 = 8\pi^3/\mho$, where $\mho$ is the volume of a rectangular crystal whose sides are L_1, L_2, L_3. The volume of the first Brillouin zone for a simple cubic structure is that of a cube of edge $2\pi/a$. Hence the total number of energy levels in this zone is $(8\pi^3/a^3)(\mho/8\pi^3) = \mho/a^3$,

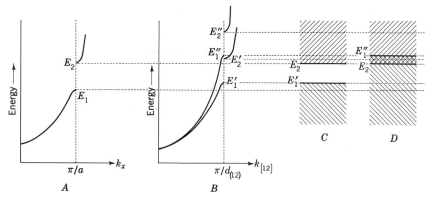

Fig. 9. Allowed-energy zones along two directions in a simple cubic crystal. (*A*) Along an a axis; (*B*) along the [12] direction. Two possible pairs of values at the zone boundary $k = \pi/d_{(12)}$ are designated by E_1', E_2' and by E_1'', E_2''. The subscript 1 refers to the highest allowed values in the first zone, and the subscript 2 to the lowest allowed value in the second zone; (*C*) allowed zones separated by forbidden-energy region; (*D*) allowed zones overlap.

exactly equal to the number of lattice points in the crystal. Since each energy level can be occupied by two electrons of opposite spin, up to two valence electrons per lattice point can be accommodated in the first zone. In a simple cubic structure, there is one atom per lattice point, so that up to two electrons per atom can be accommodated in the first zone. Elements having more than two valence electrons must distribute their electrons among the available states in the first and the second zones. The way that these available states are actually occupied depends, of course, on the relative allowed energies in the two zones, that is, on whether the zones overlap or not.

The successive occupation of available states for both possible cases is illustrated in Fig. 10. The first two diagrams (Fig. 10*A* and *B*) show the states in the first zone partially occupied by electrons. The energy contour enclosing the occupied states is called the *Fermi surface*, since it denotes the maximum energy that the electrons can have, that is, the Fermi energy. Figure 10*C* shows what happens when the number of electrons equals exactly twice the number of available energy levels, that is, when it just equals the total number of available quantum states in the first zone. (Each energy level contains two quantum states corresponding to two opposite spin quantum numbers.) Since each energy level can accommodate two electrons, the first zone is completely filled if the zones do not overlap, as shown at the bottom of Fig. 10*C*. If the zones do overlap, then the quantum states at the bottom of the second zone have energies that are lower than the energies of the states in the corners of the first zone, and these states are occupied first, as shown in

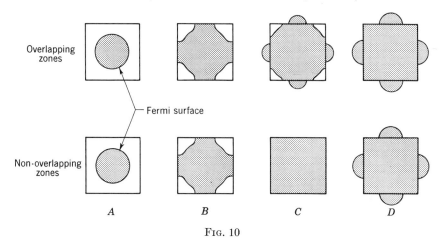

FIG. 10

the top drawing in Fig. 10*C*. The consequence of overlapping zones, therefore, is that it is impossible to complete the filling of one zone without beginning to fill the next zone. From this it follows that one or more of the zones are always partly empty whenever the zones overlap. This result is very important in explaining why some materials are conductors and others are not, as discussed in a subsequent section.

Density of states. The occupation of available states in a zone by the valence electrons in any material obviously depends on the number of states contained in each zone. As already discussed, each Brillouin zone contains one energy level per lattice point and each energy level represents two quantum states differing in their spin quantum numbers only. In a crystal, the lowest available energy levels are occupied first, in keeping with the fundamental law of nature that any system tends to a state of lowest energy. The actual energy of the system depends, however, on the distribution of available states in k space, that is, the number of quantum states per unit energy range in each zone. To determine this number, consider first the free-electron model. The energy of a free electron as a function of k is given by Eq. (5), and the Fermi surfaces for free electrons are spheres of radius $k = (k_x^2 + k_y^2 + k_z^2)^{\frac{1}{2}}$. The total number of energy levels up to some energy E, therefore, is determined by dividing the volume of a sphere of radius k by the volume in k space corresponding to a single level, which was determined in the previous section to be $8\pi^3/\upsilon$. Remembering that each energy level can contain

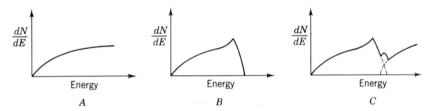

Fig. 11.   The density of available states (A) according to the free-electron theory; (B) in a single Brillouin zone; (C) when two zones overlap.

two electrons having opposite spins, the total number of electrons that can be accommodated is given by

$$N = 2 \times \tfrac{4}{3}\pi k^3 \times \frac{\upsilon}{8\pi^3}$$

$$= \frac{8\pi}{3} \frac{(2mE)^{\frac{3}{2}}}{h^3} \upsilon \tag{19}$$

where E is the kinetic energy of the electrons as determined by (5). The number of available states having energy levels lying between E and $E + dE$ is then found by differentiating (19):

$$\frac{dN}{dE} = 8\pi m \frac{(2mE)^{\frac{1}{2}}}{h^3} \upsilon. \tag{20}$$

The quantity dN/dE is called the *density of states*† and can be given a simple meaning. First, notice that it increases with increasing crystal volume, in order to accommodate the total number of electrons present, which also increases with the size of the crystal. Next, notice that it is a parabolic function of the energy, as shown in Fig. 11A for the case of free electrons. In actual crystals, this curve is modified by the Brillouin zones, as shown in Fig. 11B. For low energies the curve follows the

† The density of states is frequently designated $N(E)$ without distinguishing between states that are actually occupied and those that are available but are not occupied. In this book, the terminology $S(E)$ is used for available states and the term $N(E)$ is reserved for those states that are actually occupied, in full accord with the Fermi distribution discussed in Chap. 6.

free-electron parabola (Fig. 9A); however, as the energy of the nearest zone boundary (Fig. 8) is approached, the energies level off and more states have nearly the same energy value at the boundary. Following this, the corners of the zone are filled, causing the curve to fall as the number of available states decreases, until the curve falls to zero when the zone is completely filled. If two zones overlap, then these curves overlap and the resultant curve is obtained by a superposition of individual curves, as shown in Fig. 11C.

With this interpretation, the quantity dN/dE has the same meaning given to the quantity $S(E)$ in the preceding chapter. It will be recalled that the density of occupied states at any temperature can be determined by multiplying the available density of states by the probability of their occupation as determined by the Fermi-Dirac function. The way this is done is illustrated in Fig. 12, which shows the Fermi-Dirac function

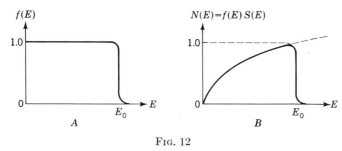

F$_{\text{IG}}$. 12

plotted in Fig. 12A and the product $f(E)S(E) = f(E)\,dN/dE = N(E)$ in Fig. 12B. Consequently, the number of electrons in a crystal of volume $\mathcal{V}$, whose energies lie between E and $E + dE$, is determined according to (20) by

$$N(E)\,dE = \frac{dN}{dE}\,f(E)\,dE = S(E)f(E)\,dE$$
$$= 8\pi m\,\frac{(2mE)^{\frac{1}{2}}}{h^3}\,\mathcal{V}\,\frac{dE}{e^{(E-E_0)/kT} + 1}. \tag{21}$$

Use is made of this relation in subsequent discussions, particularly in connection with electron emission in Chap. 11.

Applications of zone theory

Energy levels in atoms and solids. Before proceeding to a discussion of the properties of materials that can be explained with the aid of the zone theory, it is worthwhile to consider briefly the relation between the allowed- and forbidden-energy levels for electrons in isolated atoms and in solid materials. As shown in Chap. 3, the electrons surrounding a nucleus in an isolated atom can have only certain allowed energy values; that is, they must occupy the states available in certain allowed energy levels in accord with the Pauli exclusion principle.   As also shown in that chapter, when two atoms approach each other sufficiently closely to form a bond, the energy levels of the outermost, or valence, electrons split up. In the case of the hydrogen molecule, this leads to four 1s quantum states for the two electrons.   It is possible to generalize this result and show that *the joining of atoms to form molecules does not alter the total number of quantum states with a particular quantum number*, regardless of the size of the molecule.   Thus, if a crystal is thought of as an infinitely large molecule, the total number of quantum states of one kind that it contains is equal to the sum of such states in the individual atoms comprising it.

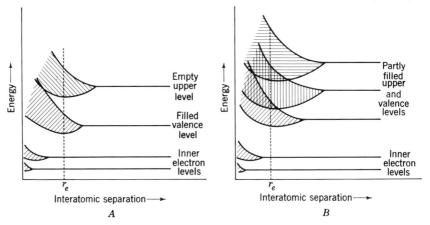

FIG. 13

The actual splitting of a level occurs whenever the electron orbitals belonging to that level overlap.   Thus, when two or more atoms are brought near each other, the outermost levels split first and the inner levels only when the atoms have interpenetrated each other sufficiently for overlap to occur.   Suppose a crystal consists of n identical atoms, each having two possible 1s states; then there exist 2n such states in the crystal.   The energies of these states differ slightly; in a crystal weighing only 1 mg there are enough atoms to make the number of such states approximately equal to 10^{19} and the energy separations between them of the order of 10^{-19} eV.   Thus the levels in a crystal split into a quasi-continuous distribution of states.   Suppose next that the n atoms are arrayed in a proper lattice array but that initially they are too far apart to interact with each other.   As their interatomic separations are progressively and uniformly decreased, the various levels split up as shown in Fig. 13.   The actual width of the levels in a crystal then can be determined by noting their width at the equilibrium separation r_e for that crystal. Since the small energy differences between the states in a level cannot be readily measured, it is convenient to think of them as forming a quasi-continuous *energy band*.

Two different cases are considered in Fig. 13. In Fig. 13*A*, the energy bands do not overlap at the equilibrium separation, whereas in Fig. 13*B*, certain levels overlap. The inner levels in both cases have not yet split at r_e because their respective electron orbitals do not overlap at this distance. Since the valence electron orbitals do overlap at this inter-atomic distance, the corresponding levels form a band. The quasi-continuous set of states in such levels is commonly called the *valence band* of the crystal.

The above results are not unlike those derived in the first part of this chapter. In fact, the zone theory and the band theory lead to essentially identical results in all crystals, and they are distinguished only by the models used in their development. In the zone theory, the starting point is the free-electron model with certain restrictions imposed on it when it is assumed that not all the electrons are truly free. Conversely, the band theory begins by assuming that all the electrons are tightly bound to their respective nuclei and considers what happens as the atoms approach each other sufficiently closely for the neighboring nuclei to exert an influence on the valence electrons. For this reason, this second procedure is commonly called the *tight-binding approximation*. As an example of the type of results obtained by applying the method of tight binding, consider the energy bands as a function of interatomic distance for diamond and sodium shown in Figs. 14 and 15, respectively. Note that the two bands in Fig. 14 first overlap and then separate. This behavior is a characteristic of the diamond structure since it does not occur in either the body-centered or face-centered cubic structures. Moreover, it turns out that the two bands thus formed each contain four *sp* states per atom, so that the states in the lower band are normally completely filled while those in the upper band are completely empty.

Conductors and nonconductors. One of the notable shortcomings of the free-electron model is its inability to explain why all materials containing atoms, and therefore valence electrons, are not conductors. The above discussion makes it abundantly clear that the answer is provided by the zone theory, or the band theory of solids. After the allowed- and forbidden-energy regions have been determined, the occupation of the available energy levels is decided by the density-of-states function $N(E) = S(E)f(E)$. It is most convenient to depict this by drawing the energy-band model of the crystal and locating the Fermi level in such a diagram. Then, at absolute zero, all the states lying below E_0 are filled, all above are empty. It should be remembered that

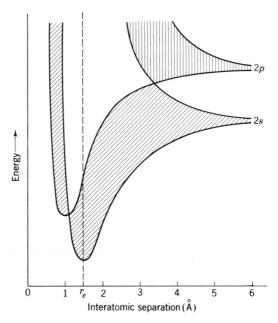

Fɪɢ. 14.   Energy bands of diamond. The upper shaded band is empty, while the lower one is completely filled.   (*After Slater.*)

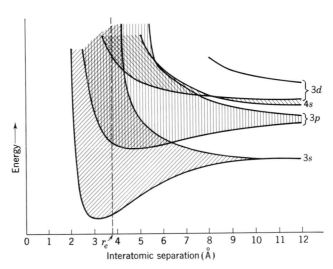

Fɪɢ. 15.   Energy bands of sodium.   Note how the bands overlap at the equilibrium interatomic separation.   (*After Slater.*)

the maximum and minimum energy values in each band depend on direction in the crystal, as already shown in Fig. 9. For convenience in representation, the band widths of the allowed-energy regions in a crystal are usually represented by delimiting each band by the maximum and minimum energy values in that band regardless of direction. This leads to the two possibilities already noted; namely, either the bands overlap or they do not overlap. When the density-of-states function is next considered, it turns out that the Fermi level lies either within an allowed band or between the bands in the forbidden-energy region. This leads to essentially three possible cases illustrated in Fig. 16. If the two bands are separated by a forbidden-energy region, then the Fermi level may lie either within an allowed band (Fig. 16A), so that one of the bands is only

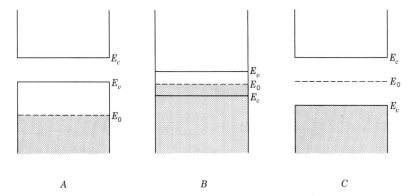

FIG. 16. Energy-band diagrams of three possible cases. Only the uppermost filled and lowermost empty bands are shown since the quantum states lying in other bands do not play an active role in determining physical properties. The top of the valence band is denoted by E_v, and the bottom of the empty band by E_c. Diagrams A and B correspond to conductors, and C to an insulator.

partly filled, or between the bands (Fig. 16C), so that the bands below E_0 are completely filled and those above are completely empty. Finally, when the bands overlap (Fig. 16B), one or both of the bands are only partly filled regardless of where the Fermi level lies.

In order for conductivity to be possible, there must be available in the same band energy states that are occupied and others that are empty. This is clearly the case for the two diagrams shown in Fig. 16A and B. Consider the case of sodium having one valence electron per atom occupying one of the two allowed $3s$ states. In a sodium crystal, there are two $3s$ quantum states per lattice point but only one atom per lattice point, so that the $3s$ band is half empty. Although this is sufficient to make sodium a conductor, it turns out that the $3s$ band overlaps the $3p$ band in a sodium crystal, as shown in Fig. 15. In fact, it has been found that the valence band overlaps the next highest band in all metal crystals examined thus far. On the other hand, the valence band is separated from the next highest empty band in diamond (Fig. 14), so that it corresponds to the diagram shown in Fig. 16C. Conductivity is not possible in diamond, therefore, because the electrons occupying states in the valence band would have to gain an energy of about 7 to 8 eV to be able to occupy the nearest allowed empty states. Thus Fig. 16C represents the energy-band diagram of an insulator. It is possible, however, that the forbidden-energy region in a crystal is quite narrow. At a sufficiently high temperature, some of the electrons can gain enough thermal energy in this case to transfer to states lying in the lowermost empty band. When an external electric field is then applied to the crystal, these electrons can gain additional energy from the field, and conduction is possible. The lowermost empty band, therefore, is called the *conduction band*. The currents produced in such crystals are necessarily small, so that they are distinguished from metallic conductors and true insulators by being called *semiconductors*. Note also that the conductivity of a semiconductor increases as the temperature increases since more valence electrons can transfer to empty states lying in the conduction band. This is exactly opposite to the temperature dependence of conductivity in metals, whose resistivity increases with increasing temperature, and provides a very convenient means of distinguishing metals from semiconductors experimentally. The conductivity in semiconductors is discussed further in the next chapter.

When metals are considered, it is usually more convenient to use the zone theory because the valence electrons in metals are relatively more free. In order to see what limitations on conductivity are imposed by this theory, consider a Brillouin zone that is only partly filled, as shown in Fig. 17. In the absence of an external electric field, each electron moves with a velocity determined by its energy; however, no net movement occurs, since for each electron with an energy determined by k there is a symmetrically located electron at $-k$ moving in the opposite direction. When an external electric field is applied, the distribution is displaced in the direction opposite to the field by electrons moving into adjacent quantum states in the same zone, as indicated in Fig. 17 by the dashed circle. Obviously, only the electrons occupying energy levels lying near the Fermi surface can move into the higher-energy states. Nevertheless, a net displacement of the electrons produces a net current, and a crystal with a partially filled zone is a conductor. As this process continues, it is possible that an electron ultimately occupies a quantum state on the Brillouin-zone boundary. At this point, it cannot move farther in the same direction without crossing the boundary, that is, without a transition to a quantum state lying in the next zone. Such a transition is highly improbable, so that the electron's velocity normal to the zone boundary is zero in this state. Physically, the electron can be pictured as being totally reflected by the crystallographic planes parallel to the zone boundary at this point. In terms of the zone model shown in Fig. 7, the reflected electron "reappears" in the zone at a translation-equivalent point on the opposite zone boundary. Such a successive occupation of available quantum states in the Brillouin zone is shown schematically in Fig. 18, where q_0 marks the initial quantum state of the electron when an external field is applied; q_R is the quantum state of the electron when reflection occurs; and q_F is the final state when the external field is removed (see also Exercise 8).

Crystal structure. The zone theory is also useful in explaining why certain crystal structures rather than other possible atomic arrays are

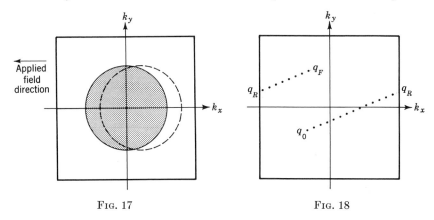

FIG. 17   FIG. 18

more stable under specified conditions of temperature and pressure. It is clear from diagrams like those in Figs. 14 and 15 that the energies of the valence electrons are markedly affected by the interatomic separation and by the structural array assumed in the calculations. Because of their complexity, such calculations have been carried out for only a few crystals. Nevertheless, it is possible to make use of the zone theory to predict structure-dependent properties of crystals. For example, the very pronounced dip in the valence band of diamond at r_e, shown in Fig. 14, reflects the very strong bonding of the four valence electrons of carbon in this structure. By comparison, the forbidden-energy region is much narrower in Ge and Si, which have the same crystal structure, so that these crystals are semiconductors. It is similarly possible to deduce the zone structure of other crystals by observing their conductivity. Consider the first long period in the periodic table. Potassium, like sodium and the other alkali metals, has one valence electron per atom and a body-centered cubic structure. Consequently, the last zone is only partly filled and potassium is a conductor. Calcium is also a conductor even though it has two valence electrons per lattice point. Thus it can be concluded that the zones overlap in calcium. Scandium, with three valence electrons per lattice point, does not require zone overlap to explain conductivity, and so forth. The valence electrons of Sc and successive elements in this row occupy the available states in the $3d$ zone (10 per atom), so that they are all conductors. The $3d$ zone is completely filled in Cu; however, the $4s$ zone is half empty. Thus copper is also a conductor. This does not mean that the zones in these metals do not overlap; only that overlap is not necessary to explain their ability to conduct electricity. The details of the observed conductivity, however, can be explained correctly only when the detailed overlap of zones is considered, as shown in the next section.

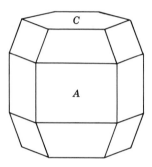

FIG. 19. First Brillouin zone of a hexagonal closest-packed metal having two atoms per lattice point of a primitive hexagonal lattice.

As a final example of the application of zone theory to the crystal structure of elements, consider the observed variation in the c/a ratio of metals crystallizing in a hexagonal closest packing. The first Brillouin zone for a primitive hexagonal lattice is a regular hexagonal prism. In the case of a hexagonal closest packing, however, it has the shape shown in Fig. 19. The reduced volume of this zone decreases the number of available states in the zone to slightly less than two per atom. Consequently, the valence electrons of divalent metals crystallizing in hexagonal closest packings must occupy states in the next-highest zone. In order to minimize the energy of this structure, the zones overlap. Because the structure is anisotropic, such overlap can occur normal to either the face marked C or A in Fig. 19, depending on which face has the lowest energy in the crystal. As already indicated in earlier discussions, a shortened interatomic distance means that the bonding between the atoms is stronger, so that the energy variation across the corresponding face of the zone is higher. This means that when the energy at the C

face E_C is higher than that at the A face in Fig. 19, then the c axis is shorter than in an ideal closest packing of like spheres and c/a is less than the ideal value 1.633. These deductions are verified by the calculated E_C and E_A values and the observed c/a ratios in several metals as shown in Table 1.

Table 1
Characteristics of hexagonal closest packings

Metal	Number of available states per atom	E_C/E_A	c/a
Be	1.731	1.196	1.585
Mg	1.743	1.137	1.625
Zn	1.799	0.863	1.861
Cd	1.805	0.859	1.890

Transition metals and alloys. As already noted, the $3d$ zone is completely filled in copper and the $4s$ zone is half empty. Nickel, lying to the left of copper in the periodic table, has one less electron per atom.

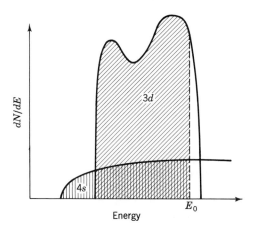

FIG. 20. Schematic representation of overlapping $3d$ and $4s$ zones of nickel. On an average, 9.4 out of 10 available $3d$ states per atom and 0.6 out of 2 available $4s$ states are occupied. (*After Slater.*)

Because of zone overlap, some of the states in the $4s$ zone actually have lower energies than those lying in the $3d$ zone Thus the zone model of nickel, shown in Fig. 20, contains a very nearly, but not quite, filled $3d$ zone and a partly filled $4s$ zone. When an electric field is applied to a nickel crystal, the electrons occupying states near the Fermi energy E_0 can transfer to nearby vacant states. It can be shown that the $4s$

electrons are primarily responsible for the observed conductivity Consider what happens, however, when an electron reaches a state at the zone boundary as discussed in connection with Fig. 18. As can be seen in the zone model in Fig. 20, there are two sets of quantum states, one in each overlapping zone, that have identical energies at the boundary. Consequently, such an electron can undergo a transition to a state in either zone. If a $4s$ electron is thus "scattered" into a state lying in the $3d$ zone, it no longer can contribute to conductivity as readily, so that the number of current carriers is decreased. Such scattering actually occurs in transition metals and accounts for their lower conductivity as compared with metals like copper. The detailed zone structure of transition

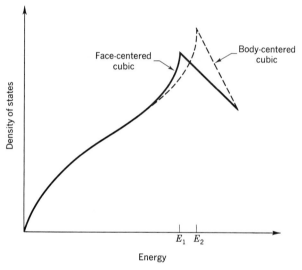

FIG. 21. Density of available states in the face-centered and body-centered cubic structures. (*After Hume-Rothery and Raynor.*)

metals like Ni, Co, and Fe similarly explains some other unusual properties that they exhibit, for example, ferromagnetism, as discussed in Chap. 12.

The zone theory also can be used to explain certain characteristics of alloys. For example, consider the Cu-Zn system (brass). Copper crystallizes with a face-centered cubic structure. The allowed density of states has a distribution like the one represented by the curve in Fig. 21, reaching a maximum value at E_1. Copper has one valence electron per atom, whereas zinc has two, so that, when zinc atoms are alloyed with copper, the additional electrons must occupy states in this zone. As successively more zinc atoms are added, the Fermi level rises by a corresponding amount. Note, however, that when $E_0 > E_1$, the density of available states starts to decrease in this structure. On the other hand, a corresponding decline is not encountered in the body-centered cubic structure until a higher energy, E_2, is reached. It follows, therefore, that a larger fraction of zinc atoms could be alloyed with copper if the crystal structure were body-centered cubic. Such a transformation from face-centered cubic (alpha brass) to body-centered cubic (beta brass), as a function of composition, actually occurs in the Cu-Zn system. The importance of the electron-to-atom ratio in alloys was first observed empirically by Hume-Rothery and later corroborated by theoretical calculations performed by H. Jones. This ratio can be used to predict quite accurately what the structure types of alloys in a larger number of intermetallic systems should be. Moreover, it turns out that, when even higher-valency atoms are added, a composition may be reached at which the number of electrons per atom exceeds the number of quantum states per lattice point available in the zone. To accommodate these extra electrons without a change in the crystal structure, it is then necessary that certain atoms be missing. This leads to so-called defect structures containing vacancies. Such a "removal" of copper atoms from their proper sites takes place in the Cu-Al and Cu-Ga systems, where, apparently, the energy of the structure containing copper vacancies is lower than it would be if the valence electrons of the missing copper atoms had to occupy higher-lying allowed-energy levels.

Superconductivity. As already discussed in Chap. 6, the resistivity of a metal decreases as the temperature decreases because the decreasing vibration amplitudes of the atoms cause the electronic mean free paths to increase. It turns out, however, that at temperatures around 4 deg K, many metals exhibit an abnormally large conductivity called *superconductivity*. This behavior is observed not only in good conductors like aluminum, but also in relatively poor conductors like lead and indium. Although superconductivity was first observed in mercury by Onnes in 1911, a quantitatively accurate theory was not developed until quite recently by Bardeen, Cooper, and Schrieffer. Consider a metal that is entirely free of any imperfections, so that its electrical resistivity at any temperature is entirely due to electron scattering by thermally vibrating atoms. At the absolute zero of temperature, the energy of the atoms is reduced to the zero-point energy $\frac{1}{2}h\nu$. This means that the ions do not have sufficient energy to scatter the electrons, and the resistivity of the metal becomes truly zero. At a slightly higher temperature, it turns out, however, that a subtle interaction between electrons and phonons can take place which effectively eliminates all electron-ion scattering. The resistivity, therefore, becomes exactly zero below a certain critical temperature. Thus, if a current is started in a ring-shaped superconductor, it continues to flow indefinitely, even after the external electric field is removed. Such an experiment has actually been performed, and the current has been circulating without detectable attenuation for years!

Even before the process of superconductivity was understood, scientists were busy searching for criteria that would determine under what conditions a material becomes superconductive. Matthias and Hulm were so engaged in 1950 when they observed that elements having a valence electron-to-atom ratio of 3, 5, or 7 invariably became superconductive. For example, technitium, having seven valence electrons per atom, becomes a superconductor below 11 deg K. Of the two elements bracketing Tc in the periodic table, molybdenum, with six valence electrons per atom, becomes superconducting only below 0.1 deg K, and ruthenium, with eight electrons per atom, becomes a superconductor below 0.5 deg K. When molybdenum and ruthenium are alloyed in equal proportion, however, the average electron-to-atom ratio is 7 and MoRu becomes a superconductor at about 10.6 deg K. Similarly, niobium nitride, having five valence electrons per atom, is a superconductor below about 15 deg. Such materials are of considerable interest because their superconductive state does not require excessively low temperatures. It turns out that a transition to the regular, or nonsuperconducting, state can be affected quite easily by either applying an alternating electric field to the superconductor or else by placing it in a strong magnetic field. Although the reasons for this are not fully understood, a number of device applications can be envisaged. An obvious one is to use the conductive $\rightleftharpoons$ superconductive transition as a switching element in, say, a computer circuit. Such applications and others are currently under active investigation.

According to the quantitative theory of Bardeen, Cooper, and Schrieffer published in 1957, the electron-phonon interaction causes the density of states of a superconductor to increase infinitely in the vicinity of the Fermi energy as shown in Fig. 22. Concurrently, a narrow forbidden-energy region, or energy gap, appears symmetrically about E_0. The existence of an infinite density of states at its edges permits continuous transitions among the available states without requiring any apparent increase in energy. The validity of this model has been demonstrated by Giaever in a series of measurements on thin sandwiches formed by two metal foils separated by a very thin insulating layer. The insulating layer provides an energy barrier for electron transfer from one metal foil to the other. As shown in Chap. 5, however, an electron may penetrate such a barrier by means of the tunnel effect of quantum mechanics. Thus, when two superconductors are separated by a thin insulating layer, their energy diagrams are as shown in Fig. 23A. Electrons occupying states near the top of the shaded region in metal 1 cannot tunnel to states of comparable energy in metal 2 because these states lie in the forbidden gap. When an external field is applied, however, the relative energy levels are shifted as shown in Fig. 23B. Now electrons can tunnel from metal 1 to metal 2 with the aid of an external field, and a measurable current is in fact observed. At temperatures one or two degrees above absolute zero, some electrons in metal 1 may actually gain

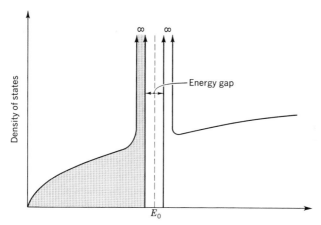

FIG. 22. Density-of-states distribution in a superconductor. The states whose energy is less than E_0 (shaded region) are normally filled, except just below the Fermi energy where the density of available states is infinite.

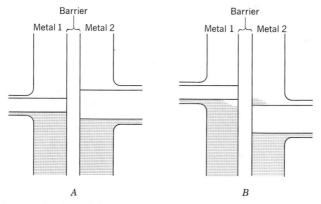

FIG. 23. Energy-band model of two superconductors separated by a thin insulator which introduces an energy barrier between them. (*A*) No voltage applied; (*B*) after voltage applied; some states in metal 1 have become empty due to tunneling. (*After Giaever.*)

enough energy to occupy states lying above the energy gap. In that case, a smaller electric field, sufficient only to raise these levels above the forbidden region in metal 2, can produce current flow across the insulator. By noting such changes in the current with applied-field strength, it is thus possible to measure the widths of the forbidden gaps in different superconductors.

Suggestions for supplementary reading

A. H. Cottrell, *Theoretical structural metallurgy*, 2d ed. (St Martin's Press, Inc., New York, 1957).

Ivar Giaever, Electron tunneling between two superconductors, *Phys. Review Letters*, vol. 5 (1960), pp. 464–466.

William Hume-Rothery, *Atomic theory for students of metallurgy* (Institute of Metals London, 1946).

William Hume-Rothery and G. V. Raynor, *The structure of metals and alloys* (Institute of Metals, London, 1956).

Frederick Seitz, *The physics of metals* (McGraw-Hill Book Company, Inc., New York, 1943).

Suggestions for further reading

Adrianus J. Dekker, *Solid state physics* (Prentice-Hall, Inc., Englewood Cliffs, N.J. 1957).

Charles Kittel, *Introduction to solid state physics*, 2d ed. (John Wiley & Sons, Inc., New York, 1956).

N. F. Mott, Recent advances in the electron theory of metals, in *Progress in metal physics*, vol. 3 (Pergamon Press Ltd., London, 1952), pp. 76–114.

N. F. Mott and H. Jones, *The theory of the properties of metals and alloys* (Clarendon Press, Oxford, 1936; reprinted by Dover Publications, Inc., New York, 1958).

John C. Slater, *Quantum theory of matter* (McGraw-Hill Book Company, Inc., New York, 1951).

Exercises

1. Making use of relation (5), explain why k is independent of direction in the free-electron model but not in the zone theory. HINT: Consider the variation of the kinetic energy of the valence electrons with position in the crystal.

2. Draw the first and second Brillouin zones, in cross section normal to the c axis, for a primitive hexagonal lattice. Assume that $((10\bar{1}0))$ planes terminate the first zone and $((11\bar{2}0))$ planes terminate the second zone. Are the Fermi surfaces likely to be more or less spherical, when these zones are partly filled, than in the case of square zones shown in Fig. 7? Why?

3. Illustrate the progressive occupation of the states in the two zones in Exercise 2 for the case when the two zones overlap along $[[\bar{2}110]]$.

4. The Fermi energy for copper is 7.1 eV. Assuming that this is the maximum kinetic energy of the electrons in copper, what is the number of electrons per cubic meter in copper as determined from Eq. (19)? How does this compare with that determined directly from the density of copper (density $= 8.92$ kg $\times 10^3$ m^{-3})? (ANSWER: 8×10^{28} m^{-3}; 9×10^{28} m^{-3}.)

5. One can derive Eq. (20) by starting with the free-electron model of Chap. 6 and noting that a spherical shell between p and $p + dp$ in momentum space contains $(8\pi p^2/h^3)dp$ states. Show that the number of available states between E and $E + dE$ in k space is then given by $S(E)\ dE$.

6. Making use of the density-of-states function $S(E)\ dE$, derive an expression for the number of electrons occupying allowed states between $E = 0$ and $E = E_0$ at absolute zero of temperature. [This should lead to Eq. (19) in the text.]

7. Why must cesium, with one valence electron per atom, be a conductor, since the body-centered unit cell of cesium contains two atoms?

8. Suppose that the electron-energy distribution in a crystal is correctly represented by the dashed circle in Fig. 17 at the instant when the external field is shut off. Since there are more electrons moving in the $+x$ direction than in the $-x$ direction at this instant, why does the current immediately become equal to zero when the external field is removed? Estimate the amount of time required for this to happen in copper. (ANSWER: 2.6×10^{-14} sec.)

9. The energy gaps in germanium, silicon, and diamond crystals are 0.72, 1.09, and 7.0 eV, respectively. All three have the same crystal structure, and the interatomic distances decrease from 2.44 Å in Si to 1.54 Å in diamond. What is the relation between the bonding in these crystals and the width of the energy gap? What does this reasoning suggest about the relative width of the energy gaps in strongly ionic crystals, for example, alkali halides? HINT: Assume that the energy-level diagram in Fig. 14 is approximately correct for Si and Ge.

10. Assume that the electron-scattering curve for an atom $f \propto (\sin \theta)/\lambda$. Show that the first-Brillouin-zone boundaries for a body-centered cubic crystal are comprised of $((110))$ planes. What are the boundary planes of the second zone? HINT: The structure-factor expression in Chap. 2 correctly expresses the reflecting power of a plane in a crystal whether x-rays or electrons are incident on it.

11. Assuming that the zone model of nickel in Fig. 20 is also correct for copper (both crystallize in the cubic closest packing), make a drawing of the $3d$ and $4s$ zones of copper and indicate the occupied states by shading. The electrons of which zone contribute to conductivity? Why?

12. It was stated in the text that the $4s$ electrons in nickel are primarily responsible for the measured conductivity. If a model like Fig. 20 also applies to face-centered cubic cobalt metal, should the conductivity in cobalt be greater or less than in nickel? Why? What do you think is the function of the unpaired $3d$ electrons in nickel?

13. By analogy to Fig. 23, draw the energy-band diagrams at 10 deg K for two unlike metals separated by a thin insulating film when:

(a) Neither metal is in the superconducting state.

(b) One metal is in the superconducting state but the other is not.

(c) Both metals are in the superconducting state. (Assume that some of the electrons occupy states above the gap in the superconductor having the narrower gap.)

Indicate the occupied states by shading, and assume that the Fermi energy lies at the same level in both metals.

14. Assume that the energy levels of metal 1 relative to those of metal 2, in Fig. 23B, shift in direct proportion to the applied voltage. Prepare a plot of the current (vs. voltage) that can be expected to flow between two superconductors separated by a thin film when the applied voltage is gradually increased.

8. theory of
semiconductors

The electrical properties of semiconductor materials are probably understood more thoroughly than those of any other class of solids. The reason for this can be attributed to the favorable physical, chemical, and electrical properties of the elemental semiconductors silicon and germanium, which have been studied so extensively, and to the inherent nature of semiconductivity. Another important fact is that the well-known and diverse practical applications have served as a stimulus to ellucidating semiconductor properties. As discussed in this and the following chapter, the applications and properties of semiconductors depend rather critically on the nature of the materials used. Moreover, it is interesting to note that many of the properties or measurements used to characterize a semiconductor can equally well be used to make it perform certain functions in a device. For example, visible light shining on a semiconductor causes a current to flow by means of a process called photoconductivity. Thus experiments using visible light can be utilized to characterize the photoconductive character of a semiconductor. Conversely, the currents produced by light in semiconductors can be used in devices designed to open supermarket doors or to dim automobile headlights when approaching another car.

The distinguishing characteristic of a semiconductor is the existence of a forbidden range of electron energies in the crystal which is so narrow that some electrons are thermally excited to states in the conduction band at room temperature. This means that normally there are some free electrons in the solid. These free electrons are markedly influenced by chemical and physical imperfections, which means that extreme chemical purity and crystalline perfection are required in semiconductor crystals. Through the development of modern semiconductor technology suitable crystals can be synthesized, so that theoretical calculations, which of necessity consider perfect crystals, can be compared directly with experimental results.

The understanding of semiconductors is based almost entirely on the energy-band model of solids. As discussed in this chapter, the energy-band model most conveniently represents the distribution of electrons in a semiconductor under various conditions. It also has contributed greatly to the understanding of other solids for which the quantitative agreement may not be as satisfactory. It should be understood, however, that the band model is an approximation, and certain semiconductor properties cannot be fully explained on this basis.

Band model of semiconductors

Energy bands. As already discussed in Chap. 7, the energy-band model of a semiconductor is quite similar to that of an insulator and consists, at absolute zero, of a filled *valence band* that is separated from the empty *conduction band* by a forbidden-energy region called the *forbidden-energy gap*. This gap is usually quite narrow in semiconductors, so that, as the temperature of the semiconductor is increased, some of the electrons occupying states near the top of the valence band can transfer to empty states near the bottom of the conduction band, after absorbing sufficient energy. This process is usually described by saying that some of the valence electrons have been "excited" to states in the conduction band. Consider the periodically varying potential energy of an electron along some row of atoms in a crystal shown in Fig. 1. The total energy that the $1s$, $2s$, $2p$, etc., electrons can have is shown by the shaded regions in Fig. 1. These energies can be compared with those calculated by the tight-binding approximation discussed in the preceding chapter. It is clearly seen in Fig. 1 that the allowed-energy values for the inner electrons are localized in the vicinity of their respective nuclei. By comparison, the $3s$ electrons do not appear to be localized at any particular atom in Fig. 1. This is the case, for example, in sodium metal, where the $3s$ wave functions of adjacent atoms overlap to such an extent that the allowed-energy levels form a quasi-continuous band extending through the entire crystal. If a foreign atom is substituted for a sodium atom, however, its electrons will have somewhat different allowed-energy levels. Consequently, the introduction of foreign atoms into a crystal structure introduces new energy levels in the vicinity of each foreign atom. Such localized levels play an important role in semiconductors, as discussed in subsequent sections.

The relative positions of the energy-band maxima and minima vary with crystallographic direction in the crystal, as described in Chap. 7, and a complete, quantitatively accurate description of a semiconductor therefore requires a three-dimensional band model. It turns out, however, that most properties of semiconductors can be understood

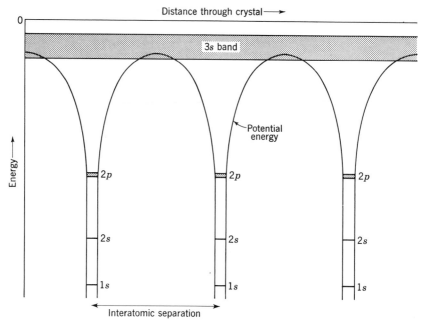

Fɪɢ. 1. Energy-level diagram of a solid showing highly localized inner electron levels and the common 3s band.

quite well by considering a one-dimensional model. As an example of the representation usually employed, consider the energy-band model of a silicon crystal along the [100] direction shown in Fig. 2A. Suppose next that the maximum energy in the valence band E_v, in this direction, is the maximum value that E_v has anywhere in the crystal. Similarly, suppose that the minimum energy in the conduction band E_c is the lowest energy in that band for all crystallographic directions. Then it is possible to define the effective energy-band model of silicon as shown in Fig. 2B, where the horizontal axis represents an "averaged" direction through the crystal. The important point here is that the width of the energy gap, $E_c - E_v$, represents the minimum width of this gap anywhere in the crystal and therefore the least amount of energy that an electron must gain to undergo a transition from the valence band to states in the conduction band. Since an insulator can conduct an electric current only after such transitions have occurred, it is clear that the model shown in Fig. 2*B* is adequate for discussing such possible transitions and the role they play in determining semiconductor properties.

Effective density of states. As already indicated in the preceding chapter, it is necessary to know the density of states in each allowed-energy band in order to determine the electron distribution under various conditions. In principle, this is easily determined because the density of allowed states is governed by the crystal structure, and their occupation

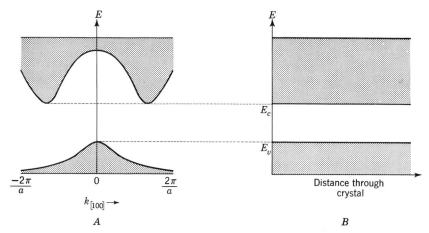

Fig. 2. Variation of energy levels with direction in a silicon crystal and the representation in terms of a conventional band model.

at any temperature is determined by the Fermi distribution function. It turns out, however, that calculations of the exact density of states $S(E)$ for actual structures are quite complex and, in fact, have not as yet been carried out for any real crystal without introducing several simplifying assumptions. Calculations such as those using the Kronig-Penney model already described, and similar models devised for real crystals, show that the electrons occupying states near the band edges behave very similarly to free electrons, except that the equations contain a term inversely proportional to d^2E/dk^2. This term measures the curvature of energy curves like the ones shown in Figs. 3 and 9 of Chap. 7. As shown there, the energy curve follows the free-electron parabola except near the band edges, that is, when the magnitude of k approaches π/d. If one defines an *effective mass* for an electron by

$$m^* = \frac{h^2/4\pi^2}{d^2E/dk^2} \tag{1}$$

then many of the equations developed in the two previous chapters can be used directly after substituting m^* for the mass of a free electron m. For example, the kinetic energy of the electron then is given by

$$E = \frac{h^2k^2}{8\pi^2 m^*}.$$

Note that differentiating the kinetic energy twice with respect to k yields Eq. (1). Electrons near a band edge respond to external forces such as an electric field as if they were free, but as though they have a different mass. The fact that m^* differs from m is a consequence of the fact that the electron is moving in a periodic potential and is not actually free.

Equation (1) shows how the structure of the crystal influences the properties of an electron through the variation of E with $\mathbf{k}$. Since m^* depends on $\mathbf{k}$, it has different values in different crystallographic directions. It is even possible for m^* to be negative, a situation which is discussed in a later section. Near a band edge the variation of E with $\mathbf{k}$ is nearly parabolic, as for a free electron, and therefore expressions derived for a free electron may be used by simply replacing the free-electron mass by m^*. In particular, the energy density of available states per cubic meter near the bottom of the conduction band is, according to Eq. (20) of Chap. 7,

$$S(E) = \frac{8\pi m^*}{h^3} \sqrt{2m^*(E - E_c)} \tag{2}$$

where E_c is the energy at the conduction-band edge.

The number of electrons per cubic meter in the conduction band now can be determined easily by multiplying $S(E)$ by the probability that a state is occupied, that is, the Fermi function $f(E)$, and integrating over all energies greater than E_c. Accordingly, the density of electrons in the conduction band is

$$n = \int_{E_c}^{\infty} S(E)f(E)\, dE. \tag{3}$$

Because $E_0 \ll E_c$, $e^{(E_c - E_0)/kT} \gg 1$, so that it is permissible to neglect the unity in the denominator of the Fermi function and (3) can be written

$$n = \frac{8\pi m^*}{h^3} \sqrt{2m^* kT}\, e^{-(E_c - E_0)/kT} \int_{E_c}^{\infty} \left(\frac{E - E_c}{kT}\right)^{\frac{1}{2}} e^{-(E - E_c)/kT}\, dE. \tag{4}$$

Let $y = (E - E_c)/kT$, so that the integral takes on the standard form

$$kT \int_0^{\infty} y^{\frac{1}{2}} e^{-y}\, dy = \frac{kT\pi^{\frac{1}{2}}}{2}.$$

Substituting back in (4),

$$n = 2\left(\frac{2\pi m^* kT}{h^2}\right)^{\frac{3}{2}} e^{-(E_c-E_0)/kT}. \tag{5}$$

Since the exponential in (5) represents the Fermi function, the quantity multiplying the exponential is called the *effective density of states* in the conduction band,

$$N_c = 2\left(\frac{2\pi m^* kT}{h^2}\right)^{\frac{3}{2}} \tag{6}$$

and has the value $2.5 \times 10^{25} \mathrm{m}^{-3}$ for $m^* = m$ and $T = 300$ deg K. Finally, the electron density is simply

$$n = N_c e^{-(E_c-E_0)/kT}. \tag{7}$$

The form of (7) is identical with that for N_c levels all located at the energy E_c. This calculation, in effect, replaces the actual wide conduction band by an equivalent number of discrete levels located at the bottom of the conduction band. This makes it possible to determine the electron density easily with (7) rather than by carrying out an integration according to (3) each time. It should be noted that it was assumed in the derivation that m^* is independent of E, which is not strictly true. As pointed out above, however, $E \propto k^2$ near the bottom of the band, so that m^* is constant. At energies above E_c, where m^* may vary, the contribution to the integral in (4) is small because of the exponential factor in the integrand. Therefore little error is introduced by this procedure. It should also be noted that (7) is valid only when E_0 is far enough below E_c to allow unity in the denominator of the Fermi function to be neglected. When E_0 approaches close to E_c, these results do not apply, the electron distribution is said to be *degenerate*, and the complete Fermi function must be used. Fortunately, this situation does not arise in most semiconductor problems, and Eqs. (6) and (7) are generally used.

It is clear from the introductory discussion of conduction-band levels that the quantum states in the conduction band are empty at the absolute zero of temperature and those in the valence band are completely filled. At higher temperatures some electrons are thermally excited and can cross the forbidden-energy gap by transitions from states in the valence band to unoccupied ones in the conduction band. This leaves behind unoccupied states in the valence band, which are called *holes* because they represent electron vacancies in an otherwise filled band. The number of holes per cubic meter in the valence band may be found by a procedure identical with that used to determine n (Exercise 2). The result for the hole density is

$$p = N_v e^{-(E_0 - E_v)/kT} \tag{8}$$

where

$$N_v = 2 \left(\frac{2\pi m^* kT}{h^2} \right)^{\frac{3}{2}} \tag{9}$$

is the effective density of states in the valence band. Although (8) is formally similar to (7), note the difference in the exponents and the fact that N_v may not be numerically the same as N_c since the effective masses in the two bands may differ.

Intrinsic semiconductors. An energy-level scheme like that shown in Fig. 2 applies to a perfect crystal structure which contains no chemical impurities and in which there are no atoms displaced from their proper sites. The properties of such a solid are thus characteristic of the ideal structure, and such a material is called an *intrinsic semiconductor*. Although it is not possible to achieve perfect structures in real crystals, it is possible in many cases to approach this ideal and, experimentally, to observe intrinsic behavior. In an intrinsic semiconductor the number of electrons occupying states in the conduction band is equal to the number of holes in the valence band. In order to calculate the actual hole and electron densities, the position of the Fermi level E_0 first must be determined. Since $n = p$, this can be done by equating (7) and (8). The result is

$$N_v / N_c = e^{-(E_c + E_v - 2E_0)/kT} \tag{10}$$

which may be solved for the Fermi energy

$$E_0 = \frac{E_c + E_v}{2} + \frac{3kT}{4} \ln \frac{m_p^*}{m_e^*}. \tag{11}$$

Here m_e^* is the effective mass of electrons in the conduction band, and m_p^* is the effective mass of electrons in the valence band, often called the effective mass of the holes.

If $m_e^* = m_p^*$, then the Fermi level lies midway between the valence and conduction-band edges, as indicated in Fig. 3. In any event, E_0 does not deviate appreciably from the center of the gap because the logarithm in the second term of (11) is small. For convenience, therefore, the Fermi level is usually assumed to be midway between the bands in an intrinsic semiconductor.

Conduction by electrons and holes. Any electrons occupying states in the conduction band are free to respond to an applied electric field. The energy they gain in the direction of the field results in motion that constitutes an electric current. The conductivity of a material is defined by Ohm's law to be the current density J per applied field, so that

$$\sigma = \frac{J}{\mathcal{E}}. \tag{12}$$

The current density due to n electrons per cubic meter moving at a resultant or average velocity $\bar{v}$ is simply $J = ne\bar{v}$, which gives, for the conductivity due to electrons in the conduction band,

$$\sigma = ne\mu. \tag{13}$$

Here, $\mu = \bar{v}/\mathcal{E}$ is the average velocity of an electron per applied field, and is called the *mobility* of the electron. Actually, of course, the

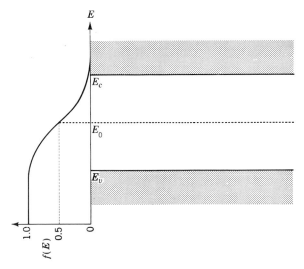

FIG. 3. Position of the Fermi level and the Fermi distribution for an intrinsic semiconductor.

electrons in the solid are in rapid thermal motion in the absence of an applied field, but their average velocity is zero in equilibrium. In the presence of an electric field, a drift velocity is added to the randomly directed thermal velocity, and this drift velocity is observable as a mobility. The properties of the structure which determine mobility are considered in a later section.

The above discussion applies to electrons in a partially filled band, such as the conduction band. It is possible, however, for electrons to move in the nearly filled valence band also. Here again, in the absence of an applied field, the net current is zero, or

$$J = ne\bar{v} = ne\left(\frac{1}{n}\sum_{i=1}^{n} v_i\right) = e\sum_{i=1}^{n} v_i = 0 \tag{14}$$

where the average velocity $\bar{v}$ is written explicitly as the summation of the individual velocities divided by the density of electrons. Next, suppose that the valence band is completely filled. The application of an electric field should make the electrons acquire an additional drift velocity, but this requires that the electrons gain additional energy. This is impossible, however, since there are no quantum states available into which an electron can transfer after acquiring the additional energy. Thus, in a completely filled band, the electrons cannot respond to an electric field, so that no conductivity results. In this case, (14) is always true, since the average velocity is zero even in an applied field.

In an intrinsic semiconductor some quantum states are unoccupied in the valence band because some electrons have been excited to states in the conduction band. The presence of such empty states in the valence band leads to conductivity because electrons in the valence band can transfer to these states and hence acquire a drift velocity due to the field. This effect can be shown in the following way: Consider the current density in (14) and focus attention on the jth electron.

$$J = e\sum_{i=1}^{n} v_i = e\sum_{\substack{i=1 \\ i\neq j}}^{n} v_i + ev_j = 0. \tag{15}$$

Rearranging terms in (15),

$$e\sum_{\substack{i=1 \\ i\neq j}}^{n} v_i = (-e)v_j. \tag{16}$$

Now the left side of (16) is the current density due to all electrons in the valence band except for the jth one. The right side represents the current density due to one electron, *but of opposite electrical charge.* Previously, this vacant quantum state has been termed a *hole.* Equation (16) demonstrates that holes in the valence band can be treated as positively charged particles fully analogous to the negatively charged electrons in the conduction band.

The use of holes to represent the behavior of electrons in a nearly filled band introduces considerable conceptual simplification, as is already evident from Eqs. (8) and (9). This is its great virtue. The concept is actually quite rigorous, as can be seen in Eq. (1). The effective mass of an electron is negative at the top of a band because the variation of energy with wave number k has a negative curvature at the top of a band. Thus electrons at the top of a band move "against" the field since they have negative mass. The electric current which results is quite equivalent to a positive electron (with positive mass) moving with the field. While it is common to speak of hole currents and hole densities, the origin of the concept must be kept clearly in mind, for a hole is actually just a convenient representation of the behavior of the large number of electrons in the valence band.

The conductivity produced by the holes is given by an expression identical with (13). Therefore the total conductivity of a semiconductor crystal is the sum of the conductivity by electrons and holes,

$$\sigma = ne\mu_e + pe\mu_h \tag{17}$$

where μ_e and μ_h are the electron and hole mobilities, respectively.

Since $n = p$, the conductivity of an intrinsic semiconductor can be obtained by combining Eqs. (7), (8), and (11). Introducing

$$E_g = E_c - E_v,$$

where E_g is the width of the forbidden-energy gap, the result is (for $N_c = N_v$)

$$\sigma = N_c e(\mu_e + \mu_h)e^{-E_g/2kT}. \tag{18}$$

Equation (18) shows that the conductivity of an intrinsic semiconductor increases exponentially with temperature. Such a conductivity variation is a characteristic of semiconductors and distinguishes them from metallic conductors. The width of the forbidden-energy gap can be determined from (18) by experimentally measuring the temperature dependence of conductivity. Taking the natural logarithm of both sides,

$$\ln \sigma = \ln N_c e(\mu_e + \mu_n) - \frac{E_g}{2k}\frac{1}{T}. \tag{19}$$

This is the expression for a straight line, since the first term on the right does not vary appreciably with temperature. Thus, a plot of $\ln \sigma$ versus $1/T$ yields a straight line of slope $-E_g/2k$, and E_g can be evaluated directly.

Extrinsic semiconductors

Impurity levels. The quantum states of an intrinsic semiconductor discussed in the previous section apply to a perfect crystal structure of the solid. When foreign, or *impurity*, atoms are incorporated into the structure, the available quantum states are altered and this introduces significant changes in the properties of the semiconductor. Since these properties now depend strongly upon the impurity content, the solid is called an *extrinsic* semiconductor. Because the number of electrons and holes normally present in an intrinsic semiconductor is small (Exercise 3), small additions of impurity atoms are sufficient to produce major changes.

Consider a single crystal of silicon. Each silicon atom has four valence electrons with which it forms four electron-pair bonds with four Si neighbors. If a pentavalent atom, for example, phosphorous, arsenic, or antimony, substitutes for a silicon atom, only four of its electrons are required to complete the covalent bonding in this structure. The extra electron is constrained to remain in the neighborhood of the impurity atom, however, because it is attracted by the extra positive charge on the nucleus. Thermal energies are sufficient to overcome this binding even at quite low temperatures, so that the electron is excited to empty states in the conduction band, where it behaves like a nearly free electron. Chemical additions from the fifth column of the periodic table thus are able to "donate" electrons to the conduction band, and they are called *donor* impurities for this reason. Since this increases the density of conduction electrons in the crystal, it is called an *n*-type extrinsic semiconductor.

The extra electron of a donor atom is primarily influenced by the excess positive charge of the ion's nucleus. Its motion near the donor atom, therefore, is quite analogous to that of the electron in a hydrogen atom. In both cases, the electron moves in the electric field of a single positive charge. In the present situation, however, this motion takes place in a medium having a relatively large dielectric constant, so that the force is correspondingly reduced. In fact, the energy state of the electron bound to the donor can be calculated by making use of the expression for the ionization energy of a hydrogen atom, already derived in Chap. 3, decreased by the square of the dielectric constant κ_e. The ionization energy of the donor thus is

$$E_D = \frac{m_e^* e^4}{8\kappa_e^2 \epsilon_0^2 h^2}. \tag{20}$$

Note that, in order to get a quantitatively correct result, the effective mass of the electron in the solid must be used. Since the first ionization potential of the hydrogen atom is 13.6 eV, the dielectric constant of silicon is 11.7, and the effective mass in silicon is approximately $0.5m$, Eq. (20) predicts $E_D \simeq 0.05$ eV. It is shown in the next chapter that this is in good agreement with experimentally observed values.

The quantum state of this electron, called a *donor level*, accordingly is located 0.05 eV below the bottom of the conduction band for pentavalent substitutional impurities in silicon. Since the donor level has a specific energy (as distinguished from a band), it is called a *discrete* state, and since it exists only in the vicinity of the donor atom, it is also said to be *localized*. One such discrete, localized donor level exists for each impurity atom, and these are usually indicated by short lines on an energy-level diagram, as shown in Fig. 4*A*.

By comparison, when a trivalent atom like boron, aluminum, gallium, or indium substitutes for a silicon atom, one of the covalent bonds in the structure cannot be satisfied. This unsatisfied electron-pair bond can be completed by the transfer of another valence electron from a nearby silicon atom. The loss of a valence electron by this silicon atom is represented in the energy-band model by the appearance of a hole in the valence band. The foreign atom is thus said to have accepted a valence electron from the crystal, and such impurities are termed *acceptors*. The inclusion of acceptors in a crystal, therefore, creates holes in the valence band and leads to *p*-type extrinsic conductivity.

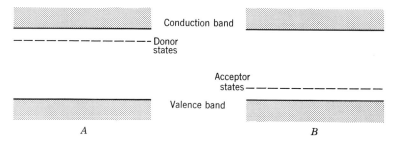

FIG. 4. Representation of discrete, localized impurity levels in an energy-level diagram for the case of *A*, donor, and *B*, acceptor, impurities.

The motion of the hole near the acceptor impurity is analogous to that of an electron near a donor. The hole has an effective positive charge, while the acceptor has an effective negative charge, since it has one less nuclear charge than the other atoms in the solid. Thus Eq. (20) can be used also for calculating the ionization energy of an acceptor level, $E_A \simeq 0.05$ eV in silicon, similarly to the value for a donor level. Acceptor states are, of course, also discrete and localized and are located 0.05 eV above the top of the valence band, as shown in Fig. 4*B*.

Other types of chemical impurities produce different kinds of impurity levels, depending on their valence-electron structure and on how they are incorporated into the crystal structure. Metals such as iron, nickel, cobalt, and copper in silicon and germanium crystals produce discrete localized levels far removed from the band edges, and are called *deep* levels. Zinc and gold atoms give rise to three discrete states per impurity atom, each at a different distance from the band edge. On the other hand, oxygen may be chemically bound in the structure in such a way as to give rise to no discrete states. In addition to chemical impurities, point imperfections may result in localized levels; for example, a vacancy-interstitial pair called a Frenkel defect behaves in many respects like an acceptor level. It is because of the generation of such states by chemical and crystalline imperfections that semiconductor crystals normally must be produced under conditions assuring extreme purity and perfection.

Position of Fermi level. The distribution of electrons (and similarly of holes) among the allowed quantum states contained in the valence and conduction bands and in the various localized levels is determined by the Fermi function. Once the position of the Fermi level is known, the occupancy of the states may be easily calculated. Since a crystal may contain both acceptors and donors, location of the Fermi level is not usually as straightforward as was the case for an intrinsic semiconductor. The principles governing its position, however, are exactly the same.

The Fermi distribution was derived in Chap. 6 for the case when the crystal is in thermodynamic equilibrium. This means that the Fermi level is so located that the crystal as a whole is neutral; that is, it contains equal numbers of positive and negative charges. Now holes and ionized donors are positive, while electrons and ionized acceptors are negative, because an ionized donor has lost its electron while an ionized acceptor has gained an electron, that is, lost its hole. The condition for charge neutrality is

$$n + N_a^- = p + N_d^+ \tag{21}$$

where N_d^+ and N_a^- are the densities of ionized donors and acceptors, respectively. Inserting the Fermi function in (21),

$$N_c f(E_c) + N_a f(E_a) = N_v[1 - f(E_v)] + N_d[1 - f(E_d)] \tag{22}$$

where N_a and N_d are the acceptor and donor concentrations, respectively, and E_a and E_d are the energy values of the acceptor and donor levels. Note that E_a and E_d are not the energy separation between the level and the nearest band edge, although conventionally the zero of energy is taken at the top of the valence band, so that $E_a = E_A$, $E_d = E_g - E_D$, and $E_v = 0$.

Normally, all the energy values in Eq. (22) are known except for E_0 [in $f(E)$], which then can be determined as a function of temperature. Because of the complexity of the equation, the solution must be carried out numerically except in very simple cases. Consider a germanium crystal containing 10^{22} donors/m^3 (a concentration of approximately one part per million) and no acceptors. A direct solution of Eq. (22), when $N_a = 0$, shows that E_0 lies 0.52 eV above the valence band at room temperature (Exercise 7). Thus the density of electrons according to Eq. (7) is

$$n = 2.5 \times 10^{25} e^{-(0.72-0.52)/kT} = 2.5 \times 10^{25} e^{-7.85} = 10^{22} \text{ electrons/m}^3 \tag{23}$$

and the density of holes in the valence band (8) is

$$p = 2.5 \times 10^{25}e^{-0.52/kT} = 2.5 \times 10^{25}e^{-20} = 5 \times 10^{16} \text{ holes/m}^3. \quad (24)$$

This distribution is indicated diagrammatically in Fig. 5, where the parabolic densities of states in the conduction and valence bands, together with their representation in terms of the discrete equivalent density of states, are shown on the left. The discrete donor levels are also shown by a sharp line. The center of the figure gives the probability of occupancy of the states as represented by the Fermi function. The

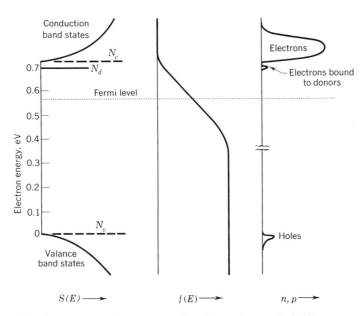

Fig. 5. The density of available states, Fermi function, and resulting electron and hole concentrations for an *n*-type germanium crystal. The hole and electron concentrations are not to scale.

densities of holes and electrons are indicated on the right in an approximate way since it is difficult to show 10^{22} electrons and 10^{16} holes on the same scale.

Figure 5 illustrates that nearly all the extrinsic electrons are in the conduction band. The number remaining bound to donors, as well as the number of holes in the valence band, is negligible in comparison. The properties of the semiconductor are dominated, therefore, by the electrons donated by the one part per million of arsenic atoms added to the crystal. In an *n*-type semiconductor, the electrons occupying states in the conduction band are called the *majority carriers* because they are responsible for most of the current-carrying ability of the semiconductor. Similarly, the holes are called the *minority carriers* in this case. Obviously, in a *p*-type semiconductor, the roles of the carriers are reversed.

The exact position of the Fermi level depends upon temperature. Results of calculations based on (22) for *n*-type and *p*-type germanium are shown in Fig. 6 for different impurity concentrations. The Fermi level remains near the center of the forbidden gap at all temperatures for an intrinsic semiconductor, as shown in an earlier section. In extrinsic material containing relatively large impurity concentrations, however,

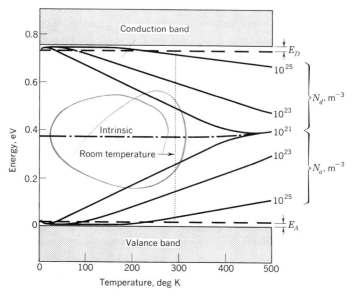

Fɪɢ. 6.   Variation of the Fermi level with temperature for both *n*-type and *p*-type germanium of various impurity concentrations.   The Fermi level always tends toward the center of the forbidden gap at high temperatures.

it remains in the region of the impurity levels up to fairly high temperatures. At sufficiently high temperatures the semiconductor must eventually become intrinsic. This happens when the number of electrons thermally excited from states in the valence band to those in the conduction band exceeds the number contributed by the impurities. The similarity between situations in *n*-type and *p*-type materials is evident in Fig. 6.

When the Fermi level lies between the donor levels and the conduction band, the *n*-type semiconductor is in the extrinsic range of conduction. After the Fermi level has crossed the energy level of the donors, it is said to be in the *exhaustion range*, since essentially all the donor electrons are occupying states in the conduction band. When $n \simeq p$, it is, of course, in the intrinsic range.   In practical devices, it is usually necessary to operate in the extrinsic, or exhaustion, ranges in order to take advantage of the properties purposely introduced by chemical impurities.   Consequently the temperature at which intrinsic conductivity becomes dominant determines the highest permissible operating temperature for such semiconductor devices.   According to (18), the width of the forbidden gap effectively determines this temperature.   In germanium ($E_g = 0.72$ eV), it is about 100 deg C, while devices using silicon ($E_g = 1.1$ eV) can be operated up to about 200 deg C.

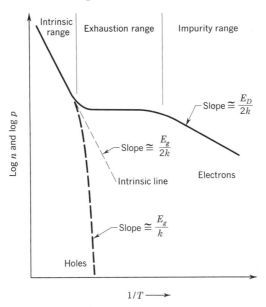

Fig. 7.   Variation of the hole and electron concentrations in an n-type semiconductor with the reciprocal absolute temperature.

Carrier concentration.   Once the position of the Fermi level is known as a function of temperature, the carrier concentrations can be calculated easily from Eqs. (7) and (8).   The form of these expressions makes it convenient to plot the logarithm of the carrier density as a function of reciprocal absolute temperature.   A straight line results in such a plot whenever the Fermi level does not change much with temperature, and the slope of this line determines the energy difference between the Fermi level and the conduction levels.

An example of the expected variation for an n-type semiconductor is shown in Fig. 7.   At low temperatures, the Fermi level is about midway between the donor levels and the conduction band and the electron concentration increases exponentially as determined by (7). The slope of the line can be used to calculate the depth of the donor levels. In the exhaustion range, the electron concentration is approximately constant since all donors are ionized. Finally, in the intrinsic range the slope determines the width of the forbidden-energy gap. Similarly, at low temperatures, the hole concentration is very small but increases rapidly as the temperature is raised. Initially the curve has a slope which corresponds to nearly the full forbidden gap, while at high temperatures it follows the intrinsic line. In addition to making possible the calculation of the various energy differences, such experimental data can be used to measure the net impurity content. This is clearly equal to the electron concentration on the horizontal part of the curve where the semiconductor is in the exhaustion range.

It is usual for semiconductors to contain both donor and acceptor impurities simultaneously. A moment's reflection shows that donor electrons fill up any available acceptor levels since the crystal must attain the lowest possible energy state consistent with its temperature. This conclusion also can be reached directly from Eq. (21). Therefore the difference between acceptor and donor concentrations, $N_a - N_d$, determines whether the crystal is n-type or p-type. For most purposes, the semiconductor can be considered to contain $N_a - N_d$ acceptors or $N_d - N_a$ donors, whichever is larger. In certain problems, particularly those having to do with electron scattering, the existence of ionized-impurity levels must be considered. Since it is easier to fabricate samples with $N_a \simeq N_d$ than with $N_a = N_d = 0$, crystals with intrinsic behavior are not as difficult to prepare as might be expected. Such specimens are said to be *compensated*.

Hall effect. It is quite fortunate that a simple experimental procedure is available to measure the concentration of the majority carriers in semiconductors. This is the *Hall effect*, which is observed when a magnetic field is applied at right angles to a conductor carrying a current, as shown in Fig. 8. The magnetic field gives rise to an electric field in a direction mutually orthogonal to the direction of the current and the magnetic field. The reason for this effect is apparent when the forces on a current carrier are considered. The electric field $\mathcal{E}$, which produces the current I, causes a force of magnitude $e\mathcal{E}$ to act on the electron. In the presence of the magnetic field, a magnetic force proportional to the magnetic-field strength B and the electron's average velocity, of magnitude $\bar{v}B$, also acts on the electron. This force is at right angles to the directions of B and $\bar{v}$, and therefore each electron is deflected toward one side of the conductor. When the electrons reach the surface of the conductor, an electrical charge is built up there which, in turn, produces an additional electric field. Under equilibrium conditions, the sideways force on the moving carriers due to this field just balances that arising from the magnetic field and the electrons can again move freely down the conductor.

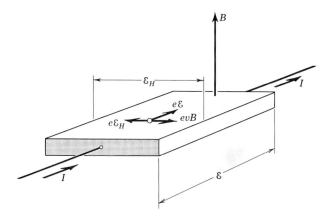

Fig. 8. The forces acting on a current carrier in a conductor placed in a magnetic field leading to the observable Hall field $\mathcal{E}_H$.

The magnitude of the transverse Hall field, $\mathcal{E}_H$, is found by equating the sidewise forces,

$$e\mathcal{E}_H = e\bar{v}B. \tag{25}$$

The average velocity can be expressed in terms of the current density J and the conduction electron density n from the relation $J = ne\bar{v}$. The result is

$$\mathcal{E}_H = \left(\frac{1}{ne}\right)JB = RJB \tag{26}$$

where the *Hall constant*

$$R = \frac{1}{ne} \tag{27}$$

gives the carrier concentration directly. The experimentally observed quantity is the Hall voltage V_H, which is obtained directly from (26) since $\mathcal{E}_H = V_H/w$, and $J = I/wt$, where w is the sample width and t its thickness. Therefore (26) becomes

$$R = \frac{1}{ne} = \frac{V_H t}{IB}. \tag{28}$$

In (28), e is the magnitude of the electronic charge, and the algebraic sign of the Hall voltage tells whether the carriers are holes or electrons. A single-crystal germanium sample prepared for Hall-effect measurements is shown in Fig. 9. The experimental data obtained from such specimens containing several different impurity concentrations are shown in Fig. 10. The carrier density for these curves has been calculated with the aid of (27). The similarity of these experimental data to Fig. 7 is evident, and the net donor concentration can be determined, therefore, directly from the curves.

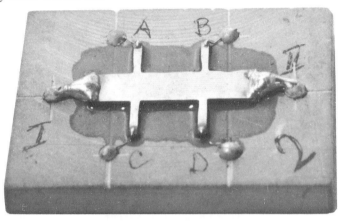

FIG. 9. Single-crystal germanium sample prepared for Hall-effect and conductivity studies. Current flow is down the main body of the crystal, and the side arms are used to prevent the side contacts from disturbing the current flow pattern. (*Courtesy of M. Epstein.*)

Combining Eqs. (13) and (27),

$$\mu = R\sigma \tag{29}$$

so that the carrier mobility can be determined experimentally directly from conductivity and Hall-effect measurements. The mobility in Eq. (29) is usually referred to as the *Hall mobility*.

The Hall constant given by (27) applies when the current is predominantly carried by majority carriers, and may be used to determine their concentration and mobility. In order to measure the mobility of both holes and electrons, n-type and p-type specimens of any semiconductor must be studied. In the case of intrinsic and near-intrinsic samples, where both holes and electrons play an important role, (27) must be replaced by

$$R = \frac{p - nb^2}{(p + nb)^2 e} \tag{30}$$

where $b = \mu_n/\mu_p$. Because of the form of the numerator in (30), the Hall voltage is generally smaller for intrinsic specimens than for extrinsic samples. Also, the sign of the Hall constant depends on the relative mobilities of holes and electrons. It is possible for the sign of the Hall voltage to change with temperature under certain conditions, for example, at the onset of intrinsic conductivity in a p-type semiconductor in which the electron mobility is greater than the hole mobility.

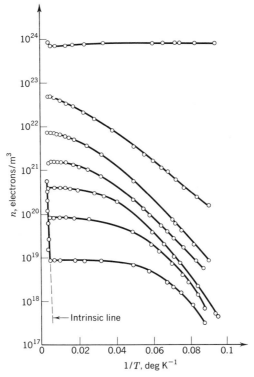

FIG. 10. Experimental determination of carrier density in germanium crystals having different donor concentrations. (*After P. P. Debye and E. M. Conwell.*)

Mobility of current carriers

Drift velocity. The conductivity of a solid has previously been expressed $\sigma = ne\mu$, where the mobility μ is the average velocity acquired by a carrier in a unit electric field. This concept must be examined more closely. First of all, the electrons and holes in a semiconductor are in rapid random motion because of their thermal energies. It is the additional velocity introduced by an external electric field which constitutes the electrical current observed. The net velocity due to the field is called the *drift velocity*. This situation is represented diagrammatically in Fig. 11. Secondly, an electron in a perfect crystal structure experiences an acceleration $-e\mathcal{E}/m^*$ in a field $\mathcal{E}$, which implies that its drift velocity continuously increases. Clearly, this cannot be true, because, according to Ohm's law,

$$\sigma = \frac{J}{\mathcal{E}} = \frac{ne\bar{v}}{\mathcal{E}} = \text{constant.} \tag{31}$$

This means that the average velocity $\bar{v}$ must be a constant. Therefore it is necessary to assume that the electron loses energy in collisions with the crystal structure so that it has a constant average velocity, or mobility $\mu = \bar{v}/\mathcal{E}$.

Suppose that the average time between such collisions is τ and that at each collision the electron loses all the energy it gained from the field

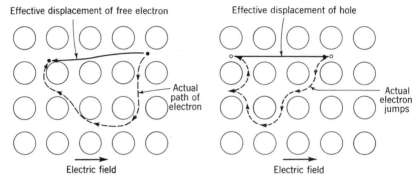

Effective displacement of free electron

Effective displacement of hole

Actual path of electron

Actual electron jumps

Electric field   Electric field

FIG. 11. Effective displacement of electrons and holes. Note that the electron follows a random path, similarly to that of a free electron in a metal, whereas a hole moves by discrete atom-to-atom jumps of a valence electron in the opposite direction.

subsequent to the previous collision. For simplicity, assume that τ is independent of the electron velocity. Then the number of collisions per second is $1/\tau$, and the rate of change of velocity is $-\bar{v}/\tau$. Under steady-state conditions, this rate of change must be equal to the acceleration due to the field, or

$$\frac{e\mathcal{E}}{m^*} = \frac{\bar{v}}{\tau} = \frac{\mu\mathcal{E}}{\tau} \tag{32}$$

and

$$\mu = \frac{e\tau}{m^*}. \tag{33}$$

The parameter τ is called the *relaxation time*.

In a perfect crystal structure, the conduction electrons move in an ideally perfect, triply-periodic potential field. In order to explain the origin of the collisions discussed above, it becomes necessary to postulate that deviations from the ideally perfect lattice array occur in crystals. There are several kinds of imperfections that are known to exist in crystals, including substitutional and interstitial atoms, missing atoms or vacancies, linear imperfections called dislocations, and others. In addition to these built-in defects, the thermal vibrations of atoms, at temperatures above absolute zero, produce instantaneous displacements which also disrupt the ideal periodicity in a crystal. It will be recalled from the discussion of specific heat that such atomic displacements can be described alternatively by the movement of phonons through the crystal, so that they are called *transient* imperfections. All these imperfections disrupt the periodicity of the atomic array and cause the electrons to be scattered. The two most important causes of scattering in semiconductors are:

1. Atomic vibrations (phonons)
2. Ionized-impurity atoms.

The effect that these imperfections have on the mobilities of electrons (or holes) can be determined most easily by calculating the corresponding relaxation time τ. In performing such calculations, it is necessary to consider the energy values that the electrons can have and it is convenient to average the relaxation times over the entire energy distribution. In general, τ may be a function of both energy and velocity, since the two quantities are independent according to the zone theory, and the actual averaging process can become quite complicated.

Scattering by phonons. As already stated, an electron moving through an imperfect lattice array of atoms can be scattered by the irregularities produced in the periodic potential, for example, phonons. This is often called lattice scattering, although it should be clearly understood that it is not the lattice array of atoms that causes the scattering, but rather the thermally induced deviations from the ideal lattice array, so that it is called *phonon scattering* in this book. The extent of this scattering depends on the degree of thermal vibration, so that it can be expected that the so-called lattice mobility is inversely proportional to temperature. An actual calculation of the mobility μ_L bears out this conclusion.

$$\mu_L = 3.2 \times 10^{-9} \frac{C_{ll}}{\mathcal{E}_1} \left(\frac{m}{m^*}\right)^{\frac{5}{2}} T^{-\frac{3}{2}} \qquad \text{m}^2/\text{V-sec} \qquad (34)$$

where C_{ll} is an average elastic constant, usually chosen as intermediate between that for [100] and [110] longitudinal acoustic waves, dyn/m^2

$\mathcal{E}_1$ is a measure of change in position of the energy-band edge due to change in volume of the unit cell, eV

T is absolute temperature, deg K.

The important feature of (34) in most semiconductor studies is the $T^{-\frac{3}{2}}$ temperature dependence. The elastic constant C_{ll} and the dependence of $\mathcal{E}_1$ on the unit-cell volume expresses, of course, the basically mechanical nature of the atomic vibrations.   As shown in the next section, Eq. (34), which is derived on the basis of a relatively simple model for the thermal vibrations, gives reasonably good agreement with experimental results.

Impurity scattering.   Ionized impurities in the structure also scatter the current carriers because the Coulomb field associated with each ionized donor or acceptor atom produces an irregularity in the periodic potential.   The field of the impurity atom is modified by the dielectric constant of the solid, and the resulting mobility depends upon temperatures because the electron velocity is a function of temperature.   The magnitude of this mobility is given by

$$\mu_I = \frac{3.2 \times 10^{17} x^2}{N_I \ln (1 + x^2)} \, T^{\frac{3}{2}} \qquad \text{m}^2/\text{V-sec} \tag{35}$$

where $x = 1.8 \times 10^5 T / N_I^{\frac{1}{3}}$, and N_I is the number of ionized impurities per cubic meter, and T is the absolute temperature.   Note that the mobility increases with temperature and is approximately inversely proportional to the density of ionized impurities present.

The scatterings by phonons and by ionized impurities are assumed to be essentially independent processes, so that the actual mobility is given by

$$\frac{1}{\mu} = \frac{1}{\mu_I} + \frac{1}{\mu_L} \tag{36}$$

or, according to (34) and (35), by

$$\frac{1}{\mu} = aT^{-\frac{3}{2}} + bT^{\frac{3}{2}} \tag{37}$$

where a and b are constants for a given material.   If a and b have comparable magnitudes, then, at low temperatures, the first term in (37) dominates and the mobility varies as $T^{\frac{3}{2}}$.   Conversely, a $T^{-\frac{3}{2}}$ variation is expected at elevated temperatures, where the second term becomes large.   This is borne out by the actual experimental measurements on the more impure silicon specimens shown in Fig. 12.   By comparison, the purest specimens exhibit phonon scattering over the entire temperature range.   These experimental mobilities are calculated by inserting experimentally measured values of the Hall constants and conductivities in Eq. (29).

In order to interpret fully experimental data such as those in Fig. 12, it is often necessary to account for scattering by other impurities that are electrically neutral.   The mobility due to such scattering is

$$\mu_N = \frac{m^* e^3}{2 \epsilon_0 \kappa_e h^3 N}$$

where N is the density of neutral impurities.   The effect of μ_N can be included by adding its reciprocal on the right in (36), and it is most important in highly compensated semiconductors. That the effect of both types of impurity scattering on mobility is inversely proportional to the impurity concentration is indicated in Fig. 13, where the theoretical curve has been corrected to include the effects of phonon scattering. It

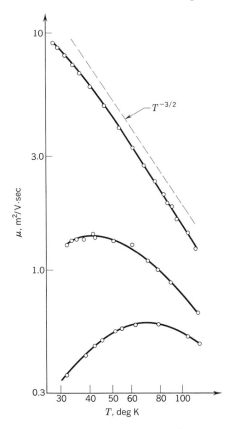

Fig. 12. The pure sample (upper curve) shows the $T^{-\frac{3}{2}}$ trend resulting from phonon scattering, while the less pure sample (lower curve) displays a $T^{\frac{3}{2}}$ dependence at low temperatures characteristic of ionized impurity scattering. (*After D. Long and J. Myers.*)

can be seen from Figs. 12 and 13 that the mobility is large at low temperatures in pure samples and is markedly reduced at high temperatures or in very impure specimens.

Diffusion constant. As is discussed in Chap. 10, it is possible to produce nonuniform concentrations of carriers in semiconductor crystals. The diffusive motion of carriers in response to such gradients is analogous to the motion of gas molecules in response to pressure gradients and is

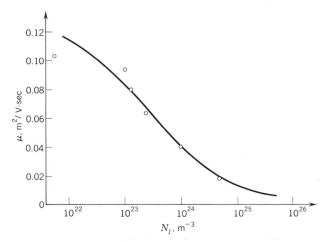

Fig. 13. Decrease in mobility as a function of ionized impurity concentration for silicon. The circles represent experimental values, while the curve represents the theoretical dependence. (*After G. Brackenstoss.*)

important in the operation of semiconductor devices. A close relationship exists between diffusion currents and conduction currents because of the many collisions the carriers experience which tend to randomize their velocities in both cases. From the usual definition of the diffusion constant D, the electric-current density caused by diffusion due to a concentration gradient dn/dx is

$$J_D = -eD\frac{dn}{dx}. \tag{38}$$

If such diffusion results in an electric field $\mathcal{E}$, steady-state conditions are reached when the conduction current $J_c(=\sigma\mathcal{E})$ produced is equal to J_D.

$$J_c = J_D$$

$$ne\mu\mathcal{E} = -eD\frac{dn}{dx}. \tag{39}$$

The presence of the field $\mathcal{E}$ produces a potential difference $e\mathcal{E}x$ over a distance x, so that the variation of n with x is

$$n = Ce^{-e\mathcal{E}x/kT} \tag{40}$$

where C is a constant. Equation (40) is an example of the application of Boltzmann statistics to the energy distribution of carriers. By combining (39) and (40), the relation between mobility and the diffusion constant is found to be

$$D = \frac{kT}{e}\mu \tag{41}$$

which is known as the *Einstein relation* and is further discussed in Chap. 12. Equation (41) is very useful in determining D because μ can be readily obtained experimentally from Hall-effect and conductivity measurements, as already shown.

Minority-carrier lifetime

The equilibrium concentration of holes and electrons in a semiconductor can be temporarily changed by external means. For example, if light photons, whose energy $h\nu$ is greater than the forbidden-energy gap,

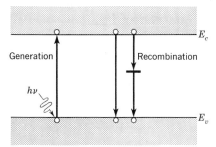

Fig. 14

are absorbed in the crystal, the energy of the photon may raise an electron from the valence band to the conduction band as indicated in Fig. 14. This process is called *generation* and produces an excess carrier concentration which persists so long as the radiation is incident. When the radiation is removed, the excess concentration decays back to the equilibrium value by means of *recombination* transitions. Such recombination may take place either directly by a band-to-band transition or by transitions to localized levels as shown in Fig. 14.

In an extrinsic semiconductor, the minority carriers can be markedly affected by such external agents but the concentration of majority carriers remains essentially unchanged because the majority-carrier density is large to begin with. Therefore attention is focused on the minority-carrier density even though both hole and electron concentrations must change simultaneously to preserve charge neutrality. Let dt/τ represent the probability that a hole will recombine in a time interval dt. Then the rate at which holes disappear is p/τ in an n-type semiconductor. Thus the rate of change of the excess hole density is

$$\frac{dp}{dt} = g - \frac{p}{\tau} \tag{42}$$

where g is the generation rate due to photon absorption. When the generation ceases, that is, when $g = 0$, (42) reduces to

$$\frac{dp}{dt} = -\frac{p}{\tau}$$

from which $p = p_0 e^{-t/\tau}$ (43)

as may be verified by direct substitution. Here p_0 is the excess hole density during generation. Equation (43) shows that the carriers decay exponentially with time after the light is turned off at a rate governed by τ, called the *minority-carrier lifetime*. A convenient experimental method of determining τ is to illuminate the semiconductor with a flash of light and measure the change in conductivity as a function of time after the light pulse. Note that τ has been previously used to denote the relaxation time, and the two should not be confused.

Recombination processes. The value of τ corresponding to direct band-to-band recombination (Fig. 14) can be calculated using quantum-mechanical perturbation theory and is found to be of the order of several tenths of a second for germanium. Experimentally observed lifetimes are always less than about 10^{-2} sec, however, which means that direct recombination is not the major effect in germanium. Direct band-to-band transitions are unlikely because the electron must lose an amount of energy corresponding to the forbidden-energy gap by the emission of photons or of phonons, and simultaneously its momentum also must be dissipated. Since the phonon energy is of the order of $kT \simeq 0.03$ eV at room temperature, many phonons must be created simultaneously in order to take up the electron's energy, and this makes the process improbable. Direct recombination accompanied by photon emission is also improbable, because, in order to conserve momentum, the free electron and free hole must collide with oppositely directed velocities.

In view of the above, it is much more likely that an electron occupying a state in the conduction band returns to a valence-band state in several sequential transitions, as illustrated in Fig. 14. An equivalent way of stating this overall process is that the discrete level first captures an electron and subsequently captures a hole. Such discrete levels, which are usually located near the center of the forbidden-energy gap, are called *recombination centers* because they aid the recombination between electrons and holes. The recombination center is effective because the amount of energy that must be dissipated in each transition is smaller than in the case of direct recombination and because it can transmit the carrier's momentum to the crystal structure.

The probability that any localized level can capture an excess carrier is expressed in terms of its *capture cross section*. A recombination center is a level that has a large capture cross section for both electrons and holes.

Depending upon the atomic origin of the level, it often happens that the cross section for capture of one type of carrier may be much larger than that for the other carrier. In this case the level is called a *trap* because it tends to trap the carrier for which it has the larger capture cross section without capturing the other kind of carrier. Since a given crystal may contain impurities which give rise to both hole and electron traps as well as recombination centers, it is possible for carriers to undergo several different kinds of transitions, which are indicated in Fig. 15.

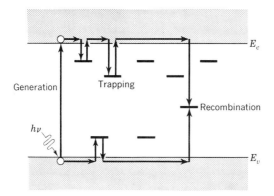

Fɪɢ. 15

As discussed in the next section, the surface of a crystal contains many imperfections. Consequently the energy-band model near the surface contains a large concentration of localized levels, and recombination transitions are more likely to occur at the surface. It is convenient, therefore, to distinguish the *bulk lifetime* τ_b from *surface lifetime* τ_s. They can be combined to give the observed lifetime by

$$\frac{1}{\tau} = \frac{1}{\tau_b} + \frac{1}{\tau_s}. \tag{44}$$

The bulk lifetime is determined by the density of recombination centers distributed throughout the crystal volume. Surface lifetime depends upon the density of recombination centers on the surface and also on the sample dimensions, since carriers must diffuse to the surface in order to recombine. It is often true for pure samples that the bulk lifetime is so large that all carriers produced in the crystal can reach the surface and therefore surface recombination dominates. In this case it is possible to change the lifetime of the crystal by various surface treatments which alter the density of recombination centers at the surface. For example, abrading the surface greatly reduces the lifetime while careful chemical etching increases it.

The ability of a recombination center to catalyze recombination depends upon whether it is already occupied or not. At equilibrium, this is determined by the position of the Fermi level. Since the occupancy of allowed states is a function of temperature, it follows that the minority-carrier lifetime changes with temperature. Experiments have shown that different types of imperfections can serve as recombination sites and the lifetime is extremely sensitive to the density of such imperfections. For example, the addition of as little as 3×10^{20} atoms per cubic meter of copper to highly purified germanium (a concentration of less than one part in 10^8) can reduce the lifetime from 2×10^3 to 40 μsec.

Diffusion length. If excess carriers are generated inhomogeneously in a semiconductor, for example, by illuminating only a small region of the crystal, they tend to diffuse away from their point of origin because the concentration gradient is largest there. These excess carriers have a finite lifetime, so that they ultimately disappear by recombination. The average distance a carrier diffuses before recombination L is related to the diffusion constant D and the lifetime τ by

$$L = (D\tau)^{\frac{1}{2}}. \tag{45}$$

The quantity L is normally called the *diffusion length*. From Eqs. (41) and (45) it can be seen that, in crystals containing carriers having high mobility and long lifetime, the carriers can travel larger distances than, for example, in crystals in which these parameters are smaller. The diffusion length is of particular importance in the operation of a transistor, as shown in Chap. 10.

Carrier fluctuations. Transitions like those illustrated in Figs. 14 and 15 for excess carriers also take place under equilibrium conditions. The distribution of carriers among the various allowed quantum states occurs under conditions of dynamic equilibrium in which such transitions take place continuously and allow the crystal to adjust to changes in its environment. Such transitions lead to the average densities of electrons and holes in the conduction and valence-band states, as calculated from Eqs. (7) and (8). Each transition temporarily changes the carrier density in a given level by one, however, so that small random fluctuations in the carrier density result. Since conductivity depends upon the density of carriers in the conduction-band states, the conductivity of a semiconductor is not exactly constant but fluctuates slightly. The conductivity variations can be detected by applying a constant potential to the crystal and measuring the ensuing fluctuations in the current. Since there are very many random transitions per second, the current fluctuations constitute a wide band of frequencies. It is convenient to represent the strength of this *current noise* in terms of an average noise current S_i. An analysis of electronic transitions in a nearly intrinsic n-type semiconductor shows that this quantity is given by

$$S_i = \frac{4i^2}{n} \frac{\tau}{1 + (2\pi f \tau)^2} \tag{46}$$

where i is the average current and f is the frequency at which the noise is detected. An experimental curve illustrating this relation is shown in Fig. 16. These results demonstrate the importance of carrier transitions in producing carrier fluctuations. Since all types of transitions cause

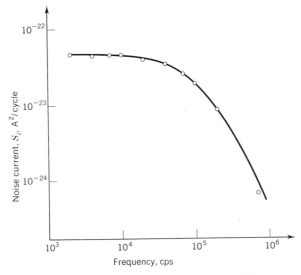

FIG. 16. Current-noise power spectrum for germanium. The curve is a plot of Eq. (46). (*After J. E. Hill and K. M. van Vliet.*)

such fluctuations, the trapping processes illustrated in Fig. 15 can be investigated by examining the current noise in semiconductors.

Surfaces

Surface states. The triply-periodic potential in a crystal abruptly terminates at its surface, and this introduces departures from the energy-band scheme appropriate inside the crystal. Furthermore, foreign atoms and/or an oxide layer adsorbed on the surface can give rise to discrete levels which also influence the properties of the solid in a region near the surface. This results in a large density of localized quantum states at the surface whose energy levels are distributed in the forbidden gap. Such levels are called *surface states* and may be present in densities as great as 10^{19} m^{-2}, or about one for each surface atom in the crystal.

The electrical properties of the semiconductor are influenced to a considerable depth in the crystal by electrons occupying surface states
because any excess charge in these states must be compensated by changes
in the free-carrier concentration in the crystal in order to maintain
charge neutrality. By comparison, surface effects are not as important
in metals, because the large number of free electrons present can easily
compensate any surface charge.

Space-charge layers. The charge density localized in surface states
is influenced by the origin of the states. Several situations can be
distinguished, depending on the sign and magnitude of the surface charge

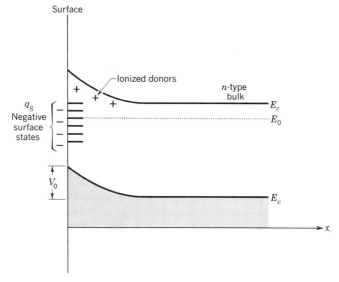

Fig. 17. Energy-level diagram for charged surface states which induce a region
immediately below the surface depleted of conduction electrons.

and on the sign of the majority carrier in the bulk of the semiconductor.
Figure 17 illustrates the case of negatively charged surface states on an
n-type crystal. The immobile negative charge q_S repels free electrons
from the surface, leaving positively charged ionized donors to neutralize
the effect of the surface charges. This behavior is illustrated in an
energy-level diagram by the upward bending of the bands near the
surface because the potential energy of the electron is now greater in this
region than inside the crystal. There is, therefore, an electric field built
up near the surface, and the difference in potential energy at the surface
compared with that in the bulk is V_0, as indicated. Since the density of
free carriers is depleted near the surface, this region is called a *depletion
layer*. A similar situation arises for positively charged surface states on a
p-type semiconductor, except that the bands bend downward.

In the simple case illustrated in Fig. 17, it is possible to determine the shape of the potential variation with depth. This is done by using the one-dimensional form of Poisson's equation, which relates the electric potential $V(x)$ to the electric-charge density $\rho(x)$,

$$\frac{d^2V}{dx^2} = \frac{\rho(x)}{\epsilon_0\kappa_e} \tag{47}$$

where κ_e is the dielectric constant. In the present case $\rho(x) = eN_d$ since the distribution of ionized donors is uniform. If the positive x direction is directed into the crystal, Poisson's equation for the additional electron potential energy becomes

$$\frac{d^2V}{dx^2} = -\frac{eN_d}{\epsilon_0\kappa_e} \tag{48}$$

where, as usual, the minus sign takes account of the electron's negative charge. The solution of (48) for $V(x)$ is

$$V(x) = \frac{eN_d}{2\epsilon_0\kappa_e}\left[\left(V_0\frac{2\epsilon_0\kappa_e}{eN_d}\right) - x\right]^2 \tag{49}$$

as may be verified by differentiating (49) twice and substituting in (48). The form of (49) shows that the potential has a parabolic dependence on x near the surface. The depth of the space-charge layer is determined by setting $V(x) = 0$ and solving for x, which then becomes the depth at which the electric field due to surface-state charges is canceled by the charge arising from ionized donors.

If the negative charge in the surface states of an n-type crystal is large, the bands are bent upward to such an extent that a large hole density is generated in valence-band states as indicated in the top drawing of Fig. 18. These free holes, together with ionized donors, neutralize the surface charge, and the shape of the potential again can be determined by using Poisson's equation. The result is that a p-type surface is produced on an n-type crystal, and it is termed an *inversion layer*. An inversion layer is also possible on a p-type semiconductor, as illustrated in the bottom drawing of Fig. 18. When the surface states on an n-type crystal are positively charged, the charge is neutralized by the free carriers in the conduction band, and this produces an *accumulation layer*. Negatively charged surface states on p-type material have a similar effect. Exercise 12 considers the energy-band model for accumulation layers.

Slow and fast states. It is found experimentally that two classes of surface states exist. States arising from foreign atoms located at the crystal-air or crystal-oxide interface are called *fast states* because their occupancy can change rapidly with changes in the bulk carrier concentration; that is, if the carrier concentration in the crystal is altered

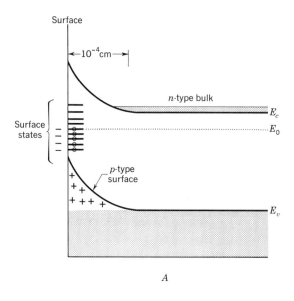

A

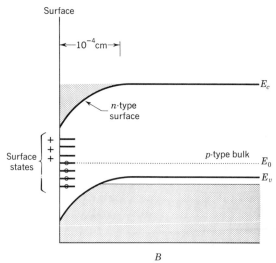

B

FIG. 18.   Effect of strongly charged surface states.   (*A*) *p*-type surface produced on an *n*-type crystal; (*B*) *n*-type surface produced on a *p*-type crystal.

(for example, by photon absorption), the density of charge in the fast surface states can readjust in times of the order of 10^{-6} sec to attain a new equilibrium.   Such levels are believed to be responsible for recombination transitions at the surface.   Transitions to surface states arising from foreign atoms located in the oxide layer or at the oxide-air interface are less probable, and the charge in these states may require many seconds to readjust. Experiments have shown that the fast surface-state density is of the order of 10^{16} m^{-2} on crystals of silicon and germanium, while the slow state density is between 10^{15} to 10^{17} m^{-2}, depending upon the thickness of the oxide layer and the type of ambient atmosphere.

Contacts

It is necessary to make electrical contact to semiconductor crystals by means of metallic conductors in order to carry out electrical measurements or to fabricate useful semiconductor devices. The junction between the metal and the semiconductor determines the current which can flow into the semiconductor. The properties of the junction can be understood in terms of energy-level diagrams similar to those discussed in the previous section.

Rectifying contacts. It is shown in Chap. 11 that, when two dissimilar metals are brought together, electrons transfer from one to the other until the Fermi levels of the two metals line up. The same thing happens when a metal and a semiconductor are brought into contact. Because of charge transfer between the two solids, a potential barrier forms at the interface. This barrier extends to an appreciable distance in the semiconductor similarly to space-charge layers due to surface states.

The energy-band models of a metal and an n-type semiconductor before they are brought together are shown in Fig. 19A. The quantities E_{W_1} and E_{W_2} are called the *work functions* of the solids and are the difference between the initial positions of the Fermi level and the height of the *surface potential barrier*, as discussed more fully in Chap. 11. In Fig. 19A the work function of the metal E_{W_1} is greater than that of the semiconductor E_{W_2}, so that, upon contact, electrons flow from the semiconductor to the metal until the negative charge on the metal builds up sufficiently to reduce the flow and make the Fermi level continuous. This illustrated in Fig. 19B and results in a space-charge layer in the semiconductor. An analogous situation occurs for a metal-semiconductor contact when the semiconductor is p-type and when the work function of the metal is smaller than that of the semiconductor, as shown in Fig. 20A. Since $E_{W_1} < E_{W_2}$, electron flow takes place from the metal to the semiconductor, and this again results in a space-charge layer in the semiconductor (Fig. 20B).

In both cases, a potential barrier forms at the metal-semiconductor junction, whose height is determined by the difference between the work functions. The form of the barrier in these simple cases is similar to that given by Eq. (49). The potential barrier impedes the flow of charge carriers across the junction more in one direction than in the other, as discussed in a later section, and such contacts are said to be *rectifying*.

Although rectifying contacts are used in semiconductor devices, it is necessary to avoid them when the electrical properties of semiconductor crystals are being measured, lest the current flow be more a property of the junction than of the bulk crystal.

Ohmic contacts. If the work function of an n-type semiconductor is greater than that of the metal, electron flow from the metal to the semiconductor makes the surface of the semiconductor more n-type, as

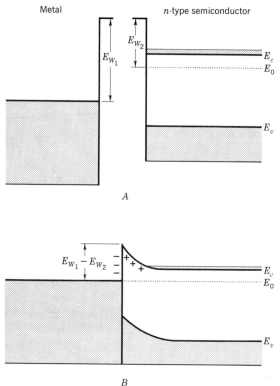

Fig. 19. Energy-level diagram of the junction between a metal and an n-type semiconductor when the work function of the metal is larger than that of the semiconductor. (*A*) Before contact; (*B*) after contact.

indicated in Fig. 21*A*. The reverse is true for a p-type semiconductor when its work function is smaller than that of the metal (Fig. 21*B*). These cases are examples of *ohmic* contacts since no barrier exists for the flow of electrons in either direction for n-type material or for holes in either direction for p-type specimens, as can be seen in the energy-level diagrams of Fig. 21.

These considerations imply that the height of the barrier is determined by the relative work functions of the metal and semiconductor. It is often observed experimentally, however, that the height of the barrier does not depend on the work function of the metal.   The reason for this is that surface states on the semiconductor may be more important in determining the size of the potential barrier.   If the surface-state density is large enough, the barrier is regulated primarily by occupancy of the surface states.   In this case, contact with a metal causes electron

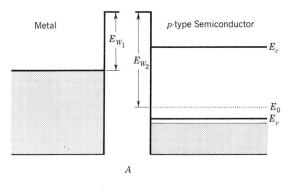

A

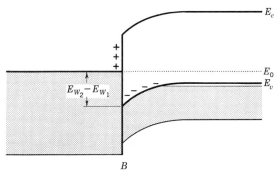

B

Fig. 20.   Energy-level diagram showing the space-charge layer that forms at a metal-semiconductor junction for a *p*-type semiconductor when $E_{W1} < E_{W2}$.   (*A*) Before contact; (*B*) after contact.

exchange between the metal and the surface states rather than the bulk semiconductor, so that the surface space-charge layer is not changed much.   Specifically, this means that, in order to obtain ohmic contacts, it is necessary to remove as many of the surface states as possible before applying the metal contact.

The rectifier equation.   The rectifying characteristics of metal-semiconductor junctions can be determined by calculating the currents flowing across the barrier in the two directions and subtracting them from each other to obtain the net current.   The case of an *n*-type semi-

conductor is considered in Fig. 22.   In equilibrium (Fig. 22*A*) a barrier,
$E_{W_1} - E_{W_2}$, exists at the interface, as discussed above.   Some electrons
in the metal have sufficient energy to overcome this barrier and flow into
the semiconductor.   Similarly, some electrons in the semiconductor have
sufficient energy to reach the metal and, at equilibrium, the two electron
currents are equal.   If an external potential is then applied between the

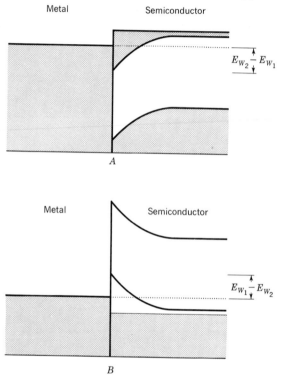

FIG. 21.   Energy-level diagrams of ohmic contacts to (*A*) *n*-type and (*B*) *p*-type
semiconductors.

metal and the semiconductor, say, from a battery, to make the semiconductor positive as shown in Fig. 22*B*, the height of the barrier is
increased to $E_{W_1} - E_{W_2} + eV$, where V is the external potential.   Since
the barrier height is increased, few electrons now can flow from the
semiconductor to the metal.   The height of the barrier viewed from the
metal is unchanged, however, so that the current from the metal to the
semiconductor is the same.   Thus there is a net current flow from the
metal to the semiconductor.   If the polarity of the external potential is
reversed (Fig. 22*C*), the barrier height is reduced and the current from
the semiconductor is increased.   This results in a net current into the

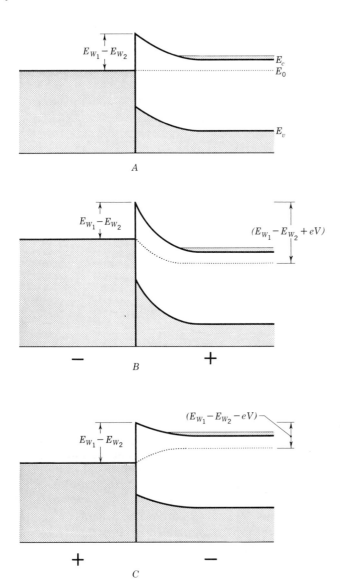

FIG. 22. Change in height of the space-charge barrier with polarity of the applied potential at a metal-semiconductor junction for an *n*-type crystal. (*A*) Zero bias; (*B*) reverse bias; (*C*) forward bias.

metal. Since the electron current from the metal is unaffected by the external potential, the rectifying characteristics of the contact depend only on the current flowing from the semiconductor. When this current is reduced, the contact is said to have *reverse bias*, and when it is increased it has *forward bias*.

The current flowing from the metal is determined by thermionic emission of electrons across the barrier $E_{W_1} - E_{W_2}$, as described in Chap. 11. Using results from that chapter, the current is

$$I_1 = BA_0 T^2 e^{-(E_{W_1} - E_{W_2})/kT} \tag{50}$$

where B is the area of the contact, and A_0 is a constant. Similarly, the current from the semiconductor is due to thermionic emission across the barrier $E_{W_1} - E_{W_2} - eV$, so that

$$I_2 = BA_0 T^2 e^{-(E_{W_1} - E_{W_2} - eV)/kT} \tag{51}$$

where, following the usual convention, the sign of V is taken to be the polarity of the metal. The net current is then

$$I = I_2 - I_1$$
$$= I_s(e^{eV/kT} - 1) \tag{52}$$

where I_s is called the *saturation current*,

$$I_s = BA_0 T^2 e^{-(E_{W_1} - E_{W_2})/kT}. \tag{53}$$

Equation (52) is called the *rectifier equation* and shows that, when V is negative, the reverse current is essentially constant and equal to I_s, while the forward current increases exponentially when V is positive. The variation of I with V as given by (53) is shown in Fig. 23.

The large ratio of forward to reverse current at a metal-semiconductor contact makes it a useful semiconductor device. Usually, the rectifying contact is in the form of a metal point resting on the semiconductor, while the other electrode is a soldered ohmic contact. The electric polarity applied to the point contact to obtain forward bias constitutes a convenient experimental procedure for determining the sign of the majority carriers in a semiconductor (Exercise 13).

Injection. Although the above discussion has concentrated on the role of the majority carriers, it is possible for the current flow across the junction to be carried partly by minority carriers. As shown in Fig. 22C, the energy difference between the Fermi level in a metal and the top of the valence band in a semiconductor represents a potential barrier for transitions from states in the valence band to those in the metal. Under certain conditions, a fraction of the current across the junction may be due to such transitions, which can be thought of as an *injection* of holes from the metal into the valence band of the semiconductor. Thus it is possible to disturb electrically the equilibrium concentration of minority carriers in a semiconductor by means of a biased metal contact.

The injected holes diffuse away from the contact a distance that depends upon their lifetime. The extra positive charge represented by the nonequilibrium concentration of holes is compensated by the charge of additional majority carriers which enter through the ohmic contact. The total majority-carrier concentration, however, does not change

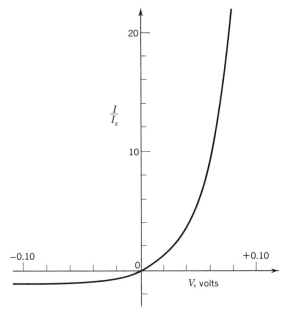

FIG. 23.   Plot of rectifier equation (52) showing increased current flow for forward bias.

much, because it is large to begin with, whereas the change in the minority-carrier concentration is relatively large. The phenomenon of minority-carrier injection is the basis of transistor action in semiconductors and is considered in more detail in Chap. 10.

Thermoelectricity

Peltier effect.   When a current flows across the junction between two dissimilar materials, heat is generated (or absorbed) at the junction. The generated heat is proportional to the current, and the phenomenon is called the *Peltier effect*. The effect is particularly interesting at metal-semiconductor junctions because it is much larger than in the case of a junction between two metals. Consider an *n*-type crystal with two ohmic contacts (Fig. 24) with a potential applied across the contacts. Although space-charge layers do not exist because the contacts are ohmic, only those electrons with energies greater than $E_c - E_0$ can get into the semiconductor (compare with Fig. 21A). Consequently, metal 1, at the left-hand contact, loses the electrons occupying its highest energy states because electrons flow from left to right in the diagram in response to the applied potential. At the right-hand contact, these electrons are deposited into metal 2, so that the "hottest" electrons are moved from metal 1 to metal 2 by virtue of the contact effects and the current flow. As a result, metal 1 is cooled and metal 2 heated by the amount of energy transferred per electron, which clearly equals $E_c - E_0$ plus the kinetic

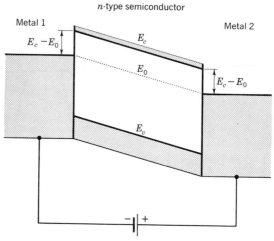

FIG. 24. Energy-level diagram of an n-type semiconductor having ohmic contacts at both ends, with an applied potential across the contacts.

energy of the carriers moving from a hot to a cold region. This is expressed in terms of the *thermoelectric power* Q_n of an n-type semiconductor, which is defined as

$$-Q_n T = (E_c - E_0) + 2kT \tag{54}$$

where the minus sign follows the usual polarity convention. Similarly, the thermoelectric power of a p-type semiconductor is

$$Q_p T = (E_0 - E_v) + 2kT. \tag{55}$$

Equations (54) and (55) show that the large values of thermoelectric power found in semiconductors basically result from the fact that the average potential energy for conduction electrons (or holes) is larger than the Fermi energy, in contrast to the situation in metals. Practical applications of the Peltier effect are discussed in Chap. 10.

Seebeck effect. If two contacts to a semiconductor are maintained at different temperatures, a potential difference can be observed between them. This is called the *Seebeck effect* and arises from the more rapid diffusion of carriers "heated" at the hot junction. These carriers diffuse to the cold junction, so that such a contact acquires a potential having the same sign as the diffusing majority carriers. For example, the cold junction of a p-type semiconductor develops a positive potential. This polarity convention is consistent with that used in Eqs. (54) and (55). The magnitude of the thermoelectric emf developed between two contacts at a temperature difference of T degrees is given by $Q_n T$ for an n-type semiconductor.

Experimental measurements of the variation of the thermoelectric power with temperature can be used to locate the position of the Fermi level directly, as shown by (54) and (55). Furthermore, the polarity of the thermoelectric emf can be used to determine the sign of the majority carrier. The familiar thermocouple is an example of the utilization of the Seebeck effect in metals.

Intrinsic thermoelectric power. In an intrinsic semiconductor, both the electrons and the holes diffuse in the temperature gradient, so that their thermoelectric potentials tend to cancel. If hole and electron mobilities are equal, $Q_i = 0$, but if they are unequal, the net thermoelectric power is a weighted average,

$$(\mu_n + \mu_p)Q_i T = \mu_n Q_n T + \mu_p Q_p T. \tag{56}$$

This expression can be simplified by assuming that the Fermi level is in the center of the forbidden gap, so that $(E_c - E_0) = (E_0 - E_v) = E_g/2$, which is sufficiently accurate for most purposes. The result, using (54) to (56), is

$$Q_i T = \frac{1 - b}{1 + b}\left(\frac{E_g}{2} + 2kT\right) \tag{57}$$

where $b = \mu_n/\mu_p$ is the mobility ratio.

The intermediate situation, where $n \neq p$ but both are important, can be treated in the same fashion. A comparison of the experimental and predicted values of the thermoelectric power in silicon is shown in Fig. 25. In the lower temperature range, the variation of the Fermi-level position with temperature is clearly indicated. As the samples become intrinsic, cancellation of the two carrier types reduces the thermoelectric power drastically, since $\mu_n \simeq \mu_p$ for silicon. The thermoelectric power of the p-type sample changes sign as the temperature increases, and this indicates that $\mu_n > \mu_p$.

Since the thermoelectric power of an extrinsic semiconductor depends upon the location of the Fermi level, it may also be expressed in terms of the majority-carrier concentration, using Eq. (7), say, for an n-type crystal. Thus there is a close relation between thermoelectric power and the Hall effect. This makes it possible to use one kind of measurement to determine either quantity, if for some reason it is experimentally difficult to carry out the other measurement. This can happen, for example, when only very small samples are available and it is difficult to

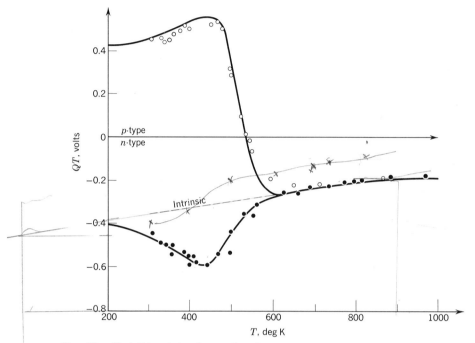

FIG. 25. Variation of the thermoelectric power of *n*- and *p*-type silicon with temperature. The circles represent experimental data, while the curves are the result of theoretical calculations. (*After T. H. Geballe and G. W. Hull.*)

attach the four contacts required for Hall-effect studies. Thermoelectric-power measurements, nevertheless, can often be made on such samples since only two contacts are required.

Suggestions for supplementary reading

William Shockley, *Electrons and holes in semiconductors* (D. Van Nostrand Company, Inc., New York, 1950).

E. Spenke, *Electronic semiconductors* (McGraw-Hill Book Company, Inc., New York, 1958).

A. van der Ziel, *Fluctuation phenomena in semiconductors* (Academic Press Inc., New York, 1959).

A. W. Wilson, *Semiconductors and metals: an introduction to the electron theory of metals* (Cambridge University Press, London, 1939).

Jay N. Zemel (ed.), *Semiconductor surfaces* (Pergamon Press, New York, 1961).

Suggestions for further reading

Adrianus Dekker, *Solid state physics* (Prentice-Hall, Inc., Englewood Cliffs, N.J., 1957).

Charles Kittel, *Introduction to solid state physics* (John Wiley & Sons, Inc., New York, 1957).

Frederick Seitz, *The modern theory of solids* (McGraw-Hill Book Company, Inc., New York, 1940).

Exercises

1. Using the definition of effective mass, show that $m^* = m$ for a truly free electron. HINT: Remember that the momentum of a free electron is $(h/2\pi)k$.

2. Derive Eq. (8) for the hole concentration by writing the expression for the number of *unoccupied* states in the valence band and integrating. Assume $E_v/kT \simeq \infty$.

3. Compute the concentration of holes and electrons in intrinsic silicon at room temperature if $m_e^* = 0.7m$, $m_p^* = m$, and $E_g = 1.1$ eV. Compare this concentration with the number of available states in the conduction band. (ANSWER: $n = p = 6.7 \times 10^{15}$ m^{-3}; $N_v = 2.5 \times 10^{25}$ m^{-3}, $N_c = 1.5 \times 10^{25}$ m^{-3}.)

4. Write down an expression for the product np. Is there anything significant about this relation that can be of practical use in determining carrier concentrations?

5. The conductivity of intrinsic germanium at room temperature is 2.2 Ω^{-1}-m^{-1}. What is the average mobility of holes and electrons? What is the approximate upper limit to the net impurity concentration to observe intrinsic behavior at room temperature? For germanium, $E_g = 0.72$ eV. (ANSWER: $\mu \simeq 0.3$ m^2/V-sec; $N_d - N_a < 10^{19}$m^{-3}.)

6. Justify each term in Eq. (22).

7. Locate E_0 at room temperature in a germanium crystal containing 10^{22} donors/m^3. Assume $E_g = 0.72$ eV and $E_D = 0.012$ eV. (ANSWER: $E_0 = 0.52$ eV.)

8. Plot E_0 as a function of T over the range 100 to 600 deg K for a silicon crystal containing 10^{21} acceptors/m^3. Assume $E_g = 1.1$ eV and $E_A = 0.05$ eV.

9. Plot the concentration of holes and electrons as a function of temperature for the silicon crystal in Exercise 8.

10. Calculate the relaxation time for scattering of electrons in germanium at room temperature given that $m_e^* = 0.2m$ and $\mu_e = 0.36$ m^2/V-sec. If the thermal velocity of an electron at room temperature is 10^5 m/sec, what is the average distance an electron travels between collisions? Compare this distance with the interatomic spacing. (ANSWER: $\tau = 4.1 \times 10^{-13}$ sec; distance = 410 Å.)

11. Compute the width of the depletion layer on the surface of an n-type germanium crystal if the departure of the surface potential is $V_0 = 0.2$ V and $N_d = 10^{21}$ donors/m^3. Compare this with the interatomic spacing. (ANSWER: $x_0 = 1.9 \times 10^4$ Å.)

12. Sketch energy-level diagrams similar to those of Fig. 18 but for accumulation layers on both an n- and p-type semiconductor.

13. Derive an expression analogous to the rectifier equation (52) for a p-type semiconductor, based on Fig. 20. Pay particular attention to the polarity of the applied potential.

14. Develop a relation between the thermoelectric power and Hall constant for an n-type extrinsic crystal.

15. By scaling Fig. 6 for p-type germanium containing $N_a = 10^{23}$ acceptors/m^3, plot the thermoelectric power as a function of temperature.

9. semiconducting materials

The semiconductor principles discussed previously are used in this chapter to describe the properties of typical semiconductor materials. The variety of electrical properties exhibited by elemental and compound semiconductor crystals can be described with the aid of the energy-band model and, in particular, by noting the width of the forbidden-energy gap. Since these features depend critically upon the crystal structure, the several types of structures common in semiconductors are of interest. Slightly imperfect structures containing chemical impurities and crystal imperfections are also considered because many properties of semiconductors are determined by such imperfections. By far the best-understood semiconductor materials are the elements silicon and germanium, because their favorable intrinsic properties and consequent commercial importance have fostered an intensive study of them. Compounds of elements from group III and group V of the periodic table, which have derivative structures of silicon and germanium, are also becoming commercially important, although the difficult technical problems concerned with preparation of sufficiently perfect crystals have not yet been completely solved. In addition to these, there exists a wide range of other materials, including oxides and even organic compounds, which have interesting and useful semiconductor characteristics. As discussed in the introductory chapters, many elements and compounds can adopt more than one crystal structure. Because the structure determines the band model, it is possible that the same element (or compound) behaves like a metal when it crystallizes in one form and like a nonmetal when it adopts another structure, for example, white and gray tin. Similarly, one polymorph may be an insulator and the other a semiconductor, for example, diamond and graphite. In all such cases, the structures of the semiconductor phases are described in this chapter.

Elemental semiconductors

Group IV materials. The elements carbon, silicon, germanium, and gray tin, from group IV of the periodic table, all crystallize with the

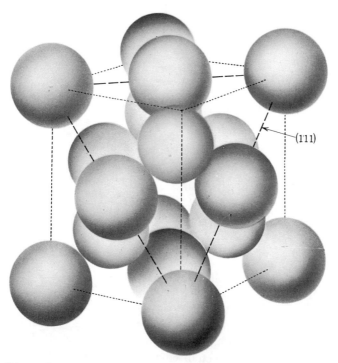

$(\bar{1}11)$

FIG. 1. Diamond structure. Note that the top-front corner atom has been omitted for clarity.

diamond structure (Fig. 1), in which each atom is tetrahedrally coordinated by four like atoms. The bonding between atoms is predominantly covalent, and the structure leads to an energy-band model having a forbidden-energy gap already described in Chap. 7. The covalent bonding is due to sp^3 hybrid formation and is strongest in diamond crystals and weakest in gray tin. This can be inferred directly from the width of the forbidden-energy gap (Fig. 2), which increases uniformly in proceeding from tin to diamond and is inversely proportional to the interatomic bond length.

The large forbidden-energy gap of diamond indicates that it is an insulator at room temperature (Exercise 1), but intrinsic semiconduction can be experimentally observed at temperatures of the order of 1000 deg C. Natural diamonds are quite impure, but are insulators, because of

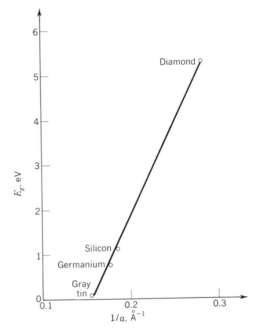

Fig. 2. Variation of the width of the forbidden gap with the reciprocal of the size of the unit cell for group IV semiconductors.

impurity compensation and gross inhomogeneities in impurity distribution, so that conducting paths through the crystal do not exist. Several p-type stones have been discovered, however. Silicon, which has a much smaller forbidden-energy gap, has yet to be produced sufficiently pure so that intrinsic conductivity can be observed at room temperature. Specimens having a room-temperature conductivity of 10^{-2} Ω^{-1}-m^{-1} are not unusual, however, and this is sufficiently pure for most research and device purposes. By comparison, intrinsic germanium is readily available, because there is a relatively large density of intrinsic carriers present at room temperature. A summary of the conduction properties of diamond-type semiconductors is given in Table 1.

The larger carrier mobilities in germanium as compared with silicon, shown in Table 1, are illustrative of the tendency for the valence electrons to resemble free electrons as the strength of the forces binding valence electrons to their parent atom decreases. This trend is consistent with that of the width of the forbidden-energy gap. In tin this binding is so weak that the stable form above 13.2 deg C is metallic. Gray tin is a laboratory-research curiosity, but its properties generally follow what is expected from an extrapolation of those observed in silicon and germanium.

Table 1
Properties of group IV semiconductors

Element	E_g, eV	σ_{RT}, $\Omega^{-1}\text{-m}^{-1}$	μ_e, $\text{m}^2/\text{V-sec}$	μ_h, $\text{m}^2/\text{V-sec}$
Diamond	5.3	10^{-12}	0.18	0.12
Silicon	1.1	5×10^{-4}	0.14	0.048
Germanium	0.72	2.2	0.39	0.19
Gray tin	0.08	10^6	0.20	0.10

Chemical doping. A most important property of semiconductors is extrinsic conductivity, which can be adjusted over a wide range by additions of chemical impurities to the crystal. As discussed in Chap. 8, the number of conduction electrons or holes is proportional to the number of substitutional foreign atoms in the structure. The process of inten- tionally introducing impurity atoms to obtain a desired conductivity is called *doping*. The commonly used doping impurities in silicon and germanium are atoms from columns III and V of the periodic table.

Group V elements such as phosphorous, arsenic, and antimony can substitute directly for a germanium atom in the crystal. Since these atoms have five valence electrons and only four are required to complete the covalent bonding in the structure, the extra electron is available to contribute to conductivity. Similarly, the group III elements, boron, aluminum, gallium, and indium, can also enter the structure substitu- tionally. These trivalent atoms require an additional electron to complete their bonding and, in accepting an electron from an adjacent germanium atom, create a mobile hole which can act as a carrier. Since one current\carrier is produced for each substitutional impurity atom, the room-temperature conductivity is proportional to the impurity content and can be adjusted between wide limits, as shown in Fig. 3 for silicon. The irregularity of the data at impurity concentrations greater than 10^{24} m^{-3} is due to reduced carrier mobility caused by ionized-impurity scattering. If a given crystal contains both donor and acceptor impuri- ties, the net number of extrinsic carriers is the difference between the two, as already discussed in Chap. 8. It is usually desirable to have a minimum total impurity content consistent with the desired conductivity in order to keep the carrier mobility as large as possible.

According to Eq. (20) of Chap. 8, these substitutional impurities give rise to discrete energy levels which are located an energy distance

$$E = \frac{m_e^* e^4}{8\kappa_e^2 \epsilon_0^2 h^2} \tag{1}$$

from the band edge. This relation predicts a value of 0.01 eV for impurity levels in germanium and a value of 0.05 eV for those in silicon.

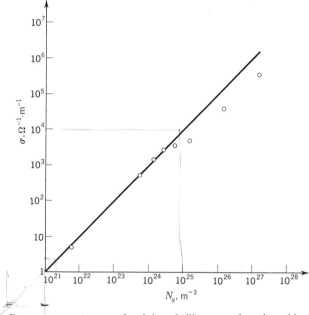

FIG. 3. Room-temperature conductivity of silicon as a function of impurity content. The departure from the straight line at high concentrations is due to a reduction in mobility caused by ionized impurity scattering.

The energy separation between the band edge and the impurity level is greater in silicon because the effective mass of carriers is larger in silicon than in germanium. Experimentally determined locations of impurity levels in the forbidden-energy gap for several kinds of atoms (Fig. 4) lie very close to the values calculated from (1). Slight departures from the predicted energy separations result from the inner electronic structure of the different atoms. This is ignored in deriving (1) because the impurities are treated as if they are similar to a simple hydrogen atom.

The same general considerations involving completion of covalent bonds apply approximately to other substitutional impurities in silicon and germanium. For example, zinc, from column II of the periodic table, can accept two electrons, and copper, from column I, can accept three. Therefore these substitutional impurities are represented on the energy-band model by two and three localized energy levels associated with each impurity atom, respectively. The locations of such levels for a number of substitutional impurity atoms in germanium are indicated in Fig. 5. Note that gold, which should be a triple acceptor according to its position in the periodic table, actually has four levels, the lowest of which is a donor level. This is an example where simple covalent bonding does not occur and a more detailed understanding of the bonds between the gold atom and surrounding germanium atoms is needed. Similarly, the

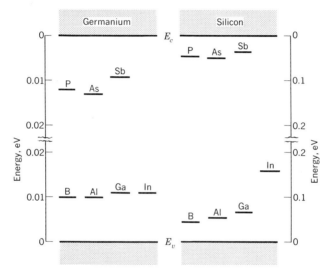

FIG. 4. Location of group III and group V substitutional impurity levels in silicon and germanium.

model does not predict too well the energy locations of many multiple-level impurities in silicon, and in practice, the number and location of the levels must be determined experimentally. The greater effective nuclear charge of these impurity atoms than that of group III or V elements implies that their carriers are bound much more tightly, and this is reflected in the greater energy depth of their levels. Because the levels are located nearer the center of the gap, they make effective trapping and recombination centers, and minute traces of these impurities drastically reduce the minority-carrier lifetime.

Silicon carbide. As discussed in the following chapter, practical semiconductor devices are limited in their high-temperature performance by the onset of intrinsic conduction. Therefore materials having a wide forbidden-energy gap are of interest, and, from this point of view, diamond is very attractive. Diamond semiconductor devices do not seem to be commercially feasible because they are very expensive, but crystals of the compound silicon carbide can be produced, and its properties constitute a useful compromise between those of silicon and diamond. The cubic form of SiC has the same structure as sphalerite and is a derivative structure of diamond, as discussed in a following section on

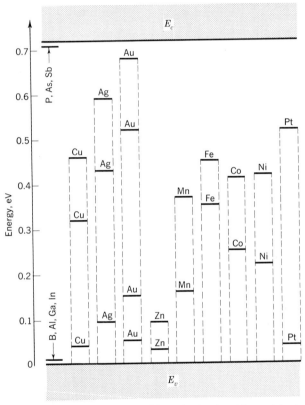

FIG. 5. Location of deep levels introduced by substitutional impurities in germanium. Note that several levels may be associated with a single impurity atom.

intermetallic semiconductors. In addition to the cubic form, SiC exists in a number of polymorphous hexagonal modifications, of which the alpha (hexagonal) form is the more common one. The properties of cubic β-SiC are analogous to silicon and germanium. The widths of the forbidden-energy gap of α-SiC and β-SiC are 2.8 and 1.9 eV, respectively. As illustrated in Fig. 6, the onset of intrinsic conductivity begins at temperatures of the order of 500 deg C, and satisfactory device operation has been observed up to this temperature.

Aluminum is commonly used to obtain *p*-type SiC crystals, while *n*-type material is produced by nitrogen doping. The donor level is 0.08 eV below the conduction band, but the acceptor level is 0.25 eV above the valence band. This means that the acceptors are not all

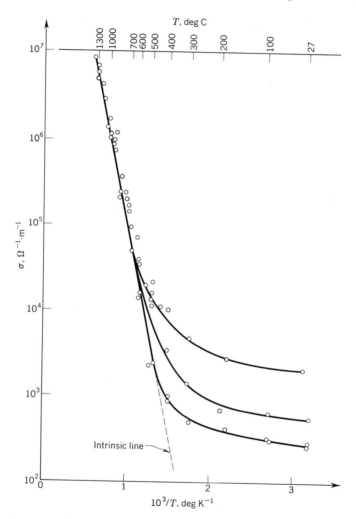

FIG. 6.   The conductivity of cubic silicon carbide crystals of different impurity content as a function of temperature.   (*After R. G. Pohl.*)

ionized at room temperature as they are in silicon and germanium, so that *p*-type SiC is in the transition range under normal conditions. This causes a significant temperature variation in the electrical properties of devices fabricated from SiC.

Graphite, Se, and Te.   The structure of the other common polymorph of carbon, graphite, is shown in Fig. 7.   This hexagonal structure
is comprised of stacked layers, each composed of carbon atoms forming
interconnected hexagonal rings.   The interatomic distances in the plane
are all equal (1.42 Å), while the planes are spaced considerably farther
apart, 3.40 Å.   The electrical conductivity of graphite is therefore highly
anisotropic, being 1,000 times
smaller in a direction perpendicular
to the planes as compared with that
along the planes.   The major features of the energy-band model of
graphite are determined by the two-
dimensional "crystal" composed of
a single (0001) plane.   Calculations
show that the conduction and valence bands just touch at the
corners of the Brillouin zone, so
that graphite can be considered to
be metallic along certain directions
in the crystal.   It is possible, however, to obtain both n-type and
p-type specimens by impurity
doping since a forbidden gap does
exist along other directions.   The
calculated band structure, modified
to take account of the true three-
dimensional crystal, correctly predicts the observed carrier densities,

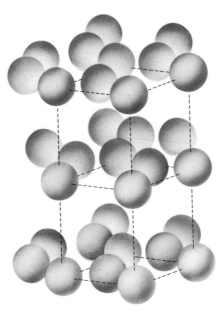

FIG. 7.   Graphite structure.

$n = p = 4 \times 10^{24} \text{ m}^{-3}$, and mobilities, $\simeq 1 \text{ m}^2/\text{V-sec}$, at room temperature.

Selenium can exist in several polymorphous modifications, including an
amorphous state, and it is interesting that all forms are semiconductors.
The stable form is hexagonal, and the structure basically consists of spiral
chains in which each atom has two covalently bonded nearest neighbors.
Amorphous selenium is essentially an insulator.   Many of its properties
can be analyzed using an energy-band model, even though crystalline
periodicity is not present, other than, possibly, over the short chain
lengths.   The forbidden gap appears to be 2.3 eV and is the same as in
liquid selenium.   This is believed to be due to preservation of the chain
structure even in the liquid state.   The forbidden-energy gap in the
crystalline form is about 1.8 eV, and carrier mobilities are very low,
about $10^{-4} \text{ m}^2/\text{V-sec}$.   Tellurium has a structure similar to that of
hexagonal selenium, but intrinsic Te is n-type, with a forbidden gap of
0.35 eV.

Intermetallic III-V compounds

General properties. Compounds formed by elements of the third and fifth columns of the periodic table having the formula $A^{III}B^V$ crystallize with the sphalerite structure shown in Fig. 8. As can be readily

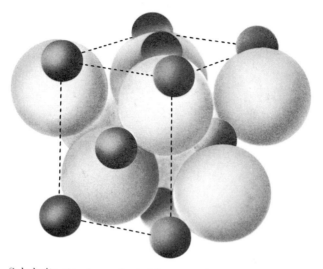

Fig. 8. Sphalerite structure adopted by many intermetallic semiconductors.

seen, by comparison with Fig. 1, this is a simple derivative structure, derived from the diamond structure by replacing alternate carbon atoms by A and B atoms and leading to a suppression of certain symmetry operations. Since each group III atom is tetrahedrally surrounded by group V atoms and vice versa, it is reasonable to assume that, on the average, each atom has four valence electrons. This suggests that the bonding has a covalent character and that the semiconducting properties of these compounds are similar to those of the corresponding group IV element. That this is approximately the case can be seen from the comparison of their crystal data:

	Si	AlP	Ge	GaAs	Sn	InSb
Unit-cell edge	5.42	5.42	5.62	5.63	6.46	6.48 Å
Interatomic separation	2.34	2.34	2.44	2.44	2.80	2.80 Å
Energy gap	1.1	3.0	0.72	1.34	0.08	0.16 eV

The three intermetallic compounds listed were selected because their constituent atoms bracket, in the periodic table, the chemical element preceding them in the above table.

The major technical importance of III-V compounds is that they provide a wider selection of forbidden-energy gaps, mobilities, etc., as compared with those of the limited number of elemental semiconductors available. The range of values that can be obtained is illustrated in Table 2. In general, the forbidden gap of a compound is wider than that of the analogous group IV element, as may be seen by comparing AlP with Si and GaAs with Ge. Also, electron mobilities are large and much greater than the hole mobilities. The temperature variation of lattice mobility does not usually follow the $T^{-\frac{3}{2}}$ law, the exponent taking on values between -1.5 and -2.0. All these effects are attributed to the 5 to 10 per cent ionicity of the bonding in these compounds, which arises from the higher electronegativity of elements in the fifth column.

Table 2
Properties of III-V compounds

Material	E_g, eV	μ_n, m²/V-sec	μ_p, m²/V-sec
AlP	3.0	. . . . .	. . . . .
AlAs	2.3	. . . . .	. . . . .
GaP	2.25	0.045	0.002
AlSb	1.52	0.040	0.020
GaAs	1.34	0.85	0.045
InP	1.27	0.60	0.016
GaSb	0.70	0.50	0.085
InAs	0.33	2.3	0.010
InSb	0.18	8.0	0.070

A major difficulty in fabricating suitable single-crystal specimens of these compounds is that not only must the chemical purity be of the same order as in the elemental semiconductors, but also the stoichiometric ratio of the constituents must equal unity to the same precision. A vacant lattice site normally occupied by a group III atom effectively acts like an acceptor level, while a group V vacancy has the characteristics of a donor level. Thus, even though the foreign-atom concentration may be negligible, the crystal must be exceedingly perfect for extrinsic conductivity to be limited.

Gallium arsenide. Gallium arsenide is of particular commercial interest because the forbidden-energy gap is larger than that of silicon; yet the impurity ionization energies are small, so that the material is in the exhaustion range at room temperature. In addition, carrier mobilities are large, and the compound is chemically stable in normal atmospheres, in contrast to the aluminum compounds, which disintegrate with time. The behavior of group IV impurities in GaAs (as well as in other

III-V compounds) is of interest because they can enter the structure substitutionally for either gallium or arsenic. A substitutional Ge atom in an As site adds an acceptor level, while in a Ga site it produces a donor level. In some III-V compounds, the group IV elements tend to favor one or the other site and hence act predominately like either a donor or acceptor. For example, Ge is a donor in InAs, whereas in GaAs both sites are preferred nearly equally and the donor and acceptor levels tend to compensate. By suitable heat-treatments during crystal growth, it is possible to induce preferential substitution on either site, so that germanium and silicon can act either as donors or acceptors in GaAs.

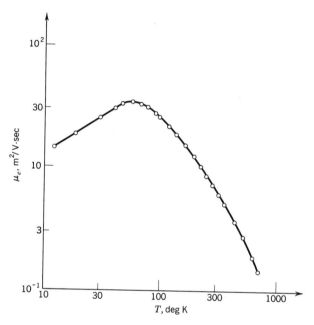

Fig. 9. Electron mobility in InSb versus temperature. (*After H. J. Hrostowski et al.*)

Indium antimonide. The electronic mobility in the intermetallic compound InSb is very large. Since the mobility is limited by phonon scattering at room temperature, it becomes even larger as the crystal is cooled (Fig. 9), and values approaching 50 m²/V-sec have been reported. The influence of impurity scattering is evident at low temperatures, even though impurity concentrations as low as 10^{19} m^{-3} have been achieved. The large electronic mobility in InSb is reflected in the narrow forbidden-energy gap in the energy-band model, which implies that the valence electrons are nearly free. This is also evident in the small effective mass and effective density of states in the conduction band, as illustrated in

Exercise 7. Furthermore, the small forbidden gap means that the Fermi level cannot be far from the bottom of the conduction band, and, in fact, intrinsic InSb is just on the verge of being degenerate even at temperatures as low as 80 deg K.

InSb is a strongly *n*-type intrinsic semiconductor because of its large electronic mobility. This results in a reversal in the sign of the Hall

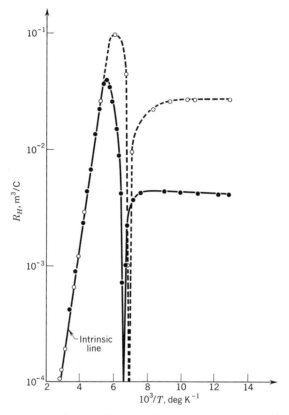

Fig. 10. Hall constant of *p*-type InSb crystals. R_H becomes negative (the sign is suppressed in the plot) in the intrinsic region because of the high electron mobility. (*After H. J. Hrostowski et al.*)

effect of *p*-type crystals, with increasing temperature, as required by Eq. (30) of Chap. 8. In typical experimental measurements (Fig. 10) the polarity of the Hall voltage is positive at low temperatures and negative at higher temperatures (above the zero in Hall constant). The horizontal slope of the curves at low temperature shows that the impurity levels are completely ionized. Actually, the impurity ionization energy is very small, about 10^{-3} eV, so that, for all practical purposes, the donor and acceptor levels can be considered part of the conduction and valence bands, respectively.

Isomorphous systems. Several pairs of III-V compounds exhibit complete solid solubility because they have identical structures and closely similar interatomic distances. In these isomorphous series, the semi-conducting properties vary smoothly with composition from those characteristic of one compound to those characteristic of the other. A number of systems have been investigated, among them In(As,Sb), In(P,As), Ga(P,As), (In,Ga)Sb, (In,Al)Sb, (Ga,Al)Sb, Ga(As,Sb), and (Ga,In)As.

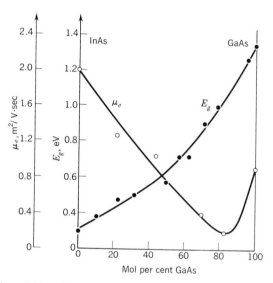

FIG. 11. Variation of E_g and μ_e with composition for (Ga, In) As crystals. (*After M. S. Abrahams, R. Braustein, and F. D. Rosi.*)

Experimental measurements on (Ga,In)As (Fig. 11) show that the forbidden-energy-gap variation is not quite linear with changing Ga/In ratio, which implies that the energy-band models of the two stoichiometric compounds differ in minor details. The mobility passes through a minimum because of scattering caused by atomic randomness, like that occurring in metal alloys, as discussed in Chap. 7. Isomorphism between the intermetallic compounds further multiplies the number of semiconductor crystals available for device application. Homogeneous single crystals, however, are difficult to fabricate while maintaining chemical purity and the desired atomic ratios to the degree necessary in semiconductor technology. As an example of the flexibility inherent in this approach, nevertheless, consider the case of (Ga,In)As. It is possible to fabricate a single crystal having a forbidden gap of 0.33 eV at one end and 1.34 eV at the other by introducing a concentration gradient of Ga (or In) across the crystal.

Compound semiconductors

The properties of the III-V compounds discussed in the preceding section resemble those of the group IV elements because their atomic bonding and structures are similar. The widths of the forbidden-energy gaps in the III-V compounds tend to be larger, however, and this is attributed to the partially ionic character of the bonding. This trend continues in the properties of $A^{II}B^{VI}$ compounds. Here, the average valency per atom is also 4, so that covalency is likely, but the increased electronegativity of the group VI elements introduces considerable ionicity. Accordingly, II-VI compounds such as CdS can be considered as semiconductors,

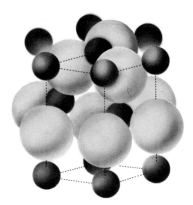

Fig. 12.   Wurtzite structure (hexagonal) adopted by CdS and CdSe.

but more nearly resemble insulators. This trend reaches an extreme in the I-VII compounds, the alkali halides, which are normally insulators having forbidden-energy gaps greater than 5 eV.

Cadmium sulfide, selenide, and telluride. Although CdS and CdSe are dimorphic, they normally crystallize with the hexagonal modification of zinc sulfide, wurtzite, shown in Fig. 12. By comparison, CdTe crystallizes only with the cubic sphalerite structure (Fig. 8). The forbidden-energy gaps of CdS, CdSe, and CdTe are 2.45, 1.74 and 1.45 eV, respectively, which correspond to photon energies in the visible wavelength region. As discussed in the next chapter, absorbed light photons can produce a change in the conductivity of a crystal by creating electron-hole pairs when the photon energy is equal to or exceeds the forbidden-energy gap. Photoconductivity measurements on these compounds (Fig. 13) illustrate this effect. Such crystals, therefore, are extensively used to detect and measure visible light radiation.

Pure or compensated crystals of CdS are essentially insulators in the dark, having conductivities of about 10^{-10} Ω^{-1}-m^{-1}. Excess cadmium in the crystal produces n-type conductivity, and electron mobilities are about 2×10^{-2} m^2/V-sec at room temperature. CdSe crystals have similar properties, except for the width of the forbidden gap. Both n-type and p-type CdTe crystals can be produced by doping with indium or gallium (donors) or silver or antimony (acceptors). Donor levels lie within 0.003 eV of the bottom of the conduction band, while acceptor

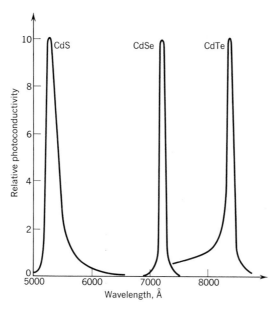

FIG. 13. Relative photoconductivity of CdS, CdSe, and CdTe as a function of wavelength. Maximum response occurs at photon energies corresponding to the forbidden gap. (*Courtesy of R. J. Robinson.*)

depths vary from 0.3 to 0.5 eV, depending upon the particular impurity atom.

The preparation of insulating crystals in which the concentration of both intrinsic and extrinsic carriers is negligible appears to be extremely difficult because very small concentrations of foreign atoms result in extrinsic carriers, (Exercise 1). Actually, a tendency toward self-compensation acts to equalize the concentration of traces of donor and acceptor atoms in the crystal and results in intrinsic conductivity. During crystal growth, some foreign donor atoms are unavoidably incorporated into the structure since absolute purity is a practical impossibility. If acceptor levels are also included, electrons from donor levels fill up the acceptor states, and the energy of the entire crystal is lowered by an amount proportional to the energy difference between the donor and acceptor levels. The lowest possible energy configuration of the crystal is reached if the number of acceptor-type foreign atoms (or an equivalent departure from stoichiometry) is sufficient to compensate for the unavoidable donor impurities. Since the energy difference between the location of the donor and acceptor levels increases with the width of the forbidden-energy gap, self-compensation during crystal growth is most effective in insulating crystals. This is fortunate, since it is in just these compounds that the number of extrinsic electrons must be very small.

Lead sulfide, selenide, and telluride. The IV-VI compounds lead sulfide, selenide, and telluride are of interest because their properties are useful in radiation detection and thermoelectric applications. The band gaps are 0.37, 0.27, and 0.33 eV, respectively, which makes the crystals most sensitive to radiation in the infrared region and suggests that the carrier mobilities are reasonably large. The structure of these crystals is that of sodium chloride or the halite structure shown on page 98. As in other compounds, the bonding is partially ionic, and this causes the mobility to vary with temperature as $T^{-\frac{5}{2}}$, instead of the $T^{-\frac{3}{2}}$ dependence expected from simple phonon scattering. Both hole and electron mobilities in all three compounds are of the order of 0.1 m²/V-sec at room temperature, and the electrons are somewhat more mobile.

Although chemical doping can be used to produce extrinsic conduction, departures from stoichiometry are equally effective. Excess lead produces n-type material, while lead deficiency results in p-type crystals. This phenomenon, which is common to all compound semiconductors, is strikingly illustrated in Fig. 14, which shows the Hall constant and resistivity at room temperature of several PbS crystals treated at various ambient sulfur-vapor pressures. The different sulfur-vapor pressures are specified in terms of the temperature of a sulfur bath in an enclosed volume containing the PbS crystals. After the concentration of sulfur in the PbS crystals attains equilibrium with the ambient sulfur atmosphere, the crystals are cooled to room temperature and examined. At low sulfur-vapor pressures the structure is deficient in sulfur and the crystals are n-type. High sulfur-vapor pressure produces an excess in the crystal and p-type conductivity. At an intermediate point between these two extremes, exact stoichiometry is attained and the resistivity is at a maximum. The minute density of vacancies required for useful extrinsic properties is quite stable at room temperatures. In PbS, the donor and acceptor ionization energies are 0.03 and 0.001 eV, respectively.

Lead telluride is a useful semiconductor for the thermoelectric applications discussed in Chap. 10. The important semiconductor parameters for thermoelectric devices are the electrical conductivity and thermoelectric power. Both depend upon the location of the Fermi level in the forbidden-energy gap, so that it is possible to achieve an optimum chemical doping for device applications.

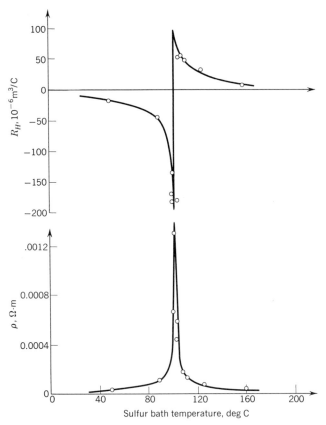

Fig. 14. Hall constant and resistivity of several PbS crystals equilibrated at different sulfur-vapor pressures. The stoichiometric composition is that for which the resistivity is intrinsic, and therefore largest. (*After R. F. Brebrick and W. W. Scanlon.*)

According to Eq. (55) in Chap. 8, the thermoelectric power of an extrinsic p-type semiconductor is given by

$$QT = (E_0 - E_v) + 2kT. \qquad (2)$$

If $E_0 - E_v$ is expressed in terms of the conductivity σ, (2) becomes (see Exercise 11)

$$\frac{Q}{k} = (2 + \ln N_v e \mu_p) - \ln \sigma. \qquad (3)$$

Since the first term in (3) is constant, a plot of Q versus $\ln \sigma$ is a straight line. The room-temperature thermoelectric power of a series of PbTe samples is shown as a function of conductivity in Fig. 15. The solid curve is a plot of (3), while the data points are experimental. The conductivity of these crystals was controlled by departures from stoichiometry produced by the heat-treatment procedure discussed above. Note that the range of conductivities obtainable by this means is quite large. Since the experimental data agree with (3), μ is essentially independent of the impurity content. This means that the mobility is determined by phonon scattering even when very large impurity concentrations are present. This result is typical of compounds having an appreciable ionic component in their bonding.

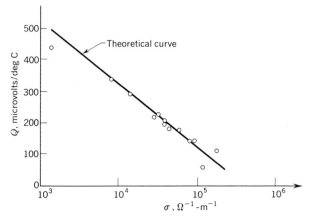

FIG. 15. Room-temperature thermoelectric power of PbTe as a function of conductivity. The solid line represents the predicted variation given by Eq. (3).

Other compounds. In thermoelectric applications it is desirable to use a conductor which has a small thermal conductivity. The concentration of free electrons is much smaller in semiconductors than in metals, so that the thermal conductivity is determined primarily by atomic vibrations, or phonons, as discussed in Chap. 6. Compounds such as Bi_2Te_3, Sb_2Te_3, Bi_2Se_3, etc., from columns V and VI of the periodic table, have low thermal conductivity because of their relatively complicated layer-like crystal structure (hexagonal) and because of the disparity in the atomic masses of the constituent atoms. Both features tend to impede the transfer of atomic vibrations through the crystal by causing the phonons to be scattered, and therefore reduce thermal conductivity. Bi_2Te_3, the most promising thermoelectric compound known, has a forbidden gap of 0.16 eV and electron and hole mobilities of 0.11 and 0.049 m^2/V-sec, respectively, making it not unlike PbTe with regard to its semiconductor behavior. The thermal conductivity is only

2 W/deg C-m (W = watts), however, compared with a value of 4 for PbTe.

A very large number of other semiconducting compounds have been prepared and examined. In most cases the characteristics of a given compound may be inferred by comparing it with one of the comparable materials discussed in the previous sections. GeTe, for example, is similar to PbTe, except for a somewhat larger forbidden gap. A group of materials, called ternary compounds, composed of three kinds of atoms, has recently received attention. A partial listing of compounds that have been prepared is given in Table 3. The chemical formulas of these compounds suggest that they are analogous to the isomorphous series (In,Ga)As (Fig. 11), and, to some extent, similar semiconducting properties are found. The ternary compounds in Table 3 differ from the solid solutions, however, in that they are stoichiometric and have specific crystal structures.

Table 3
Ternary-compound semiconductors

AgS_2S_2	$CuInSe_2$	$AgCrTe_2$
AgS_2Se_2	$CuInTe_2$	$CuFeSe_2$
$AgSbTe_2$	$AgGaSe_2$	$AgFeTe_2$
$AgBiS_2$	$AgGaTe_2$	$CuAsSe_2$
$AgBiSe_2$	$AgInSe_2$	$CuSbS_2$
$CuGaSe_2$	$AgInTe_2$	$CuBiS_2$
$CuGaTe_2$	$AgCrSe_2$	$AgAsSe_2$

Oxides

The crystal structures of metal oxides usually consist of a hexagonal or cubic closest packing of anions in which cations occupy either octahedral or tetrahedral voids. The strong electronegativity of oxygen suggests that the bonding is predominantly ionic. Because the ionic bond is very strong and does not involve the sharing of electrons by two or more atoms, intrinsic electronic conductivity in ionic compounds is usually negligibly small. The transition-metal oxides, on the other hand, can become electronic conductors by taking on compositions that deviate from stoichiometry. One way this can happen is through loss of oxygen by the crystal, with the result that the structure contains an excess of metal ions. To preserve charge neutrality, excess electrons occupying quantum states in the forbidden-energy gap remain in the crystal. These electrons can make transitions to states in the conduction band and become free carriers.

Simple oxides: ZnO and MgO. In zinc oxide, the ions have closed outer shells and the energy bands arise from the filled $2p$ levels of the O^{2-} ion (valence band) and the empty $4s$ levels of the Zn^{2+} ion (conduction band). Zinc oxide crystallizes with the wurtzite structure, and carefully prepared single crystals have room-temperature conductivities between 1 and 10^2 Ω^{-1}-m^{-1}, even though the forbidden gap is about 3.3 eV. The extrinsic conductivity is n-type and is attributed to interstitial zinc atoms which act as donors. This is not surprising since the structure contains large voids which can easily accommodate interstitial atoms. Electronic mobility is 2×10^{-2} m^2/V-sec at room temperature and is limited by phonon scattering. The conductivity of ZnO can be changed markedly by heating in a reducing or oxidizing atmosphere. If the crystal is caused to lose oxygen, the ions escape as neutral atoms (O_2), leaving two electrons per atom behind. They, in turn, combine with a Zn^{2+} ion in an interstitial position to produce a neutral zinc atom. One electron is then easily ionized (0.05 eV) into the conduction band from a zinc atom in this position.

The most refractory of the alkaline earth oxides, magnesium oxide, has the halite structure shown on page 98. Because the bonding in MgO is predominantly ionic, the forbidden gap is very large, approximately 7.3 eV. In the temperature range 600 to 1000 deg C, extrinsic n-type conductivity ($\simeq 10^{-7}$ Ω^{-1}-m^{-1} at 600 deg C) having a thermal activation energy of 2.3 eV is observed. The donor centers present are the result of excess magnesium in the structure in the form of oxygen vacancies instead of interstitial Mg atoms. The halite structure has very small unoccupied voids (tetrahedral), so that interstitials are less likely than in wurtzite, and direct experimental evidence agrees with the presence of vacant O^{2-} sites. Many discrete levels distributed throughout the forbidden-energy gap in MgO have been detected by optical measurements, but their association with specific chemical impurities remains largely speculative. Photoconductivity by holes is observed at room temperature, with a very low value of hole mobility, 2×10^{-4} m^2/V-sec.

Transition-metal oxides. The energy-band model of the transition-metal oxides is unique in that the energy bands important for semiconductivity are related to inner electronic levels of the atoms in the structure rather than to the outermost valence-electron levels, as is the case in all previous structures discussed. A filled and empty band originating from the $2p$ anion levels and $4s$ cation levels, respectively, are present as in ZnO. The $3d$ levels of the cations are only partially filled, however, so that, if the wave functions of $3d$ electrons on adjacent ions overlap sufficiently to form a band, metallic conduction characteristic of a partially filled band is expected. Several transition-metal oxides, TiO, Ti_2O_3, and V_2O_3, do have metallic properties, but on the other hand, others such as Cr_2O_3, Mn_2O_3, Fe_2O_3, NiO, etc., are insulators when stoichiometric.

In titanium and vanadium oxides, the overlap of $3d$ wave functions is barely sufficient to develop an energy band, and therefore the band is narrow when compared with that of the $2p$ and $4s$ band, for example. Because of the small width of the band, the effective mass of the carriers is large and the mobility is small (Exercise 6). The carrier density in this band can be inferred from the electronic configuration of the metal ion (Table 4). There are no $3d$ electrons in TiO_2 when it is stoichiometric

Table 4
Properties of *d*-band conduction

Compound	d-band electronic configuration	d-band mobility, m^2/V-sec	Type of conduction
TiO_2	d^0	10^{-4}	Narrow-band semiconductor
TiO	d^2	3.6×10^{-5}	Metallic
V_2O_3	d^2	10^{-6}	Metallic, very narrow band
Fe_2O_3	d^5	10^{-7}	Localized levels
NiO	d^8	10^{-7}	Localized levels

so that it is an insulator. Extrinsic conductivity produced by departures from stoichiometry show that the $3d$ band carrier mobility is only 10^{-4} m^2/V-sec. Therefore it is concluded that a narrow $3d$ band exists in TiO_2 and that it is normally empty. The configurations of TiO and V_2O_3 are such that the $3d$ band is partially filled in each compound, and this produces metallic conductivity. The very low mobility in V_2O_3 suggests that the band is actually very narrow, and this is indicative of the fact that the larger ionic radius of vanadium compared with titanium increases the separation between nearest-neighbor metal ions and reduces the $3d$-wave-function overlap. As a further example, the next-heavier metal oxide, Fe_2O_3, displays semiconducting properties even though the $3d$ levels of the ions are only partially filled. This means that the wave-function overlap is so small that a $3d$ band does not exist. The $3d$ electrons are localized on the individual ions and can move from one ion to the next only by a sort of hopping process. This results in a very low mobility, which is an exponential function of temperature.

NiO is similar to Fe_2O_3, and the experimental values of mobility are the same in both compounds. For conduction to occur via the localized $3d$ levels it is necessary to have some Ni^{3+} ions in the structure so that electrons may be exchanged with the normal Ni^{2+} ions. These are produced either by departures from stoichiometry or by introducing Li^{1+} ions substitutionally for Ni^{2+} ions in the crystal. The Li^{1+} ions effectively act as acceptor centers, and an equal number of Ni^{3+} ions are produced in the structure to maintain charge neutrality. The energy-band scheme representative of this situation is given in Fig. 16 for NiO, where the

$3d^8$ levels are shown as dots to suggest that bandlike conduction does not occur. Also shown are the next set of $3d$ levels, $3d^9$, which are normally unoccupied in NiO. At high temperatures these can be populated by thermal excitation of electrons from the $3d^8$ levels (which effectively produces additional Ni^{3+} ions thermally) or from the $2p$ band. The Fermi level lies approximately midway between the $3d^8$ and $3d^9$ levels at high temperatures, as in an intrinsic semiconductor, producing

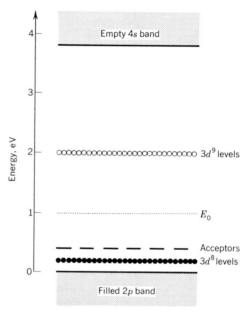

FIG. 16.   Energy-level diagram for NiO. The localized $3d$ levels are represented by circles to indicate they do not form a band.

an activation energy for conductivity of 1.0 eV. When holes are produced in the $2p$ band by thermal excitation, they dominate the conductivity, because their mobility is characteristic of normal band carriers. Mobilities in the range 0.01 to 0.1 m^2/V-sec have been measured. NiO is thus a p-type semiconductor under nearly all conditions. The properties of Fe_2O_3 can be interpreted in a similar fashion, except that the Fe ion can take on a lower as well as a higher valence state, and therefore oxidation results in p-type crystals while reduction produces n-type. Introduction of Ti^{4+} produces Fe^{2+} ions and results in n-type conductivity.

A large number of oxides having the general formula AB_2O_4 crystallize with the spinel structure discussed further in Chap. 13. The structure consists of oxygen atoms arranged in a cubic closest packing, the cubic unit cell containing 32 oxygen atoms. The metal atoms are distributed among the tetrahedral voids (A atoms) and octahedral sites (B atoms). The distribution of cations among these sites determines the electrical (and magnetic) properties of the crystal. The spinel structure is adopted by a large number of crystals having a variety of atoms present in different ratios. Nevertheless, it is possible to divide these compounds into groups according to their composition, commonly called aluminates, $MeAl_2O_4$; ferrites, $MeFe_2O_4$; titanates, Me_2TiO_4; and so forth. According to the compositional formula AB_2O_4, the total positive charge of the three cations must equal $+8$. It is possible to satisfy this condition by combinations of divalent and trivalent ions, or divalent and quadrivalent ions, etc. These are usually designated as 2-3 spinels, 2-4 spinels, and so forth. The natural distribution of cations in, say, a 2-3 spinel is to place the divalent ions in the tetrahedral sites and the trivalent ions in the octahedral sites. This is the structure adopted by many compounds, such as $ZnFe_2O_4$, for example, and is called a *normal* spinel structure. It is also possible, however, to distribute the cations in another way. In $MgFe_2O_4$, for example, the tetrahedral sites are occupied by Fe^{3+} ions and the octahedral equipoints are statistically shared by Mg^{2+} and Fe^{3+} ions. Such an arrangement is called an *inverse* spinel structure and is best represented by the structurally correct formula $B(AB)O_4$.

The structure of magnetite, Fe_3O_4, is of this type, and the structural formula, therefore, is written $Fe^{3+}(Fe^{2+}Fe^{3+})O_4$. Since Fe^{2+} and Fe^{3+} ions are present in equivalent sites, conductivity occurs by electron exchange between nearest-neighbor iron atoms as discussed above for NiO. In the case of magnetite, however, the structure already has equal numbers of ions in the two charge states, so that many electrons participate in the hopping process simultaneously. This means that the conductivity of Fe_3O_4 should be very large, as is indeed the case. Furthermore, deviations from stoichiometry actually cause the conductivity to decrease since the numbers of Fe^{2+} and Fe^{3+} ions then are no longer equal. The conductivity also decreases abruptly at about 120 deg K, and this is attributed to an ordering among the cations in the octahedral positions and a consequent decrease in the ease of electron exchange between them. The transformation causes the symmetry of Fe_3O_4 to become orthorhombic below -158 deg C.

Organic semiconductors

Molecular crystals. The conductivity of many organic compounds varies exponentially with temperature similarly to inorganic semiconductors. Although the conductivity in most organic substances is small and more nearly characteristic of insulators than semiconductors, it appears that some of the conduction mechanisms are similar to those described in the preceding sections of this chapter. A detailed study of organic semiconductors is still in its early stages, so that, for example, it is not even clear to what extent a band model can be applied to interpret experimental results. The potentially large diversity of properties inherent in the multitude of organic molecular structures is the major reason for interest in organic semiconductor materials. The structurally simpler materials that have been investigated are the conjugated and aromatic molecules, which have a common basic structure similar to benzene. In C_6H_6, the carbon atoms form a regular plane hexagon, and the six hydrogen atoms are so arranged in the same plane that all the valence angles are 120 deg. Therefore three of the four valence electrons of each carbon atom are hybridized in such a way as to give localized carbon-carbon and carbon-hydrogen bonds. The remaining valence electron on each carbon atom does not take part in the bonding, and the wave functions of these six electrons in the ring overlap, so that these electrons are mobile along the conjugated carbon atoms. The conductivity of organic materials is ascribed to the movement of these so-called π *electrons* between molecules.

Anthracene, $C_6H_4:(CH_2):C_6H_2$, is the most studied of the organic semiconductors because single crystals can be fabricated relatively easily and photoconductivity has been observed in anthracene. The crystal structure is monoclinic and consists of an array of anthracene molecules held together by van der Waals forces. Experimental measurements on pure crystals indicate hole and electron mobilities of the order of 10^{-4} m^2/V-sec, which are anisotropic because of anisotropy in the crystal structure. Conductivity is in a narrow band which arises from overlap between molecular electronic wave functions on adjacent molecules. The forbidden-energy gap is about 3.1 eV and is determined largely by the structure of the isolated molecule. Other molecular crystals composed of chemical groups which act as electron donors, together with groups which act as electron acceptors in the same molecule, can have conductivities as large as 10^4 Ω^{-1}-m^{-1}. The electron exchange between these chemical groups has the effect of producing many free carriers, so that the electrical conductivity is nearly metallic and mobilities are very small, less than 10^{-6} m^2/V-sec. Here again it appears that the conductivity can be explained by determining the wave-function overlap between adjacent molecules in the crystal, so that a suitable energy-band model can be devised.

Polymers. The semiconducting properties of polymers containing very large molecules are not well understood, partly because the physical-chemical structures are not as well known as for the simpler molecular crystals. Most plastics are insulators, but appreciable conductivity can be produced by heating or by chemical treatments. In the approximately

200 different compounds examined to date, conductivity activation energies ranging from essentially zero to several electron volts and conductivities as high as 10^3 Ω^{-1}-m^{-1} have been measured. Mobilities are 10^{-7} m^2/V-sec or smaller, and the materials are always p-type. Attempts to explain conductivity and thermoelectric-power measurements on the basis of an energy-band model have not been successful. It is felt that an electron hopping model, similar to that of the transition-metal oxides, may be more appropriate.

Crystal preparation

As discussed in previous sections, the electrical properties of semiconductors are very sensitive to the disturbing effects of chemical impurities and crystal imperfections. Indeed, many of the useful semiconductor devices described in the following chapter are operable only if they are fabricated from very nearly perfect single crystals. The techniques which have been developed to grow large single crystals of the commercially important semiconductor materials are generally applicable to other elements and compounds.

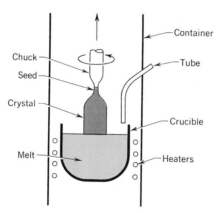

Fig. 17. Model of crystal puller.

Growth from the melt. The rate of growth of a crystal from its own liquid is determined primarily by the rate of diffusion of the heat of fusion away from the growing crystal face, which is necessary to maintain the temperature below the melting point. A widely used procedure for growing large crystals from the melt, called the *Czochralski method*, starts the growth by immersing a previously oriented seed crystal into the surface of the melt and slowly withdrawing it as growth proceeds (Fig. 17). The temperature at the crystal-melt interface is controlled by the rate of pulling, by controlling the temperature gradient in the crystal (which can be done with afterheaters surrounding the crystal), and by controlling the temperature of the melt. To start, the melt is kept slightly above the melting point, and the seed is lowered into the liquid. The tip of the seed is first melted, to make sure that the liquid wets its surface, and then the temperature is reduced, until the melt begins to freeze onto the seed. The seed is then slowly raised in order to maintain the liquid-solid interface near the surface of the melt.

It is common practice to rotate the seed during growth to promote stirring in the melt and to average out slight temperature asymmetries. If necessary, the entire apparatus can be placed in a sealed enclosure containing an inert atmosphere or high vacuum to reduce contamination or oxidation. A sealed system is necessary to grow crystals of compounds which are unstable at their melting point. For example, it is possible to pull single crystals of GaAs from the melt if the temperature of the entire system is adjusted to produce the proper arsenic-vapor pressure over the melt. The proper pressure is determined by the thermodynamic properties of the compound and is such that the crystal is in equilibrium when stoichiometric at the melting point. The existing vapor pressures are sometimes adjusted by controlling the temperature of a side reservoir containing a melt of the element.

The Czochralski method is the standard procedure used for growing silicon and germanium single crystals. Specimens up to 15 cm in diameter have been grown, although 3 cm is sufficient for most semiconductor-device applications. Lengths can be of the order of tens of centimeters, depending only upon the size of the apparatus. Linear growth rates as high as 3×10^{-4} m/sec are possible. Either n-type or p-type crystals can be grown, depending upon the composition of the melt. Crystals which change from n-type to p-type can be grown by dropping pellets of the desired impurity into the liquid through a tube (Fig. 17), at predetermined intervals.

Zone melting. By selectively heating a small region of an elongated polycrystalline rod it is possible to produce a molten section, or zone, which can be moved from one end of the sample to the other. This *zone-melting* process is often used to grow single crystals of both elements and compounds, as discussed below. If several heaters are arranged as in Fig. 18, a number of molten zones may be passed through the sample during a single excursion of the crucible, which greatly improves the efficiency of the process. Zone melting also can be used to purify semiconductor crystals, and in this application it is called *zone refining*. In equilibrium, the minor constituent or impurity has a greater concentration in the liquid phase than in the solid phase. This is a consequence of the thermodynamic properties of the solid and liquid phases and is measured by a distribution coefficient k defined by

$$k = \frac{C_S}{C_L} \tag{4}$$

where C_S is the concentration of the impurity in the solid phase, and C_L is the concentration in the liquid phase. Impurity atoms tend to concentrate in the liquid zone as it passes through the crystal because the distribution coefficient is always less than unity. Thus impurities can be swept to one end of the crystal, which is later discarded.

It can be shown that the distribution of an impurity in a rod-shaped crystal after passage of a molten zone of length z is

$$C_S(x) = C_0[1 - (1 - k)e^{-kx/z}] \tag{5}$$

where C_0 is the initial uniform concentration, and x is the distance along the rod measured from the end first melted. Equation (5) shows that

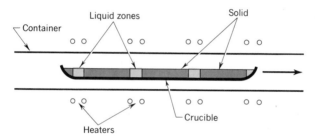

Fig. 18. Simple zone-melting apparatus. As the crucible is moved past the heaters, molten zones pass through the solid.

the starting end of the rod ($x = 0$) attains a concentration of C_0k after a single pass. After the second pass the concentration is C_0k^2, and so on. Since k may be of the order of 0.1 to 0.001 for, say, substitutional impurities in germanium, the impurity concentration is markedly reduced near the starting end of the rod in several passes. The length z can be made effectively very large by multiple passes, so that the desired purification can be obtained throughout most of the length of the crystal. Note that, if $k = 1$, the impurity concentration is not influenced by zone refining. This is the case for boron in silicon, for example. At distances far removed from the starting end ($x \to \infty$), the impurity concentration is not changed according to (5). The reason for this is that the impurity concentration in the molten zone becomes so large that as many impurity atoms are trapped during freezing as are removed by melting. Actually, this situation never occurs in practical zone refining since the ratio of the length of the crystal to the width of the molten zone is usually at least 10:1. This saturation effect proves useful in obtaining uniformly doped specimens, however, for which the process is termed *zone leveling*. Here, after the crystal is purified by zone refining, a large concentration of the desired impurity is introduced into the molten zone. The impurity is then uniformly distributed along the crystal as the doped zone traverses its length. This procedure may also be used to level out spatial fluctuations in residual impurity concentrations present in crystals prepared by other methods.

Zone melting can be used to grow single crystals by inserting a seed crystal at the starting end of a crucible containing a polycrystalline sample. A molten zone is established in the polycrystalline rod at a distance removed from the end containing the single crystal so as not to melt the oriented seed. The zone is then backed up to the seed end until the seed is wetted, and then the zone is moved slowly through the specimen. If the zone is doped, zone leveling will take place at the same time as crystal growth. In this way a polycrystalline rod, previously purified by zone refining, can be converted to a uniformly doped single crystal. This process is often used in an adaptation of the zone-melting technique, which has the major advantage of eliminating crucible contamination. This is the *floating-zone method*, in which a polycrystalline rod is held vertically between two chucks and a single molten zone passed from end to end. Surface tension in the liquid zone maintains the liquid between the two solid portions. Since the solid serves as a container for its own melt, crucible contamination is eliminated. The simple floating-zone technique is limited to crystals a few centimeters in diameter because surface-tension forces are not large enough to hold the zone together in larger diameters.

Suggestions for supplementary reading

J. J. Brophy and J. W. Buttrey (eds.), *Organic semiconductors* (The Macmillan Company, New York, 1962).

H. E. Buckley, *Crystal growth* (John Wiley & Sons, Inc., New York, 1951).

Harry C. Gatos (ed.), *Properties of elemental and compound semiconductors* (Interscience Publishers, Inc., New York, 1960).

T. S. Moss, *Optical properties of semiconductors* (Academic Press Inc., New York, 1959).

Suggestions for further reading

N. B. Hannay (ed.), *Semiconductors* (Reinhold Publishing Corporation, New York, 1959).

A. F. Joffe, *Physics of semiconductors* (Academic Press Inc., New York, 1960).

W. G. Pfann, *Zone melting* (John Wiley & Sons, Inc., New York, 1958).

Various authors, *Semiconducting compounds* (W. A. Benjamin, Inc., New York, 1961).

Exercises

1. Given that the intrinsic conductivity of Ge is 2.2 Ω^{-1}-m^{-1} at room temperature, calculate the approximate intrinsic room-temperature conductivities of Si and diamond, assuming that the carrier mobility is the same in all three semiconductors. What do these results imply regarding the chemical purity required to obtain intrinsic Si and diamond crystals? (ANSWER: 1.4×10^{-3} Ω^{-1}-m^{-1}; 8.8×10^{-39} Ω^{-1}-m^{-1}.)

2. Estimate the maximum net impurity content allowable in Ge, Si, and diamond in order to permit intrinsic behavior at room temperature. (ANSWER: 10^{19}, 10^{16}, and 10^{-19} m^{-3}.)

3. What is the most probable reason that all natural semiconducting diamonds are found to be *p*-type?

4. Assuming that a Ge crystal with an impurity concentration of 10^{23} m^{-3} becomes intrinsic at 70 deg C, determine the temperature at which the following crystals will similarly become intrinsic if they have the same net impurity content: Si, cubic SiC, diamond.

5. Determine the position of the Fermi level and the conductivity of β-SiC containing 10^{22} acceptors/m^3 at room temperature.

6. Assuming that two semiconductors have the same forbidden gap and conductivity, derive a relation between the ratio of their carrier mobilities and density of states in the conduction band. Similarly, derive a relation between the mobility ratio and their effective masses. Based on these results, what general statement can be made regarding the effective mass and density of states in the conduction band of a semiconductor having a large electron mobility?

7. Given that $m_e^* = 0.25m$ for Ge, calculate the effective masses of electrons and holes in InSb using mobility values given in the text and the result of Exercise 1. Also, determine the density of states in the conduction and valence band. (ANSWER: $m_e^* = 0.03m$; $m_h^* = 1.1m$; $N_v = 1.2 \times 10^{25}$ m^{-3}; $N_c = 1.1 \times 10^{23}$ m^{-3}.)

8. Using the results of Exercise 7, calculate the position of the Fermi level in intrinsic InSb at room temperature. Determine the hole and electron concentration and the conductivity. (ANSWER: $E_0 = 0.01$ eV.)

9. Using the results of Exercise 7, determine the location of the Fermi level at room temperature in intrinsic InSb using the full Fermi function to allow for the possibility that the electrons in the conduction band are degenerate. Compare with the result of Exercise 8.

10. Using expressions for the Hall constant given in Chap. 8 and the data for InSb in Fig. 11, determine the ratio of electron to hole mobility in InSb from the temperature at which $R_H = 0$.

11. Starting with Eq. (2), derive the relation between thermoelectric power and conductivity given by (3).

12. Given that $\mu_p = 8 \times 10^{-2}$ m^2/V-sec and $m_p^* = 0.3m$ for PbTe, plot a curve of room-temperature thermoelectric power vs. conductivity and compare with Fig. 15.

13. If the extrinsic conductivity of a *p*-type PbTe crystal is 10^5 Ω^{-1}-m^{-1} and the sample is chemically pure, what is the fractional departure from stoichiometry? Assume one carrier per vacancy and $\mu_p = 8 \times 10^{-2}$ m^2/V-sec. Which atom species is deficient?

10. semiconductor devices

The dry-disk rectifier, a device based on the semiconducting properties of copper oxide, was first sold commercially as early as 1920. Later selenium rectifiers and photoelectric devices of both selenium and copper oxide were used in many applications. It was not until the discovery of the transistor in 1949, however, that the era of semiconductor devices can be said to have really begun. The impetus given to semiconductor physics by the commercial importance of the transistor has, in subsequent years, spawned a wide variety of semiconductor devices, and additional applications are continually being uncovered.

The distinguishing feature of semiconductor devices that has prompted their replacement of vacuum tubes is that their electronic functions occur within a solid body. This implies small physical size and an essentially infinite operating life, since the active elements, electrons, cannot wear out. Another important feature is their ability to operate with very low power consumption. This is due to the relatively small numbers of current carriers required in semiconductors and the fact that they are already present and need not be constantly generated. The variety of semiconductor devices available stems from the wide range of electrical properties which can be developed in semiconductors by minor chemical additions, as discussed in the previous chapter. A drawback of such devices is that they require materials of extreme chemical purity and that

268

they are sensitive to chemical attack. In addition, they are inherently temperature-sensitive, so that the ambient scope of their operation is limited. Some of the important features of semiconductor devices are discussed in this chapter. While the transistor is undoubtedly the most important of such devices because of its ability to amplify electric signals, a wide range of other equally intriguing devices have been devised and, in the aggregate, are of equal importance. It is not the intent of this discussion to consider engineering aspects of device applications, but rather to develop an understanding of the basic operation and performance of several typical semiconductor devices.

Theory of *p-n* junctions

Structure of *p-n* junction. The junction between a *p*-type region and an *n*-type region in the same semiconductor single crystal is basic to the operation of many devices and is called a *p-n junction*. Such a structure may be fabricated in any one of several ways, which result in an abrupt change from donor impurities to acceptor impurities across a given region of the crystal. For example, if the crystal is grown from the melt as described in Chap. 9, by slowly withdrawing the solidifying crystal, a *grown junction* can be created by suddenly changing the impurity concentration in the melt from *p*-type to *n*-type. An *alloy junction* may be produced by, say, placing a pellet of indium on an *n*-type crystal of germanium and heating to the melting point of indium. The molten indium dissolves some of the germanium in contact with it, and upon cooling, the base crystal contains the indium as an impurity in the refrozen portion. The regrown region is now *p*-type, since the concentration of indium is quite large and the original *n*-type impurity is heavily overcompensated. The *p-n* junction, of course, lies between the original *n*-type crystal and the regrown *p*-type region. Alternatively, if a semiconductor crystal is heated in the presence of a donor or acceptor gas, the gas atoms diffuse part way into the crystal to form a *diffused junction*. A number of other methods have also been developed, and, in each case, the object is to produce a *p*-type region adjacent to an *n*-type region in the same single crystal.

The electrons in the *n*-type region tend to diffuse into the *p*-type region, and, in equilibrium, this must be compensated by an equal flow of electrons in the reverse direction. Since the concentration of electrons is much larger in the *n*-type material compared with the *p*-type, the electron current from the *n*-type region would be greater if it were not for the presence of a potential rise at the junction which reduces the current flow in this direction. The polarity and magnitude of this internal potential are such that they make the currents equal (Exercise 1). A precisely identical argument applies to the hole currents to and from the *n*- and *p*-type regions.

At equilibrium, the Fermi level is continuous throughout the crystal, so that the energy-level diagram appears as in Fig. 1. Notice that the polarity of the potential rise, V_0, tends to keep the electrons in the *n*-type region and the holes in the *p*-type region. Since the Fermi level is continuous throughout, the magnitude of the potential step is simply the difference between the locations of the Fermi level in the *n*-type material and the *p*-type material with respect to, say, the top of the valence band. Therefore the magnitude of the built-in potential of a *p-n* junction depends upon the width of the forbidden-energy gap, the

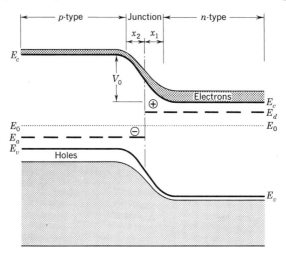

FIG. 1. Energy-band model of a *p-n* junction.

impurity concentration, and the temperature. At high temperatures, when both regions become intrinsic, the potential rise disappears, so that the *p-n* junction may be said to be nonexistent. Because many semiconductor devices are based on the properties of the *p-n* junction, improper device operation sets in at such temperatures.

The potential rise V_0 is due to a double layer of charge at the interface between the *p*- and *n*-type regions. Electrons from donor levels in the junction transfer into adjacent acceptor levels, and the two layers of ionized-impurity centers give rise to the potential barrier as shown in Fig. 1. In the case of an *abrupt* junction, one in which the transition from *p*-type to *n*-type occurs very suddenly in the crystal, the equality of charge transferred from donors to acceptors means that

$$N_d x_1 = N_a x_2 \tag{1}$$

where x_1 is the width of the junction in the n-region, and x_2 is the extent of the junction in the p-region. The magnitude of the potential rise in the n-region V_1 is determined by Poisson's equation in one dimension, which is

$$\frac{d^2V_1}{dx^2} = \frac{eN_d}{\epsilon_0\kappa_e} \tag{2}$$

where κ_e is the dielectric constant of the material. The solution of (2) gives the width x_1.

$$V_1 = \frac{eN_d}{2\epsilon_0\kappa_e} x_1^2 \tag{3}$$

A similar expression,

$$V_2 = \frac{eN_a}{2\epsilon_0\kappa_e} x_2^2 \tag{4}$$

applies to the p-region. The total width of the p-n junction d is then, from (3) and (4),

$$\begin{aligned} d = x_1 + x_2 &= \left(\frac{2\epsilon_0\kappa_e}{e}\right)^{\frac{1}{2}}\left[\left(\frac{V_1}{N_d}\right)^{\frac{1}{2}} + \left(\frac{V_2}{N_a}\right)^{\frac{1}{2}}\right] \\ &= \left(\frac{2\epsilon_0\kappa_e V_0}{e}\right)^{\frac{1}{2}}\left[\left(\frac{1}{N_d} \times \frac{1}{1 + V_2/V_1}\right)^{\frac{1}{2}} + \left(\frac{1}{N_a} \times \frac{1}{1 + V_1/V_2}\right)^{\frac{1}{2}}\right] \end{aligned} \tag{5}$$

where, of course, $V_0 = V_1 + V_2$. Using Eqs. (3), (4), and (1),

$$\frac{V_1}{V_2} = \left(\frac{x_1}{x_2}\right)^2 \frac{N_d}{N_a} = \frac{N_a}{N_d} \tag{6}$$

and the width (5) may be expressed

$$d = \left(\frac{2\epsilon_0\kappa_e V_0}{e(N_a + N_d)}\right)^{\frac{1}{2}}\left[\left(\frac{N_a}{N_d}\right)^{\frac{1}{2}} + \left(\frac{N_d}{N_a}\right)^{\frac{1}{2}}\right]. \tag{7}$$

If, say, $N_d \gg N_a$, this reduces to

$$d = \left(\frac{2\epsilon_0\kappa_e V_0}{eN_a}\right)^{\frac{1}{2}} \tag{8}$$

which shows that the junction gets narrower as the impurity concentration is increased. Furthermore, (6) shows that the potential rise is almost entirely confined to the p-type region when the n-type region is heavily doped. These results apply only to an abrupt junction, for example, an alloy junction. In other situations, where a concentration gradient of donors and acceptor impurities may extend over an appreciable distance during the transition from n-type to p-type, somewhat different expressions are obtained, but the principles are the same.

The rectifier equation. When the p-n junction is in equilibrium, the number of holes diffusing from the p-region to the n-region per second is equal to the number diffusing in the reverse direction, as discussed above, and the same thing is true of the electron currents. If an external potential is applied to the junction with a polarity such that the height of the potential step is increased (top of Fig. 2), the hole diffusion current from the p-region to the n-region is reduced, since there are few holes with sufficient energy to surmount this larger barrier. Holes diffusing from the n-region to the p-region are not affected because they do not have to climb the barrier. Thus a net hole current flows, but it is limited by the small number of holes in the n-region. If the polarity is reversed (Fig. 2, bottom), the barrier is reduced and a large hole current flows from the p-region to the n-region, since there are many holes in the p-region. The reverse flow, from n to p, again remains unaffected. The net current in this case is large, and the polarity of the applied potential is called *forward bias*. For the opposite polarity the current is small and the junction is said to be under *reverse bias*. The magnitudes of the hole currents are indicated by the lengths of the arrows I_1 and I_2 in Fig. 2 for the various conditions of bias. A similar argument applies to the electron currents as well, except that, since the charge on the electrons is negative, the actual direction of electron motion is opposite to that of the current arrows. The diagram shows that forward- and reverse-bias polarities for electrons are the same as those for holes. A p-n junction is clearly a rectifier since a larger current flows for one polarity of applied potential than for the other. A similar behavior is found for certain metal-semiconductor contacts discussed in Chap. 8. In the case of the p-n junction, however, the current is most likely to be carried by both holes and electrons. In determining the current-voltage characteristic of the junction, the same considerations apply to the current carried by holes as to that carried by electrons, and the total current is the sum of the two.

Focusing attention on the hole current first, that flowing from the n-region to the p-region is proportional to the equilibrium hole concentration in the n-region, p_n, so that

$$I_2 = C_1 p_n. \tag{9}$$

C_1 is a constant involving the junction area and semiconductor parameters and is discussed in the next section. As described above, I_2 is independent of the applied potential V. The current I_1, flowing from p to n, is proportional to the number of holes in the p-region with sufficient energy to surmount the barrier, or

$$I_1 = C_1 p_p e^{-(V_0 - eV)/kT} \tag{10}$$

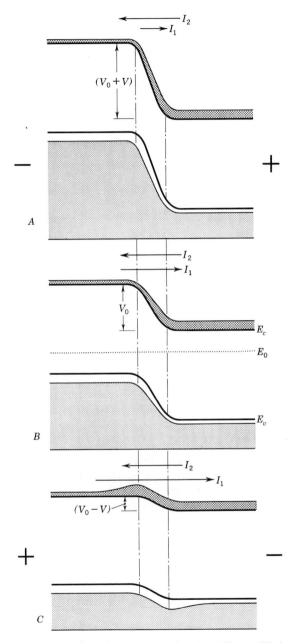

FIG. 2. Energy-band model of a *p-n* junction for (*B*) equilibrium; (*A*) reverse bias; and (*C*) forward bias.

where p_p is the equilibrium concentration of holes in the p-region, and the exponential is the Boltzmann distribution law discussed in Chap. 5. When the applied potential $V = 0$, $I_1 = I_2$, so that

$$p_n = p_p e^{-V_0/kT}. \tag{11}$$

The net hole current I_h, therefore, is

$$\begin{aligned} I_h &= I_1 - I_2 \\ &= C_1 p_p (e^{eV/kT} - 1) e^{-V_0/kT} \\ &= C_1 p_n (e^{eV/kT} - 1). \end{aligned} \tag{12}$$

An identical expression can be derived for the electron current:

$$I_e = C_2 n_p (e^{eV/kT} - 1) \tag{13}$$

where n_p is the equilibrium concentration of electrons in the p-region. The total current is the sum of (12) and (13),

$$\begin{aligned} I &= I_e + I_h \\ &= I_0 (e^{eV/kT} - 1) \end{aligned} \tag{14}$$

where $I_0 = C_1 p_n + C_2 n_p$ is called the *saturation current*. Equation (14) is called the *rectifier equation* and has the familiar form already found to apply to rectifying metal-semiconductor contacts.

The polarity of the applied potential is conventionally chosen so as to make the p-region positive for forward bias. When the junction is strongly biased in the reverse direction, the magnitude of the current is I_0 and is essentially independent of the applied potential. The current determined by (14) is compared with experimental measurements in Fig. 3, and the agreement is seen to be very good over a wide range of currents.

Minority-carrier injection. Application of forward bias to a p-n junction alters the concentrations of minority carriers near the junction. This has already been indicated schematically in Fig. 2, where the concentrations of holes in the n-region and electrons in the p-region are shown to be increased over their equilibrium values near the junction. Examination of this behavior leads easily to an interpretation of the saturation current in terms of material parameters.

The excess of hole concentration $p(x)$ on the n-type side of the junction over the equilibrium concentration deep in the n-region, p_n, may be written, using the same reasoning leading to (12), as

$$p(0) = p_n (e^{eV/kT} - 1). \tag{15}$$

These excess minority carriers diffuse away from the junction because of the concentration gradient, and as they diffuse away, they recombine in a time which is characteristic of the minority-carrier lifetime. On proceeding away from the junction, the excess hole concentration thus decreases, according to Exercise 6,

$$p(x) = p(0)e^{-x/L_p} \qquad (16)$$

where L_p is the diffusion length of holes in the n-region. The concentration of holes in the various regions is indicated schematically in Fig. 4,

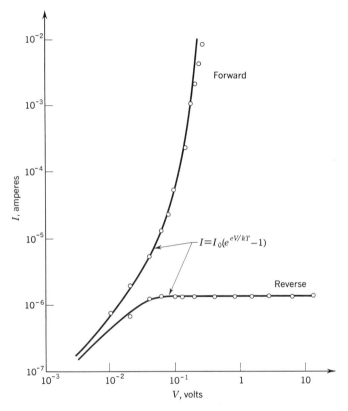

FIG. 3. Current-voltage curve of a p-n junction. The solid line represents the rectifier equation, while the circles represent experimental values.

where the exponential decrease of excess holes near the junction is shown, together with the large concentration p_p in the p-region and the small concentration p_n in the n-region, far from the junction.

The current carried across the junction by the diffusing holes is now simply given by

$$I_h = -BeD_p \left(\frac{\partial p}{\partial x}\right)_{x=0} \qquad (17)$$

where D_p is the diffusion constant for holes in the n-region, and B is the

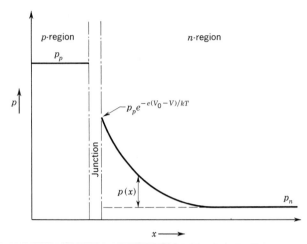

FIG. 4.   Hole concentration near a *p-n* junction.

area of the junction.   Thus, combining Eqs. (15) to (17),

$$I_h = B\left(\frac{p_n e D_p}{L_p}\right)\left(e^{eV/kT} - 1\right) \tag{18}$$

which should be compared with (12).   Here again, an analogous relation for electrons may be derived.   The total current is, as before,

$$I = Be\left(\frac{p_n D_p}{L_p} + \frac{n_p D_n}{L_n}\right)\left(e^{eV/kT} - 1\right) \tag{19}$$

where the saturation current now can be written explicitly as

$$I_0 = Be\left(\frac{p_n D_p}{L_p} + \frac{n_p D_n}{L_n}\right). \tag{20}$$

Here D_n and L_n are, respectively, the diffusion constant and diffusion length of electrons in the *p*-region.

The junction current is carried predominantly by holes according to (20), if the *n*-type region is lightly doped and if the *p*-type region is heavily doped, $n_p \ll p_n$.   By this means, large excess hole concentrations can be injected into the *n*-region and these minority carriers diffuse, on the average, one diffusion length away from the junction.   This is called *minority-carrier injection* and forms the basis for transistor action in semiconductors.   In most semiconductor devices using *p-n* junctions, either the *n*-region or the *p*-region is heavily doped in order to obtain large minority-carrier injection.

Junction capacitance.   The *p-n* junction is a double layer of opposite charges separated by a small distance and so has the properties of an electrical capacitance.   The magnitude of this capacitance per unit area is, using (8) for a heavily doped n-region,

$$C = \frac{\epsilon_0 \kappa_e}{d}$$

$$= \left(\frac{\frac{1}{2}\epsilon_0 \kappa_e e N_a}{V_0 + V}\right)^{\frac{1}{2}} \tag{21}$$

where κ_e is the dielectric constant. According to (21), the capacitance varies with the applied potential, as illustrated by the experimental values plotted in Fig. 5. Such an electrically variable capacitor is an important

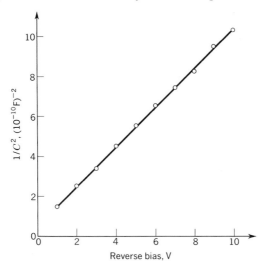

FIG. 5. Variation of the capacitance of a p-n junction as a function of reverse bias. The straight line is in agreement with Eq. (21).

semiconductor device because the ability of a p-n junction to vary its electrical capacitance rapidly in response to an impressed signal can be used in several applications. Of particular importance is the parametric amplifier circuit, which is capable of amplification at extremely high frequencies.

p-n junction devices

Rectifiers. The large forward currents and the small reverse currents generated at p-n junctions make such devices particularly useful as rectifiers. Junctions in silicon and germanium are widely used as power rectifiers to convert an alternating current to a direct current. The rectifying action can be understood with the aid of the simple circuit shown in Fig. 6. A sinusoidal alternating voltage $v(t)$ results in a pulsating, nearly unidirectional current through the load because the p-n

junction passes an appreciable current only in the forward direction.   In
most practical applications, the pulsating load current is further smoothed
to a true direct current by means of additional electrical components.

An ideal rectifier passes no current in the reverse direction, and for
this reason, it is desirable to make the saturation current of a p-n junction rectifier as small as possible.   Equation (20) shows that this may
be effected by reducing both p_n and n_p, which can be done by increasing

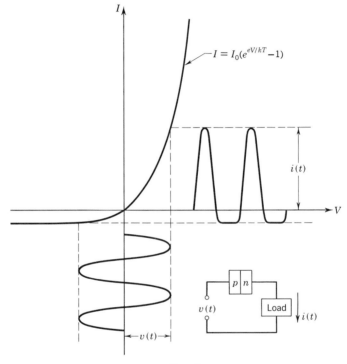

FIG. 6

the impurity content or by choosing a semiconductor with a wide forbidden-energy gap.   Furthermore, since both p_n and n_p depend exponentially on temperature, I_0 also increases rapidly with temperature and a
semiconductor having a wide forbidden-energy gap must be used for high-
temperature applications.   In high-power rectifiers, the electrical losses
in the device itself may cause appreciable temperature rises independently
of the surroundings.   For these reasons, p-n junctions in silicon are
generally preferred over germanium devices, and silicon carbide rectifiers
are used for applications requiring operation at extreme temperatures.
Junction rectifiers have been fabricated with cross-sectional areas of several square centimeters, capable of carrying many hundreds of amperes.

The current-voltage characteristics of the medium-size silicon rectifier in Fig. 7 (junction area $\simeq 7 \times 10^{-6}$ m²) show a forward current of 40 A (amperes) for an applied potential of one volt. At the same potential, the reverse current is about 0.2 μA, so that I_0 is negligible and the rectifier is very nearly ideal. The rapid increase in saturation current with temperature is evident in the figure. The ability of such devices to handle large currents is limited almost entirely by the ease with which they can dispose of the heat generated in the semiconductor by conducting it to the surroundings.

At high values of reverse bias, the current is observed to increase very rapidly as shown in Fig. 7. This is caused by the extremely large electric

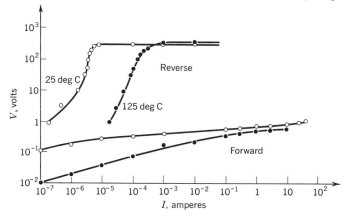

Fig. 7. Current-voltage characteristics of a medium-power silicon *p-n* junction rectifier at two temperatures. (*After M. B. Prince.*)

fields created in the junction because of the small junction width. Carriers in the junction are accelerated to such high velocity by this field that they cause impact ionizations by collisions with electrons bound to atoms in the crystal. Electrons so ionized are similarly accelerated and in turn cause other ionizations. This avalanche process leads to very large currents, and the junction is said to have suffered breakdown. The breakdown is not destructive, however, unless the power dissipation is allowed to increase the temperature to the point where local melting destroys the semiconductor. The voltage across the junction remains quite constant over a wide current range in the breakdown region, and this effect is used in devices to provide constant voltages in electronic circuits. These junctions are called *Zener diodes* because Zener first derived an expression for the magnitude of the voltage required to achieve breakdown. It is possible to increase the potential at which breakdown begins in rectifiers by increasing the junction width. This reduces the electric field and means that the impurity concentration must be decreased, according to

(8). This, in turn, is in conflict with the requirement of a small saturation current, so that each rectifier design represents a compromise which is resolved in terms of its ultimate application.

Photocells. Any free carriers generated in a reverse-biased *p-n* junction are immediately swept by the electric field toward the region where they are the majority carriers. One way to produce carriers at or near the junction is by the absorption of light photons energetic enough to excite an electron from valence-band states to conduction-band states (Fig. 8). These carriers move in response to the junction field and contribute to the junction current, which can be measured in the external circuit. A *p-n* junction, therefore, can be used as a photoelectric detector, or *photocell*. It has been found that very nearly one hole-electron pair is produced for each absorbed photon whose energy is greater than the forbidden-energy gap. Clearly, a long-wavelength limit to the response of a junction photocell exists, and it depends on the width of the forbidden-energy gap. Junctions in Ge and InSb are commonly used to respond to near- and far-infrared radiation, respectively.

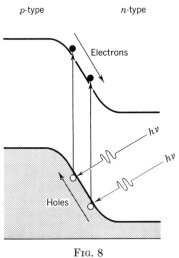

The usefulness of the *p-n* junction photocell stems from the fact that only a small saturation current flows under reverse bias, so that it is possible to detect a current produced by the excitation of relatively few carriers by absorbed photons. The same number of additional carriers produced in a regular crystal, not containing a junction, is more difficult to detect because the fractional change in conductivity is smaller. It is common practice to cool junction photocells in order to reduce I_0 as much as possible. This is particularly true when the forbidden-gap width required to achieve photoresponse is small, for example, for a far-infrared cell made of InSb.

The small reverse current of a junction photocell can be expressed conveniently in terms of the junction resistance. The resistance of a *p-n* junction, $R = dV/dI$, is found by differentiating (14),

$$1 = \left(\frac{I_0 e}{kT}\right) e^{eV/kT} \frac{dV}{dI}$$

and

$$R = \left(\frac{kT}{eI_0}\right) e^{-eV/kT}. \tag{22}$$

p-type   n-type

Electrons

$h\nu$

$h\nu$

Holes

F IG. 8

Under reverse bias, that is, when V is negative, R becomes very large. The change in conductivity, due to photo-excited carriers, is then quite large. (This matter is considered further in Exercises 9 and 10.)

Because of the internal potential rise in a p-n junction, photo-excited carriers are collected at the junction even in the absence of an external potential. These extra carriers are then observed as a current in the external circuit. This is called the *photovoltaic effect* and is used to convert radiant energy into electrical power. It has widespread application in the familiar photographic exposure meter and in the solar battery discussed in the next section.

Solar batteries. The photovoltaic effect in a p-n junction can serve as a useful electrical generator employing solar energy and has the virtue

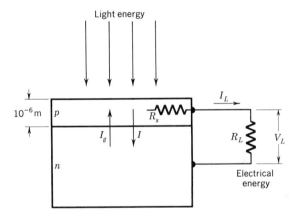

Fig. 9. Sketch of a p-n-junction solar-energy converter.

of extreme simplicity in operation. Silicon junctions are almost universally used in this application because the forbidden gap is appropriate for the wavelengths of solar radiation. CdTe and GaAs are also useful materials, particularly for higher-temperature applications, but the quality of the junctions produced in these compound semiconductors is not as good as those possible in silicon.

Holes and electrons produced in the junction region are swept to the p-type and n-type sides, respectively, as shown in Fig. 8. This produces a current I_g across the junction and also acts to charge the p-type region positively and the n-type region negatively. Thus, if there are no external connections to the junction, this forward bias causes a forward current I to flow. Under this condition, the forward current just balances I_g. When the p- and n-type sides are connected externally through an electrical load, as in Fig. 9, a portion of I_g flows in the external circuit and the so-called solar battery acts as a converter of light energy to electrical energy.

The current I_g consists of that flowing across the junction and that flowing through the external load I_L,

$$I_g = I + I_L \tag{23}$$

and, from (14), $$I_g = I_0(e^{eV/kT} - 1) + I_L. \tag{24}$$

For reasons to be discussed below, the p-type region is made very thin, and this introduces an internal series resistance R_S in the solar battery

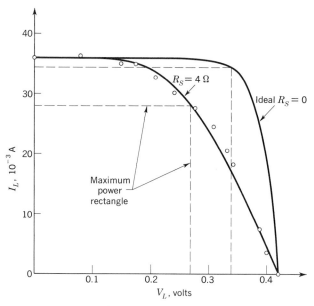

Fig. 10.   Current-voltage curves of a 1.7 cm² silicon solar cell in bright sunlight. The circles represent experimental data, while the solid lines represent Eq. (26).

which tends to reduce its efficiency.   Referring to Fig. 9, the voltage drops in in the circuit are

$$V = I_L R_S + V_L. \tag{25}$$

Equations (24) and (25) may be combined to give the relation between the load current and load voltage for a given I_g,

$$V_L = \frac{kT}{e} \ln\left(1 + \frac{I_g - I_L}{I_0}\right) - R_S I_L. \tag{26}$$

The current-voltage characteristic of a typical silicon solar battery is compared with (26) and with an ideal battery ($R_S = 0$) in Fig. 10.   The electrical power delivered to the load is given by the rectangle bounded by V_L and I_L, and the maximum power output is represented by the rectangle of largest area that can be fitted under the curve.   The presence of R_S clearly reduces the efficiency of the solar generator.

According to (26), it is desirable to make I_0 as small as possible in order to increase V_L. This means that the quality of the junction must be as good as possible in order to attain the theoretical value given by (20) for the forbidden-energy gap of a particular semiconductor. Non-uniform junctions, unwanted chemical impurities, and crystal imperfections all act to increase I_0. It is particularly difficult to fabricate high-quality junctions of the large areas necessary to intercept the maximum amount of incident light. For these reasons only silicon solar cells have achieved practically useful conversion efficiencies so far.

The solar radiant energy is a maximum at wavelengths corresponding to green light ($h\nu = 2.5$ eV), so that satisfactory solar-energy converters must be made of semiconductors having a forbidden-energy gap in the range 1.0 to 1.5 eV in order for most of the incident photons to be energetic enough to create electrons and holes. Furthermore, the carriers must be produced at or very near the junction so that they may be separated by the junction field and contribute to I_g. Electrons and holes produced far from the junction simply recombine without contributing to the cell current. For this reason the junction is located very near the surface and the thickness of the p-type layer is conventionally about 10^{-6} m in a silicon cell. This very thin layer, through which the load current must flow as indicated in Fig. 9, is the origin of R_S. The thickness of the p-type layer represents a compromise between the value of R_S and the collection efficiency of the junction for photo-excited electrons and holes. An important source of inefficiency, optical reflection at the surface of the semiconductor, is reduced by antireflection coatings similar to those used in optical instruments. The best silicon cells have overall efficiencies (delivered electrical power divided by total incident solar energy) of 15 per cent.

Tunnel diodes. An unusual phenomenon occurs in very heavily doped junctions under small forward bias. At doping levels of 10^{25} impurities/m^3 or more, the electron or hole concentration becomes degenerate and the Fermi level actually lies in the valence band or conduction band as shown in Fig. 11A. Furthermore, the junction width becomes so narrow that quantum-mechanical effects are important. When the width approaches about 100 Å, the wave function of an electron occupying a state in the valence band on the p-type side extends into states in the conduction band on the n-type side, and vice versa. Therefore an electron has an appreciable probability of crossing the junction by what is called quantum-mechanical tunneling through the potential barrier at the interface. The tunneling current thus produced depends upon the junction width, upon the number of electrons capable of tunneling, and upon the number of empty energy states into which they can transfer. At equilibrium (Fig. 11A) the tunneling currents across the junction in the two directions are equal and the net current is zero.

For a small forward bias (Fig. 11*B*), the empty valence-band states on the *p*-type side are lined up with the electrons occupying states in the conduction band on the *n*-type side, and a large current flows from *p* to *n* (electrons tunnel from *n* to *p*). As the bias is increased beyond this point, electrons near the Fermi level cannot tunnel because there are no states available in the forbidden gap on the *p*-type side. Figure

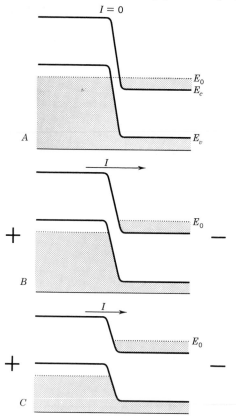

Fɪɢ. 11. Energy-band models of a heavily doped *p-n* junction (*A*) at equilibrium; (*B*) under forward bias; (*C*) under increased forward bias. The Fermi level indicated in *B* and *C* marks its equilibrium value in the bulk of the *p*-type and *n*-type portions.

11*C* shows the case for a still larger bias in which no tunneling can occur since no states at all are available. In this condition a normal forward *p-n* junction current flows. For biases between those represented by Fig. 11*B* and *C*, each *increase* in voltage results in a *decrease* in current. The junction therefore appears to have a negative resistance, which has a number of useful applications. (See also discussion of superconductivity in Chap. 7.)

The current-voltage characteristics of a *tunnel diode* (also referred to as an *Esaki diode* after its discoverer) are given in Fig. 12. The peak current corresponds to the bias condition of Fig. 11*B*, while the condition in Fig. 11*C* occurs to the right of the valley and is the normal forward current of *p-n* junction. The negative-resistance region between the peak and the valley is clearly evident. This negative resistance in a tunnel diode is important because all normal circuit components have positive resistance and hence dissipate power. If the tunnel diode is placed in a circuit and the components chosen so that the net resistance

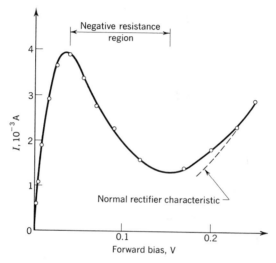

FIG. 12. Current-voltage characteristic of a tunnel diode. (*After L. Esaki.*)

is zero, no power dissipation results. Such a circuit therefore oscillates at a frequency corresponding to its characteristic resonant frequency, which is determined by its capacitance and inductance. Thus extremely simple oscillators consisting only of a tunnel diode together with a capacitor and an inductance (and a battery to bias the junction to the negative-resistance point) are possible. Such circuits can oscillate at very high frequencies, since tunneling takes place essentially instantaneously, and useful generators at frequencies as high as 10^{11} cps have been built. As can be seen from Fig. 12, tunnel diodes operate at very low potentials, which means that very small input powers are required.

Transistors

Theory of junction transistors. A junction transistor consists of two parallel *p-n* junctions juxtaposed in the same single crystal and separated by less than a minority-carrier diffusion length. Two distinct types are possible, the *p-n-p* transistor and the *n-p-n* transistor, depending upon the conductivity type of the common region. The behavior of the two is conceptually identical except for the interchange of minority-and majority-carrier types (and the polarity of the bias voltages), so that it suffices to discuss the operation of the *p-n-p* transistor. Examples of physical structures having this multiple-junction arrangement are described in a following section.

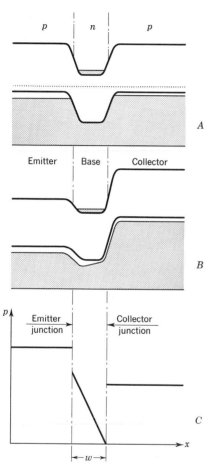

The operation of a junction transistor can be derived from the properties of a *p-n* junction. The energy-band scheme of a *p-n-p* transistor structure without applied bias voltages sketched in Fig. 13*A* is simply that of two *p-n* junctions back to back. In operation, one junction, called the *emitter*, is biased in the forward direction and the other, the *collector*, is biased in the reverse direction, as shown in Fig. 13*B*. Holes injected into the *n*-type *base region*, at the emitter junction, diffuse across to the collector junction, where they are collected by the junction field. The resulting concentration of holes in the various regions is sketched in Fig. 13*C*. Variation of the emitter-base voltage changes the injected current correspondingly, and this signal is observed at the collector junction. Since the forward-biased emitter represents a small resistance and the reverse-biased collector a large resistance, and nearly the same current flows through both, a large power gain results. Thus the transistor is basically a power amplifier capable of increasing the strength of electrical signals. For maximum amplification it is desirable that the collector current be as large a fraction of the emitter current as possible, and this

FIG. 13.   Energy-band model of a *p-n-p* transistor (*A*) at equilibrium; (*B*) under operating bias; (*C*) hole concentration for case *B*.

imposes a number of design criteria. First of all, the current through the emitter junction should be carried primarily by holes, since electrons flowing from base to emitter cannot influence the collector current. According to (19), this can be accomplished if the *p*-type emitter region is heavily doped while the *n*-type base region is lightly doped. Second, the minority-carrier lifetime and mobility of the holes in the base region should be large, so that they will diffuse an appreciable distance in the base before disappearing by recombination. (A lightly doped base is conducive to high mobility and lifetime, as discussed in Chap. 8.) Finally, the base region must be thin compared with a diffusion length, so that few carriers are lost by recombination before they reach the collector junction. The conductivity of the collector region, so long as it remains extrinsic, is less important than that of either the emitter or base regions. If convenient, it is desirable to make it lightly doped in order to increase the junction width, thus reducing the capacitance of the collector junction and also increasing the reverse breakdown voltage. In a typical grown-junction transistor the collector, base, and emitter conductivities are about 10, 10^2, and 10^5 Ω^{-1}-m^{-1}, respectively, and the width of the base region is of the order of 10^{-5} m.

A useful figure of merit for a transistor is the *current gain factor* α, which is the ratio of the collector current to the emitter current (frequently specified for constant collector potential when this is important). Alpha is the product of two terms, the emitter efficiency γ and the base transport efficiency ϵ, so that

$$\alpha = \epsilon\gamma. \tag{27}$$

The emitter efficiency is defined as the fraction of the emitter-junction current carried by holes (for a *p-n-p* transistor), while ϵ is the ratio of the collector current to the hole current injected into the base at the emitter. In terms of these definitions, the carrier currents in the various regions of a transistor are shown in Fig. 14, using the emitter current I_e as a starting point. The current I_e flows across the emitter junction; γI_e holes are injected into the base, and $\epsilon\gamma I_e$ reach the collector. The difference between the emitter and collector currents, $(1 - \alpha)I_e$, must flow into the base region from the external ciruit. Obviously, when the current gain is unity, all the emitter current appears in the collector circuit and the base current is zero.

The emitter efficiency of an isolated *p-n* junction is considered in Exercise 14, but it is important to note here that γ is larger in a transistor structure because the collector junction acts as a sink for holes in the base. Thus the hole concentration in the base falls more rapidly with distance (Fig. 13C) compared with the exponential decrease for an isolated junction (Fig. 4). The concentration gradient of holes at the base side of the emitter junction, therefore, is greater when the collector junction is nearby, and, according to (17), the forward hole current is correspondingly increased. Using (16) and (17), the hole gradient for the isolated junction is $-p(0)/L_p$. Since the variation of hole concentration in the base region of a transistor is nearly linear, the gradient in this case is $-p(0)/W$, where W is the width of the base. Thus the forward hole current in the emitter is increased by the factor L_p/W and the emitter efficiency γ is similarly larger. The emitter efficiency can easily be made to exceed 0.995 when the doping ratio between emitter and the base is large and when $W \ll L_p$.

An expression for the base transport efficiency can be derived from a solution of the diffusion equation in the base region. When the density

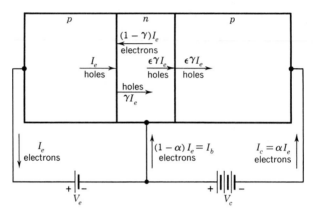

FIG. 14. Schematic diagram of the currents in a *p-n-p* transistor expressed in terms of the emitter current, injection efficiency, and current gain.

of injected holes is much larger than the equilibrium concentration and when $W \ll L_p$, the concentration is as depicted in Fig. 13C. Under these conditions ϵ is given by

$$\epsilon = 1 - \frac{1}{2}\left(\frac{W}{L_p}\right)^2. \tag{28}$$

A reasonable value for L_p is 2×10^{-4} m, so that if $W = 10^{-5}$ m, the base transport efficiency is 0.9988. The current gain α, therefore, is typically of the order of 0.994 or greater.

Minority carriers injected into the base move to the collector by diffusion under the action of their concentration gradient. The finite time taken by the carriers to cross the base limits the high-frequency usefulness of junction transistors, since α is drastically reduced when the transit time approaches the period of the high-frequency signal. Because of the higher mobility of electrons, *n-p-n* transistors are more satisfactory in high-frequency applications. By ingenious fabrication procedures, it is possible to produce extremely narrow base widths ($\simeq 5 \times 10^{-7}$ m) and obtain useful transistors at frequencies of 10^3 megacycles. It is also necessary to reduce the area of the junctions in such devices in order to minimize the adverse effects of the junction capacitances.

Transistor characteristics. A most convenient way of representing the electrical properties of a transistor is through a set of current-voltage curves of the collector junction for different values of emitter current. A representative example for a *p-n-p* transistor is given in Fig. 15. When

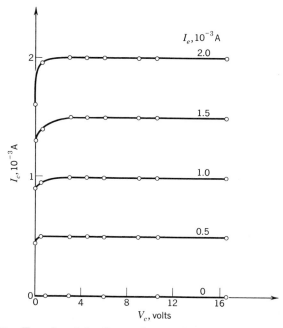

FIG. 15. Experimental collector characteristics of a *p-n-p* transistor.

$I_e = 0$, the curve is simply the reverse saturation curve of the collector junction, and the addition of an emitter current has the effect of translating this characteristic curve along the current axis. An estimate of alpha can be made by comparing the collector current I_c with I_e at high, as well as small, emitter currents. Actually, since alpha is very nearly unity for most devices, this is difficult to determine from collector characteristic curves alone unless the base current is also given. Then the collector-base current gain β may be computed directly. The relation between the collector-base current gain and α is, from Fig. 14,

$$\beta = \frac{I_e}{I_b} = \frac{\alpha}{1 - \alpha}. \tag{29}$$

The collector-base current gain becomes very large when α approaches unity, so that β is a sensitive measure of the quality of a junction transistor. Typical values range from 20 to 10^3 in practical devices.

The slopes of the collector characteristics in Fig. 15 have the dimensions of conductivity (amperes per volt). In engineering design of transistor circuits it is therefore possible to represent the reverse-biased collector junction by means of a resistance together with a constant-current generator. Similarly, the forward-biased emitter junction may be represented by a resistance according to (22). These, together with such ohmic resistances as may be important, in particular the ohmic resistance of the thin base region, can be used to describe the performance of a transistor in any electrical circuit. The values of the various resistances used to represent the device ultimately derive from the shape of the collector characteristics as exemplified in Fig. 15.

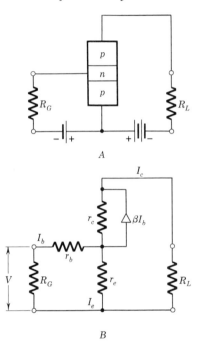

FIG. 16. Grounded-emitter transistor amplifier. (*A*) Circuit diagram of transistor amplifier; (*B*) equivalent circuit.

Grounded-emitter amplifier. An example of a transistor used as an amplifier is shown in Fig. 16*A*. This circuit configuration is called the *grounded-emitter circuit* because the emitter is common to both input and output. The grounded-emitter connection is the one most commonly used, since it provides the highest amplification per stage, but *grounded-base* and *grounded-collector* circuits are also useful for specific purposes. The properties of the transistor as an amplifier can be calculated with the aid of the equivalent circuit given in Fig. 16*B*. Here the various resistances are appropriate to the small alternating currents superimposed on the main bias currents which set the operating point of the device at the desired place on its characteristic.

The emitter and collector resistances are those corresponding to the properties of the respective junctions. The base resistance r_b is composed of two parts: that due to the ohmic resistance of the base semiconductor (which may be appreciable since the base is so thin) and that due to an electrical feedback between the collector and the base. The origin of this effect is the decrease in base width with collector voltage due to the greater width of the collection junction. This increases the base transport factor according to (28) and also tends to increase the emitter efficiency. Therefore a smaller emitter-base voltage is required to maintain a constant emitter current. This is exactly the effect that the series resistance r_b has in the circuit of Fig. 16B, since it reduces the influence of the applied input voltage. Typical values for r_b, r_e, and r_c are 500, 20, and 10^6 Ω, respectively. In the grounded-emitter amplifier, it is usually arranged that $R_G \gg (r_b + r_e)$ and $r_c \gg R_L$ in order that the input and output circuits are sufficiently isolated. Under these conditions the input power is

$$VI_b = (I_e r_e + I_b r_b)I_b = [I_e r_e + (1 - \alpha)I_e r_b](1 - \alpha)I_e. \qquad (30)$$

The output power is

$$I_c^2 R_L = \alpha^2 I_e^2 R_L \qquad (31)$$

and the power gain of the amplifier is the ratio of (31) to (30):

$$
\begin{aligned}
G &= \frac{I_c^2 R_L}{VI_b} \\
&= \frac{\alpha^2 R_L}{[r_e + (1 - \alpha)r_b](1 - \alpha)} \\
&= \frac{\beta^2 R_L}{[r_b + (\beta/\alpha)r_e]}.
\end{aligned} \qquad (32)
$$

With $R_L = 30{,}000$ Ω, $r_b = 500$ Ω, $r_e = 20$ Ω, and $\beta = 50$, the power gain is 5×10^4, so the transistor stage provides very useful amplification.

Transistors are used in a wide variety of circuit configurations, quite analogously to the vacuum tube. Various oscillator, pulse, switching, and other circuits have been devised, in addition to the three amplifier connections discussed above. Furthermore, flexibility not available in vacuum-tube circuits is provided by the availability of both *n-p-n* and *p-n-p* devices, with their reversed-bias polarities. In addition, the ruggedness, very small size, and minute electrical-power requirements of transistors further enhance their usefulness. It is quite feasible, for example, to design a solar-powered amplifier or oscillator using transistors and solar batteries which will operate unattended, essentially forever.

Transistor types. Of the several ways to produce a semiconductor single crystal containing two *p-n* junctions, three procedures are most popular. These are the grown junction, alloy junction, and diffused junction illustrated in Fig. 17. Many modifications of these three types have been devised, but they do not differ substantially from those shown in Fig. 17. The grown-junction transistor was historically the first

junction-transistor type to be fabricated. It is prepared during growth
of the semiconductor single crystal from the melt by altering the impurity
content of the melt as the solidified crystal is slowly withdrawn. To
obtain a *p-n-p* transistor, for example, the process is started with a *p*-type
melt. After crystal growth has proceeded for some time, sufficient *n*-type
impurities are added to overcome the original *p*-type impurity. When
the base region has grown as wide as desired, the melt is made *p*-type

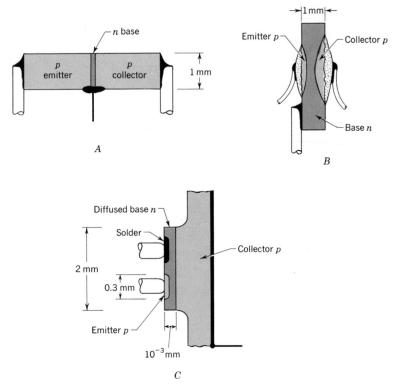

Fig. 17. Three common types of junction transistors. (*A*) Grown; (*B*) alloy; (*C*) diffused.

again by new impurity additions and the collector region grown. The
resultant single crystal has a thin planar *n*-type region perpendicular to
the direction of growth. Since a typical crystal may be one or two
centimeters in diameter, while a single transistor is about one millimeter
wide, a large number of transistors is obtained by cutting up the original
single crystal. Connections are made to the emitter, base, and collector
regions by an appropriate soldering process. It is common practice to use
a solder containing a small amount of doping impurity in order to obtain
good electrical connections. If the base-region solder joint tends to make

the germanium n-type, it produces an ohmic contact to the base region, but not to either the emitter or collector regions.

The p-n-p alloy junction transistor is produced from a thin slab of n-type single-crystal material by placing indium pellets on opposite surfaces and heating. The indium melts and dissolves the germanium beneath it. During subsequent cooling, the dissolved germanium recrystallizes upon the base material and includes indium atoms in its structure. The recrystallized material is therefore p-type, and a transistor structure results. By careful control of this process it is possible to

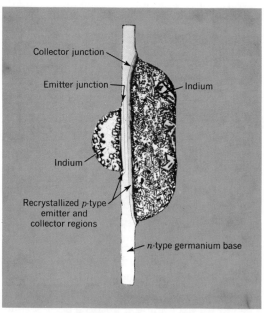

FIG. 18. Micrograph of the cross section of an alloy junction transistor. The emitter and collector junctions, revealed by chemical etching, are clearly visible. (*Courtesy of Raytheon Manufacturing Company.*)

produce quite satisfactory devices. A cross section of an actual device in which the junctions have been made visible by chemical etching is shown in Fig. 18. The emitter and collector junctions, recrystallized regions, and indium pellets are clearly visible in the micrograph.

It is necessary to make the dimensions of the transistor as small as possible in order to achieve good high-frequency performance. This can be done by forming a diffused junction. A p-type slab is placed in an atmosphere containing an n-type impurity in the gas phase. While the crystal is heated, the n-type impurity is allowed to diffuse into it to a depth of about 10^{-3} mm. The crystal is subsequently masked and etched so as to produce a small elevated region about 2 mm in diameter (Fig.

17C). The elevated region is then suitably masked so that two areas, about 0.33 mm in diameter, are exposed, and an appropriate metal is deposited on it by evaporation in high vacuum. Next the slab is maintained at an elevated temperature while the vapor-deposited metal diffuses into the n-type region. If the masking was properly done, this results in the formation of a p-type emitter region by one of the metal deposits while the other one makes ohmic contact to the base. The precise control of mechanical dimensions afforded by the gas-diffusion step and the vapor-deposition process results in very small transistors, because the active region in the device is the part lying directly beneath the emitter region in Fig. 17C. This can have a volume less than 10^{-12} m³. Useful operation at 1,000 megacycles has been attained by such designs.

Homogeneous devices

Photoconductors. Semiconductor crystals which are insulators in the dark make good photoconductors because free holes and electrons created by absorbed photons can produce a significant change in their conductivity. A convenient measure of the ability of a crystal to detect radiation, that is, its *photosensitivity*, is the ratio of the carrier photocurrent to the rate of creation of carriers.

$$G = \frac{I}{e}\frac{1}{F} \tag{33}$$

where I is the photocurrent, and F is the number of electrons and holes produced each second by the absorbed photons. The quantity G is called the *photoconductive gain* and may be interpreted, according to the form of (33), in terms of the number of carriers passing between the electrodes per second for each photon absorbed per second.

The photocurrent can be expressed as the ratio of applied voltage V to the electrical resistance R of the crystal. According to Ohm's law,

$$\frac{I}{e} = \frac{V}{eR} = \frac{VA}{e\rho L} = VA\frac{n\mu}{L}, \tag{34}$$

since $\rho = 1/ne\mu$, and L is the distance between two electrodes of area A. Also, the carrier concentration is related to the carrier lifetime τ and the generation rate such that

$$F = \frac{nAL}{\tau}. \tag{35}$$

Using (34) and (35), the gain becomes

$$G = \frac{\mu\tau V}{L^2}. \tag{36}$$

The transit time of an electron between electrodes is $t = L/\mu(V/L)$, so that from (36) the gain is the ratio of carrier lifetime to transit time

$$G = \frac{\tau}{t}. \tag{37}$$

Both (36) and (37) show that maximum photosensitivity is realized when the lifetime is large. This sensitivity is obtained at the expense of a sluggish response, since τ also determines the rapidity with which the photoconductor can respond to changes in light levels.

CdS and CdTe single crystals are widely used photoconductors because of their high sensitivity and response to visible light wavelengths, as shown in Chap. 9. The peak photosensitivity occurs at a wavelength corresponding to the forbidden-energy gap because less energetic photons are unable to create electrons and holes while more energetic photons are heavily absorbed very near the surface, so that the lifetime is small and the response is decreased. Long-wavelength infrared detectors are made from germanium crystals containing deep-lying impurity levels such as those introduced by copper, nickel, or manganese. When the crystal is maintained at a low temperature, the conductivity is small because all the carriers occupy the impurity levels. Infrared photons, whose energies correspond to the energy depth of the impurity levels, excite free carriers and produce conductivity. The photoconductive-response peak, therefore, is in the far-infrared region, at wavelengths of some tens of microns.

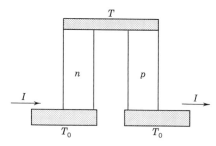

Fig. 19. Single-element thermoelectric cooler.

Thermoelectric cooling. Current flowing in a circuit containing a semiconductor-metal contact tends to pump heat from one electrode to the other because of the Peltier effect described in Chap. 8. The thermoelectric power of semiconductors is large enough to make such electronic cooling of practical interest, particularly where small size and absence of mechanical movements are desired. A single cooling unit consisting of a p-type element and an n-type element joined with ohmic contacts is sketched in Fig. 19. The series current I pumps heat from the common junction, cooling it an amount ΔT below the ambient temperature T_0. It is advantageous to use a p-type and an n-type element together, because the thermoelectric effects of the two are additive. This is so because the thermoelectric powers are of opposite sign and the current flows in the opposite direction (with respect to the cold junction) in each leg.

The Peltier cooling effect is reduced by heat conducted down the elements by normal thermal conductance, together with Joule heating in the elements due to the electric current. In the absence of external heat sources, the cooling effect is balanced by these heat losses and

$$QTI = K\,\Delta T + \tfrac{1}{2}I^2R \tag{38}$$

where Q is thermoelectric power

K is total thermal conductance

R is total electrical resistance of semiconductor elements.

The factor $\tfrac{1}{2}$ comes from an exact solution of the heat-transport equation

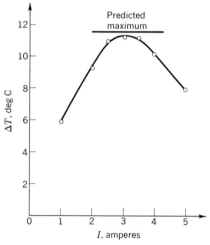

FIG. 20. Experimental variation of the temperature differential of a thermoelectric junction with current. The maximum cooling is in good agreement with Eq. (39).

for this problem. Equation (38) can be solved for ΔT and maximized with respect to the current. The result is (Exercise 15)

$$\Delta T_M = \frac{Q^2 T_M^2}{2KR}. \tag{39}$$

An optimum current is not unexpected since ΔT is small at low currents because the Peltier effect is small. At large currents, the cooling effect is large, but the Joule heating is even larger because it increases as the square of the current. The optimum point occurs between these two extremes, when the heat flow into the cold junction is due equally to thermal conductance and Joule heat. An experimental plot of the cooling effect as a function of current for a PbTe element is shown in Fig. 20. The observed maximum temperature differential is in good agreement with (39).

Considering the individual legs of a single cooling element, a figure of merit for a semiconductor may be defined, based on the expression for ΔT_M. This thermoelectric figure of merit is

$$\zeta = \frac{Q^2 \sigma}{\kappa} \tag{40}$$

where σ is the electrical and κ is the thermal conductivity. It can be seen that a useful material has a large thermoelectric power and electrical conductivity but a small thermal conductivity. A definite inverse relation between Q and σ exists, as already discussed in Chap. 9, so that it is possible to achieve the suitable compromise between the two by chemical doping. Methods used to reduce the thermal conductivity with mixed isomorphous crystals have been described in the previous chapter. Using such procedures for Bi_2Te_3, figures of merit of the order of 3×10^{-3} deg C^{-1} have been attained. These lead to maximum temperature differentials of 80 deg C, which is an indication of the potential practicality of electronic cooling. High values of ΔT_M are necessary since additional heat loads, such as would be experienced in actual devices, act to reduce the temperature differential.

It should be noted that the thermoelectric effect can be reversed, so that, when a temperature differential is maintained across the thermo-electric element, as by heating the junction, a current is caused to flow in the circuit. Such a *thermoelectric generator* is an equally attractive practical device, particularly for high-ζ materials, since the same figure of merit applies. Most often, different semiconductors are used for power generation than for cooling because they are required to withstand higher temperatures. High-temperature operation is desirable in order to increase the efficiency of the generator. This is so because a thermoelectric generator is basically a heat engine using electrons as the working fluid and is therefore limited to the Carnot-cycle efficiency maximum. To span large temperatures, it is possible to construct layered elements such that each semiconductor operates in its optimum temperature range. Typical possibilities are Bi_2Te_3 or Sb_2Se_3 (n- and p-type) operating at 25 to 250 deg C, and GeTe (p-type) operating at 250 to 550 deg C. Such a simple generator using materials with an effective ζ of 3×10^{-3} deg C^{-1} can achieve an overall efficiency of the order of 10 per cent.

Hall-effect devices. The Hall effect in semiconductors is the basis of a number of devices, the most common of which is a magnetic-field detector (Fig. 21). In this application, the high sensitivity, wide frequency response, and simplicity of the Hall element make it superior to other methods.   In Fig. 21, the Hall voltage V_H is (as derived in Chap. 8)

$$V_H = \frac{R_H I B}{t}$$ (41)

where the Hall constant $R_H = 1/ne$.   According to (41), increasing the current improves the ability of such a Hall element to measure small magnetic fields.   But the current is limited by heating effects, so that this approach has a limited value.   In order to ascertain the semiconductor parameters important for Hall devices, it is useful to express I in terms

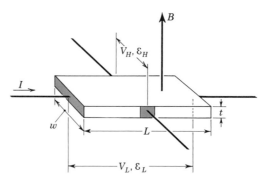

FIG. 21.   Semiconductor Hall element used to measure magnetic fields.

of the longitudinal applied voltage V_L and then rearrange the equation to express the ratio of the Hall electric field to the longitudinal field.   Thus, from (41),

$$V_H = \frac{1}{ne} \frac{V_L}{(1/ne\mu)(L/wt)} \frac{B}{t}$$

$$= V_L B \frac{w}{L} \mu$$ (42)

and, finally,

$$\frac{\mathcal{E}_H}{\mathcal{E}_L} = \mu B.$$ (43)

The sensitivity of a Hall element, therefore, depends essentially only upon the carrier mobility.   Because the mobility in InSb is large, it is widely used as a Hall-effect material.   A calibration curve for an InSb magnetic-field detector is illustrated in Fig. 22.   The linearity and wide dynamic range are clearly evident.   Furthermore, the Hall effect is independent of frequency all the way from direct current to microwave frequencies, so that the same Hall elements can be used to detect and measure radio-frequency magnetic fields.

According to Eq. (41), the Hall voltage is the product of I and B.   If the magnetic field is produced by a solenoid carrying a current I_2, B is proportional to I_2 and the output of the Hall element represents the product $I \times I_2$; that is, the Hall device acts as an electronic multiplier. A useful application of this principle is to an electronic wattmeter, if the current in the Hall element is multiplied by a current proportional to the voltage in the circuit. A microwave wattmeter can be made by orienting the semiconductor element in the electromagnetic field of a waveguide in such a way that the electric field induces a longitudinal current and the the magnetic field generates the Hall voltage by reacting with this current. Then the output voltage of the Hall element is proportional to the incident microwave power.

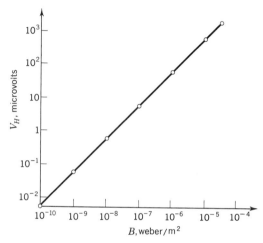

Fig. 22. Output voltage of an InSb Hall element as a function of magnetic field. (*Courtesy of M. Epstein.*)

Thermistors. The exponential temperature dependence of conductivity characteristic of semiconductors can be used in applications involving temperature measurement, and suitable thermally sensitive resistors are called *thermistors*. Because its conductivity varies so rapidly with temperature, a thermistor is actually a very sensitive electrical thermometer, capable of detecting temperature changes as small as 10^{-6} deg C. Remembering that the conductivity of semiconductors increases with temperature, in contrast to metals, it is possible to use thermistors in electrical circuits to compensate for conductivity changes and thus design systems having zero overall temperature coefficients.

Thermistors are generally fabricated from sintered oxide semiconductors, which makes them quite refractory and chemically inert. A typical composition consists of a mixture of cobalt, nickel, and manganese oxide powders sintered to yield a single homogeneous oxide semiconductor. The conductivity dependence on temperature of such an oxide semiconductor is shown in Fig. 23. The extremely wide conductivity and temperature ranges characteristic of this material are apparent. These are presumably due to the conductivity mechanisms in oxide semiconductors discussed in Chap. 9.   The particular material of Fig. 23 has a room-temperature conductivity of 0.4 Ω^{-1}-m^{-1} and an activation energy of approximately 0.3 eV.   As shown, the activation energy varies

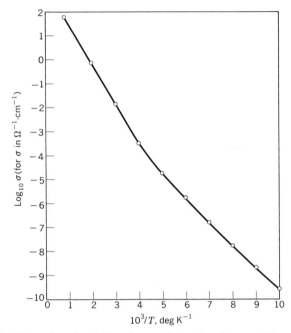

FIG. 23.   Variation of conductivity of an oxide semiconductor with temperature.

somewhat with temperature but is still remarkably constant, considering that the conductivity ranges over 12 decades.

Miscellaneous devices

Modular electronics.   The transistor and similar semiconductor devices make possible major reductions in the physical dimensions of electronic circuits because of their small physical size and minimal power requirements.   An even further reduction is possible, however, by extending semiconductor principles to the operation of the circuit as a whole, rather than limiting them to the individual components.   Indeed, in *modular electronics*, it becomes difficult to distinguish between component and circuit as such because an entire function is integrated into a single unit.   This concept is most easily illustrated with a particular example, sketched in Fig. 24.   The conventional circuit of a single-stage grounded-emitter transistor amplifier consists of the six components shown in Fig. 24A connected together with conductors. Each of these components may be fabricated individually from semiconductors, since a homogeneous bar has a specific resistance and a *p-n* junction acts as an

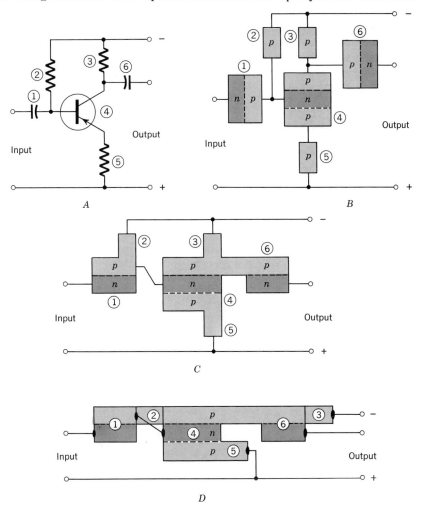

FIG. 24. Development of a modular electronics unit from a standard transistor amplifier circuit. (A) Standard circuit; (B, C) intermediate states; (D) final unit.

electrical capacitance. Thus a conventional amplifier circuit can be constructed from, say, only silicon and wires (Fig. 24B). The wires are not really necessary to connect together two silicon pieces of the same conductivity type, however, so that a further simplification is possible as shown in Fig. 24C. Finally, the configuration shown in Fig. 24D is obtained by judiciously rearranging the components. The structure represented by the last sketch is simply a three-layer semiconductor slab with suitable shapes cut (or chemically etched) away and provided with electrodes at appropriate points. Forgetting for the moment that its particular shape was derived from a conventional circuit, it resembles somewhat more a single semiconductor device than a conventional circuit. Yet this semiconductor device, within one module, is capable of performing the functions of a complete amplifier. The goal of modular electronics is to integrate components and circuits into single functional units so as to achieve substantial reductions in the number of individual units and drastically reduce their physical size. The elementary illustration shown above represents only the rudiments of this approach. Entire amplifier circuits, radios, and computer subassemblies have been constructed of single semiconductor slabs using these principles.

Space-charge-limited devices. In all the semiconductor devices discussed previously, electrical-charge neutrality is always maintained. When minority carriers are injected into the base region of a transistor from a *p-n* junction, for example, their electrical charges are compensated by the ever-present majority carriers of opposite sign. Similarly, when photons are absorbed by a semiconductor, equal numbers of holes and electrons are produced and charge neutrality exists. In insulators, on the other hand, it is possible to depart from charge neutrality under certain conditions because of the lack of free charges in the insulator. This produces several interesting effects and useful devices. Consider an insulating crystal containing two ohmic contacts with a high electric field between them. Since the crystal is an insulator, it cannot contribute to current flow. If the field is high enough, it is possible for electrons to be introduced into the conduction band from the negative electrode and thus constitute a current. The crystal may be said to be supporting current flow as a "conducting insulator," quite analogously to the vacuum in an electron tube which freely allows electrons to pass through it. Under these conditions, the current is *space-charge-limited* because the electrical charge of the uncompensated electrons alters the electric-field distribution in such a way as to limit the current. This effect is well known in electron tubes as Child's law.

Under space-charge-limited conditions, the current density J, flowing due to an applied potential V, is given by (Exercise 17)

$$J = \frac{9\epsilon_0 \kappa_e \mu}{8w^3} V^2 \tag{44}$$

where w is the crystal thickness. This expression has been termed *Child's law for solids*, and the dependence on voltage and crystal thickness is in good agreement with experimental observations. This phenomenon is of interest for semiconductor devices because of the large current densities possible (Exercise 18) and because the uncompensated electrons are free to be influenced by electric fields. Consequently, devices directly analogous to the electron tube are possible, which have superior high-frequency performance, since the carriers move in response to the applied field rather than diffuse because of a concentration gradient as in the transistor. Diodes based on space-charge-limited current behavior have been successfully fabricated, and they do show the expected excellent high-frequency characteristics. Three-electrode amplifier devices, while operational, are not yet satisfactory, primarily because of carrier trapping effects in the insulator. As the technology of producing perfect insulating crystals improves, it seems likely that devices based on space-charge-limited currents will become increasingly important.

Suggestions for supplementary reading

A. Coblenz and H. L. Owens, *Transistors: theory and applications* (McGraw-Hill Book Company, Inc., New York, 1955).

R. F. Shea, *Transistor circuits* (John Wiley & Sons, Inc., New York, 1953).

John N. Shive, *Semiconductor devices* (D. Van Nostrand Company, Inc., New York, 1959).

H. C. Torrey and C. A. Whitmer, *Crystal rectifiers* (McGraw-Hill Book Company, Inc., New York, 1948).

Suggestions for further reading

Lloyd P. Hunter, *Handbook of semiconductor electronics*, 2d ed. (McGraw-Hill Book Company, Inc., New York, 1958).

A. K. Jonscher, *Principles of semiconductor device operation* (John Wiley & Sons, Inc., New York, 1960).

M. J. O. Strutt, *Transistoren* (S. Hirzel Verlag KG, Stuttgart, 1954).

Exercises

1. Show that the potential difference across a p-n junction is equal to the difference between the position of the Fermi levels in the n- and the p-regions by finding the condition for zero net electron current across the junction. Assume that the current leaving a region is proportional to the electron concentration.

2. Using Fig. 6 of Chap. 8, determine the internal potential rise in a p-n junction of germanium at room temperature when the donor concentration is 10^{25} m^{-3} and for acceptor concentration of 10^{21}, 10^{23}, and 10^{25} m^{-3}. Repeat for $N_d = 10^{21}$ m^{-3}. (ANSWER: 0.45, 0.56, 0.67 eV; 0.22, 0.33, 0.44 eV.)

3. Using the results of Chap. 8, Exercise 8, plot the variation of the internal potential rise in a silicon p-n junction as a function of temperature from 100 to 600 deg K if the acceptor concentration is 10^{21} m^{-3} and the donor concentration is 10^{25} m^{-3}.

4. Determine the widths of the six p-n junctions considered in Exercise 2. (ANSWER: 6.3, 0.71, 0.11, 6.3, 5.5, 6.3 $\times$ 10^{-7} m.)

5. Using the same reasoning leading to Eq. (12), derive (13). Pay particular attention to the directions of current and electron flow.

6. The differential equation for the excess hole concentration in the n-region of a p-n junction is

$$D_p \left(\frac{d^2 p}{dx^2} \right) - \frac{p}{\tau_p} = 0.$$

Solve this equation for $p(x)$ to derive Eq. (16).

7. Show why $N_a \ll N_d$ at a p-n junction leads to electron injection into the p-region.

8. Compute the capacitances per unit area of a p-n junction in germanium if $N_d = 10^{25} \text{ m}^{-3}$ and $N_a = 10^{21}, 10^{23}, 10^{25} \text{ m}^{-3}$.

9. By taking the derivative of the rectifier equation, determine the resistance of a 1-mm² p-n junction in germanium under a reversed bias of 1 V, if $N_a = N_d = 10^{23} \text{ m}^{-3}$, $\tau_n = \tau_p = 10^{-6}$ sec, and $\mu_n = \mu_p = 0.1 \text{ m}^2/\text{V-sec}$. Compare with the resistance of a piece of intrinsic germanium as thick as the width of the p-n junction. Explain the difference. (ANSWER: 1.6×10^{22}, 4.1×10^{-2} Ω.)

10. Compare the response of a p-n junction in InSb and of a simple bar of intrinsic InSb with a completely absorbed photon flux of 10^{15} sec^{-1} by computing $\Delta I/I$ for an applied potential of 1V. Assume minority-carrier lifetimes of 10^{-7} sec, $N_a = N_d = 10^{23} \text{ m}^{-3}$, and that the cells operate at 77 deg K; use $\mu_n = 10 \text{ m}^2/\text{V-sec}$, $\mu_p = 10^{-2} \text{ m}^2/\text{V-sec}$.

11. Assuming that the radiant energy of the sun at the earth's surface is equivalent to 500 W/m², compute the maximum efficiency of the practical cell of Fig. 10. The cell's area is $1.7 \times 10^{-4} \text{ m}^2$. (ANSWER: 8.9 per cent.)

12. Calculate the doping required to produce an Esaki diode in Ge. Assume $N_a = N_d$ and $w = 100$ Å. (ANSWER: 10^{25} m^{-3}.)

13. Draw the energy-band diagram for an n-p-n transistor in equilibrium. Repeat for proper biases. Sketch the electron concentration in the three regions as a function of distance.

14. Assuming that the I_2 currents of Fig. 2 are negligible for both holes and electrons in a p-n junction under forward bias, derive an expression for the emitter efficiency of a hole emitter in terms of the conductivities and diffusion lengths.

15. Starting with Eq. (38), derive the expression for the maximum temperature differential of a thermoelectric cooling element.

16. Consider a thermoelectric cooling element of PbTe for which $\mu = 0.17 \text{ m}^2/\text{V-sec}$, $m^* = 0.3m$, and $K = 3$ W/deg C-m. Develop an expression for Q as a function of σ and determine the optimum value of σ by plotting ΔT_M versus σ.

17. Derive Child's law for solids by starting with Poisson's equation

$$\frac{d^2 V(x)}{dx^2} = \frac{\rho(x)}{\epsilon_0 \kappa_e}$$

and noting that the current density is

$$J = \sigma \mathcal{E} = \rho(x) \mu \frac{dV}{dx}.$$

Suitable boundary conditions are

$$V(0) = \frac{dV(0)}{dx} = 0.$$

18. Compute the space-charge-limited current density in a CdS crystal 5×10^{-5} m thick under 10 V applied potential. Assume $\mu = 2 \times 10^{-2} \text{ m}^2/\text{V-sec}$ and $\kappa_e = 2$. (ANSWER: $4 \times 10^5 \text{ A/m}^2$.)

11. electron emission

The behavior of electrons inside solid materials has been considered in other chapters. It is now of interest to consider phenomena that occur when electrons are caused to leave the solid. This process is called *electron emission*, and it is effected by several external agents, each of which can impart to the electron sufficient energy to enable it to escape. This can happen as a result of applying sufficiently high temperatures or electric fields to supply the energy for the escape of the electrons. Similarly, energetic photons or charged particles bombarding a surface may transfer energy to the electrons through collisions, resulting in electron emission. Each of these processes is strongly dependent upon the properties of the material itself, namely, the forces that normally constrain the electrons to remain inside it. Another factor that must be considered is the efficiency of energy transfer from the external agent to the electrons. Thus it turns out that some substances emit copious numbers of electrons when subjected to one type of external excitation while others respond efficiently to other ones. Primary attention is focused on metallic solids because of their commercial importance and the ready availability of relatively free electrons in such materials. The principles of electron emission, however, apply directly to all materials.

After an electron is emitted from a solid and has traveled the equivalent of a few hundred atomic diameters, it can be considered free and essentially independent of the processes which caused it to reach that point. The properties of the electron are then, to a large extent, unrelated to the characteristics of the solid from whence it came. The behavior of such electrons is exhaustively treated in many texts on electron ballistics and lies outside the scope of this book. Electron emission from solids is the basis for a large variety of important practical devices, of which the vacuum tube is undoubtedly the best known. Until the advent of the transistor, the vacuum tube was the only effective means of amplifying high-frequency voltages. As such it formed the basis of the entire radio, television, and electronics technology.

Effects at metal surfaces

Work function. Under normal conditions, the constituent electrons of a substance are confined to the material. Yet, as has been shown in Chap. 6, the valence electrons inside metals can be considered to be free

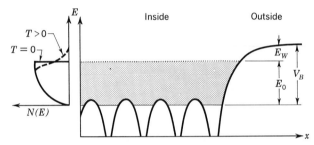

FIG. 1. Idealized potential energy of an electron along a row of atoms near the surface, and the potential-energy barrier at the surface which constrains the electrons to remain inside. The shaded area represents electron energy as given by the Fermi distribution shown on the left.

in that they are not bound to any particular atom. The potential-energy barrier at the surface which constricts electrons to the interior is shown diagrammatically in Fig. 1. The Fermi distribution function for the electrons in this metal is shown on the left. The height of the potential-energy barrier at the surface V_B is greater than that of the Fermi energy E_0 by an amount E_W called the *work function*. This is the minimum extra energy required to permit an electron to escape from the solid at the absolute zero of temperature. It is convenient to define the work function as the energy difference between the height of the surface-potential barrier and the position of the Fermi level inside the metal.

Surface-potential barrier. The various ways an electron can gain sufficient energy to surmount the surface barrier constitute the different types of electron emission. The magnitude of the work function depends on the properties of the solid in two ways. It is determined by the position of the Fermi level and by the height and shape of the surface-potential barrier. The procedure for determining the location of the Fermi level already has been described in Chap. 6. Consequently, it is sufficient to consider here only the characteristics of the barrier and its dependence upon the structure of a crystal. Consider the force acting on an electron after it has escaped some distance outside of the metal. Since the escaping electron is negatively charged, it leaves a positive charge behind in the metal. Thus, an attractive force exists between the metal and the electron which tends to return it to the metal. If the electron is many atomic diameters away from the surface, then the surface of the metal can be considered to be a plane perfect conductor for present purposes. In this model, the force on the electron is equivalent to that of a positive charge placed an equivalent distance behind the plane surface as the electron is in front. This is an example of the well-known *method of images* used in electrostatic problems. The force acting on the electron can then be expressed by

$$F = \frac{e^2}{4\pi\epsilon_0(2x)^2}. \tag{1}$$

The potential energy U associated with this force is the integral of the force from the point x to infinity, so that

$$U(x) = \int_x^\infty F(x)\ dx = -\frac{e^2}{16\pi\epsilon_0 x} \tag{2}$$

is the form of the barrier at a distance far from the surface. This approximation breaks down at closer distances because (2) predicts an infinite energy when $x = 0$, whereas the potential energy of an electron at the surface (or inside) must have a finite value. As the electron approaches the metal surface to within a few atomic diameters, it is most strongly affected by the ions closest to it. Therefore a more accurate model of the shape of the barrier is required at these distances. Such a model can be derived by considering the force on the electron due to its nearest-neighbor ions. For simplicity, consider the primitive cubic structure shown in Fig. 2. Next, let an electron approach the four ions forming one face along a line passing through the midpoint of the face. Since, as discussed above, the total effective charge on the metal is $+e$, each ionic charge is taken to be one-fourth of an electron charge. The total force on the electron is the vector sum of the force of attraction of each ion.

$$F = \frac{e^2 x}{4\pi\epsilon_0(x^2 + a^2/2)^{\frac{3}{2}}} \tag{3}$$

where a is the interatomic spacing. The potential energy is again found by integrating the force,

$$U(x) = \int_x^\infty F(x)\ dx = \frac{e^2}{4\pi\epsilon_0} \int_x^\infty \frac{x\ dx}{(x^2 + a^2/2)^{\frac{3}{2}}}.$$

This is a standard integral which is evaluated in integral tables.

$$U(x) = \frac{e^2}{4\pi\epsilon_0} \left[\frac{1}{(x^2 + a^2/2)^{\frac{1}{2}}} \right]_x^\infty$$

$$= -\frac{e^2}{4\pi\epsilon_0(x^2 + a^2/2)^{\frac{1}{2}}}. \tag{4}$$

Note that, when $x \gg a/2$, $U(x) \simeq -e^2/4\pi\epsilon_0 x$, which is in agreement with (2), apart from the factor $\frac{1}{4}$, which results from the approximation introduced by considering only the four nearest-neighbor ions in (3).

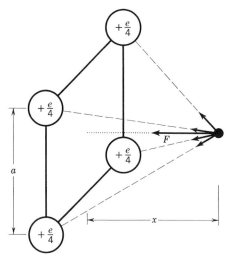

FIG. 2.   Total force on an electron very near the surface.

The shape of the potential-energy barrier determined by (4) is plotted in Fig. 3 for the case of an interatomic spacing of 1 Å.   In this diagram, the potential energy has been chosen equal to zero at infinity for convenience.   One particularly important feature seen in Fig. 3 is that the major increase in potential energy of the barrier occurs within a very few interatomic distances from the surface.

The total height of the potential-energy barrier for electrons is determined from (4) by setting $x = 0$.   Accordingly, the magnitude of the surface barrier V_B in electron volts is

$$V_B = \frac{e}{2^{\frac{3}{2}}\pi\epsilon_0} \frac{1}{a}. \tag{5}$$

Note that V_B is inversely proportional to the interatomic spacing.   This is in good agreement with experimental results for the alkali metals as shown in Fig. 4.   The slope of the predicted and observed lines in the figure differs somewhat, because of the simplifications introduced in the

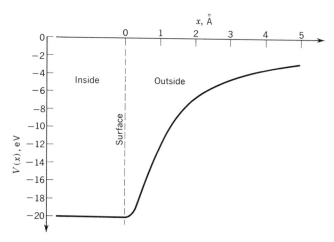

FIG. 3. Potential-energy barrier near the surface of a simple cubic structure having an interatomic spacing of 1 Å.

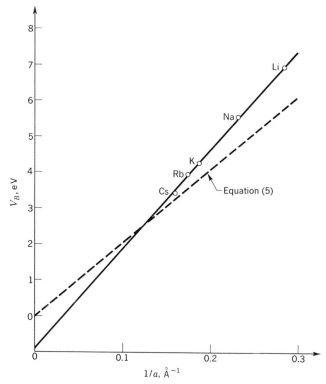

FIG. 4. Relationship between the height of the potential-energy barrier and the reciprocal of the interatomic spacing for the alkali metals compared with the simple theory.

analysis.   The comparisons in Fig. 4, however, show that the major
features of the surface-potential barrier can be understood in terms of the
attraction for the escaping electron by the ions in the metal surface it has
left behind.

Since the potential barrier depends upon the interatomic spacing of ions
in the surface layer, crystals have different work functions in different
crystallographic directions.   A polycrystalline metal, in which individual crystallites present different faces to the surface, has a hetero-
geneous work function which varies from grain to grain.   Those regions

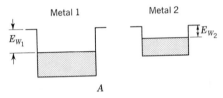

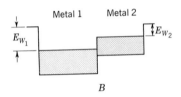

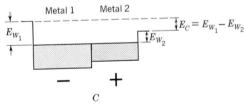

FIG. 5.   Potential-energy barriers of two dissimilar metals.   (A) Before the two
metals are joined; (B) immediately after contact; (C) after equilibrium is attained.

which have smaller work functions are more copious electron emitters, so
that the emission is not homogeneous over the entire surface.

Contact potential.   When two dissimilar conductors are brought
together, a difference of potential is produced between them.   This
contact potential is a direct result of the difference in the work functions of
the materials.   Consider the two metals shown in Fig. 5*A*, where the
potential barriers are indicated by simple vertical lines because the
distances of interest are much larger than interatomic dimensions.
When the metals are brought into contact, the potential-energy barrier
between them disappears (Fig. 5*B*).   Electrons from one metal can now

spill over into the other metal because it has unoccupied states at lower energies. This flow of electrons charges metal 1 negatively and metal 2 positively, thereby building up a retarding force which tends to reduce the electron flow. Equilibrium occurs (Fig. 5C) when the potential difference reduces the flow to zero, that is, when the Fermi levels of the two metals finally line up. As is apparent in Fig. 5, the contact-potential difference between the two metals is the difference in their work functions, or

$$E_C = E_{W_1} - E_{W_2}. \tag{6}$$

The magnitude of the contact potential E_C for combinations of most metals is usually of the order of a few tenths of a volt.

Thermionic emission

At temperatures above absolute zero the Fermi distribution predicts that some of the electrons in a metal have energies greater than that of the surface barrier (Fig. 1). Because the Fermi function is temperature-dependent, the number of electrons with sufficient energy to surmount the barrier increases as the temperature increases, so that more of them can escape. In order to determine the magnitude of this *thermionic emission*, it is necessary to calculate the number of such electrons at any temperature. Only those electrons headed toward the surface with sufficient energy to escape need be considered.

Richardson-Dushman equation. The electrons in a solid are in random motion in all directions. To determine the fraction which is headed in one direction with sufficient energy to overcome the surface barrier, start with the energy distribution of the electrons in a metal already derived in Chap. 7, Eq. (21). The number of electrons per cubic meter with energy between E and $E + dE$ is

$$N(E)\,dE = \frac{8\pi m \sqrt{2mE}}{h^3} \frac{dE}{e^{(E-E_0)/kT} + 1}. \tag{7}$$

It is convenient to convert $N(E)\,dE$ to a speed distribution. This is done by expressing the electron's energy in terms of its speed S, using

$$E = \tfrac{1}{2}mS^2. \tag{8}$$

Substituting (8) in (7), $N(E)\,dE$ becomes $N(S)\,dS$,

$$N(S)\,dS = \frac{8\pi m^3}{h^3} \frac{S^2\,dS}{e^{(E-E_0)/kT} + 1} \tag{9}$$

where the meaning of $N(S)\,dS$ is clearly the number of electrons per cubic meter having speeds between S and $S + dS$. Equation (9) can now be converted to the velocity-distribution function $N(v)\,dv$. This is the number of electrons per cubic meter with velocity components between v_x and $v_x + dv_x$, v_y and $v_y + dv_y$, and v_z and $v_z + dv_z$, where

$$S^2 = v_x^2 + v_y^2 + v_z^2 \qquad (10)$$

and v_x, v_y, and v_z are the components of the velocity vector. The velocity distribution takes into account the direction of each electron's motion, in addition to its energy or speed. Referring to Fig. 6, the speed distribution in (9) refers to all the electrons in the spherical shell bounded by S and $S + dS$. The velocity distribution, on the other hand, is concerned only with the electrons contained in the volume element

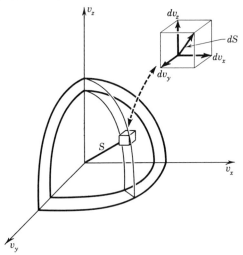

FIG. 6

$dv_x \, dv_y \, dv_z$. Therefore the velocity distribution is simply the ratio of these volumes times the speed distribution, or

$$N(v) \, dv = N(S) \, dS \, \frac{dv_x \, dv_y \, dv_z}{4\pi S^2 \, dS}$$

$$= \frac{2m^3 \, dv_x \, dv_y \, dv_z}{h^3 e^{(E-E_0)/kT} + 1} \qquad (11)$$

where
$$E = \tfrac{1}{2} m(v_x^2 + v_y^2 + v_z^2). \qquad (12)$$

The electron-velocity distribution can now be used to determine the number of electrons which will strike a unit area of the surface each second. If the surface is a plane perpendicular to the x axis, all electrons in a volume of size $v_x \times$ (unit area) will strike a unit area of the surface in one second if they have velocity components between v_x and $v_x + dv_x$. This is true quite independently of the values of the y and z velocity components. Thus the number striking a unit area of the wall per second, with energy sufficient to escape, is equal to this volume times the electron-velocity density $N(v) \, dv$, suitably integrated over the range of electron energies; that is, $N(v) \, dv$ is multiplied by the volume $v_x \times$ (unit area) and summed over all v_y and v_z values. The integration over v_x extends only from a velocity v_{0x} sufficient to surmount the barrier, where

$$\tfrac{1}{2}mv_{0x}^2 = eV_B. \tag{13}$$

Finally, multiplying by the electronic charge, the emitted current density J is

$$J = \frac{2em^3}{h^3} \int_{v_x = v_{0x}}^{\infty} \int_{v_y = -\infty}^{\infty} \int_{v_z = -\infty}^{\infty} \frac{v_x \, dv_x \, dv_y \, dv_z}{e^{(E-E_0)/kT} + 1} \tag{14}$$

where (12) is substituted for E in the exponent before integrating. Only electrons with energies greater than E_0 can be emitted, so that it is a good approximation to neglect the unity in the denominator of the integrand. Equation (14) then reduces to three simple integrations:

$$J = \frac{2em^3}{h^3} e^{E_0/kT} \left[\int_{v_{0x}}^{\infty} v_x e^{-mv_x^2/2kT} \, dv_x \right]\left[\int_{-\infty}^{\infty} e^{-mv_y^2/2kT} \, dv_y \right]$$
$$\left[\int_{-\infty}^{\infty} e^{-mv_z^2/2kT} \, dv_z \right]. \tag{15}$$

The last two brackets are standard integrals which each result in $(2\pi kT/m)^{1/2}$. Substituting $y = v_x^2$ in the first, this integral becomes

$$\frac{1}{2} \int_{v_{0x}}^{\infty} e^{-my/2kT} \, dy = -\frac{kT}{m} [e^{-mv_x^2/2kT}]_{v_{0x}}^{\infty} = \frac{kT}{m} e^{-mv_{0x}^2/2kT}.$$

Therefore (15) becomes

$$J = \frac{2em^3}{h^3} e^{E_0/kT} \left(\frac{kT}{m} e^{-V_B/kT}\right)\left(\frac{2\pi kT}{m}\right) \tag{16}$$

and, since $E_W = V_B - E_0$, it can be further simplified to

$$J = \frac{4\pi mek^2}{h^3} T^2 e^{-E_W/kT} \tag{17}$$

and

$$J = A_0 T^2 e^{-E_W/kT}. \tag{18}$$

The quantity A_0 is a universal constant which has the value

$$A_0 = 1.2 \times 10^6 \text{ A/m}^2 \text{ deg K}^2.$$

Thermionic-emission constants. Equation (18) is the *Richardson-Dushman equation* for thermionic emission. The emission current is exponentially dependent upon the work function and the inverse absolute temperature, which means that the variation of emission with these quantities is very rapid. The emitted current is a property of the material because the work function is determined by the interatomic spacing. In fact, thermionic-emission measurements can be used to determine the work function of a substance. Rearranging (18) and taking the natural logarithm of both sides results in

$$\ln \frac{J}{T^2} = \ln A_0 - \frac{E_W}{k} \frac{1}{T}. \tag{19}$$

A plot of $\ln (J/T^2)$ against $1/T$ is a straight line having a negative slope E_W/k and an intercept $\ln A_0$. An example of typical experimental data is shown in Fig. 7. The results are in good agreement with (19) and constitute experimental verification of the analysis.

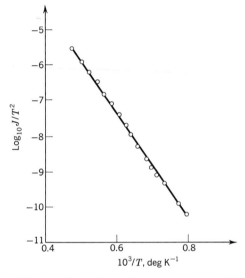

FIG. 7. Variation of thermionic emission from uranium carbide as a function of temperature plotted according to the Richardson-Dushman equation. (*After G. A. Haas and J. T. Jensen, Jr.*)

A list of experimentally determined values of A_0 and E_W for several metals is given in Table 1. Although the work functions of these metals are all of the order of a few electron volts, the variation between different materials is quite significant because of the exponential dependence of emission on work function. For example, the electronic emission from tungsten is a factor of 10^5 greater than that from nickel at a temperature of 1000 deg K, even though their work functions differ by only 0.09 eV. Other factors, such as the melting point, may be equally important for practical applications, however. In spite of its relatively large work function, tungsten is an effective thermionic emitter because it can withstand extreme temperatures, whereas, for example, cesium melts before reaching temperatures at which appreciable emission can occur.

Table 1
Thermionic-emission constants of some metals

Metal	A_0, 10^4 A/m² deg K²	E_W, eV
Ca	60	3.2
Cr	48	4.60
Cs	162	1.8
Fe	26	4.48
Mo	55	4.3
Ni	30	4.61
Pt	32	5.32
Ta	55	4.19
W	60	4.52

It should be noted that the experimental values of A_0 are considerably smaller than the predicted 1.2×10^6 A/m² deg K². This discrepancy is related to the difficulties associated with experimental measurements. One problem is the space charge built up by emitted electrons immediately outside the surface, which tends to inhibit emission of other electrons because of electrostatic repulsion. This aspect is considered in greater detail in the next section. Of major significance is the variation of work function with crystallographic direction since polycrystalline wires are generally used. Because the surface-potential barrier, and therefore the work function, depends on the interatomic spacing, different crystal faces have different work functions. The exponential dependence of emission on work function implies that a major fraction of the emission current comes from crystallites having a low work-function plane exposed to the surface. Thus the true emitting surface area may be only a fraction of the total geometrical surface area, and the value of A_0 determined from the measurements is smaller than the predicted one. Even in the case of single-crystal emitters, it is doubtful that the true surface area, considered on an atomic scale, is equal to the macroscopic area. Furthermore, (19) implies that the work function is independent of temperature. The change in work function with temperature due to thermal expansion of the structure and other thermal effects also influences the derived values of A_0.

To increase the thermionic emission from metals, it is necessary to lower the work function. A useful procedure is to adsorb a monolayer of a suitable foreign material on the surface. The case of thorium on tungsten, having considerable practical importance, is typical. The ionization potential of a thorium atom (4.0 eV) is less than the work function of tungsten (4.52 eV), so that, when the atom approaches very close to the surface, it loses an electron to the metal. The positively charged thorium ion is tightly bound to the surface by electrostatic forces, and the positive charge effectively lowers the work function because it aids electrons to escape. Such a thoriated tungsten cathode has an effective work function of about 2.6 eV compared with 4.5 eV for clean tungsten.

Schottky effect. Thermionic-emission measurements are influenced by the space charge of the emitted electrons. An accelerating electric field is usually applied to the cathode surface to circumvent this difficulty. The field causes emitted electrons to be removed from the region of the cathode and accelerated to a collector electrode, where they constitute the measured thermionic-emission current. The accelerating field also alters the potential-energy barrier at the cathode surface and increases emission. This increase and its dependence upon the electric field is called the *Schottky effect*.

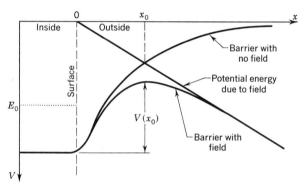

Fɪɢ. 8.   Reduction in the surface barrier due to an external electric field.

The accelerating electric field produces a constant force on the escaping electron of $e\mathcal{E}$, where $\mathcal{E}$ is the electric field, and this results in a contribution to the potential energy of $-e\mathcal{E}x$ at a distance x from the surface. The total potential energy of the escaping electron is then the sum of the surface-potential barrier as given by Eq. (4) and that due to the field $-e\mathcal{E}x$. As shown in Fig. 8, the new potential-energy curve has a maximum value, and the effective work function is reduced. The point x_0 is the distance at which the attractive force of the surface in the negative x direction is equal to the force due to the external field in the positive x direction. Equating these forces and using the image-force law of Eq. (1), since x_0 turns out to equal many interatomic spacings,

$$\frac{e^2}{16\pi\epsilon_0 x_0^2} = e\mathcal{E} \tag{20}$$

and

$$x_0 = \left(\frac{e}{16\pi\epsilon_0 \mathcal{E}}\right)^{\frac{1}{2}}. \tag{21}$$

The effective potential energy in the presence of the field is that given by (2) less that resulting from the field,

$$U(x) = -\frac{e^2}{16\pi\epsilon_0 x} - e\mathcal{E}x. \tag{22}$$

The height of the barrier at x_0 is found by substituting (21) into (22).

$$U(x_0) = -e\left(\frac{e\mathcal{E}}{4\pi\epsilon_0}\right)^{\frac{1}{2}}. \tag{23}$$

Since the work function is the difference between the barrier height and the position of the Fermi level, (23) represents the effective reduction in

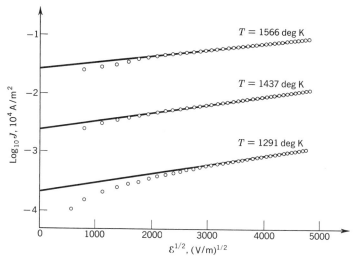

FIG. 9. Schottky plots of the thermionic emission from a uranium carbide cathode at three temperatures. (*After G. A. Haas and J. T. Jensen.*)

the work function due to the applied field. The thermionic-emission current density is then, according to (18),

$$J = A_0 T^2 e^{-(E_W - e\sqrt{e\mathcal{E}/4\pi\epsilon_0})/kT}. \tag{24}$$

Equation (24) shows that the observed thermionic emission at constant temperature increases with applied field and that a plot of $\ln(J/T^2)$ as a function of $\mathcal{E}^{\frac{1}{2}}$ is a straight line.

Experimental thermionic-emission currents from a uranium carbide cathode for several different temperatures are shown in Fig. 9. At the larger fields, the data points fall on a straight line having the Schottky slope given by (24), and extrapolation of this line to zero field gives the value of J, which is to be used in the Richardson-Dushman equation to determine E_W and A_0. At lower fields, the experimental points fall below the line predicted by (24), and this is a manifestation of space-charge effects. The good agreement between the predicted Schottky effect and experimental results constitutes experimental verification of the validity of the image-force law in determining the shape of the surface-potential barrier at distances equal to many interatomic separations. In fact, these results present the most direct experimental proof of such behavior.

Thermionic emission from semiconductors. An extreme example of the variation of work function with temperature occurs in the case of thermionic emission from semiconductors. As discussed in Chap. 8, the Fermi level in a semiconductor may shift a considerable distance toward the center of the forbidden-energy gap as the temperature increases, depending upon the concentration of donor and acceptor impurities. The work function, therefore, is a strong function of temperature because it is the difference between the height of the potential-energy barrier and the Fermi level. An interpretation of experimental results with the aid of (19) is then quite inappropriate.

The most important commercial thermionic cathode material is made from mixed crystals of barium, calcium, and strontium oxides and is an oxidic semiconductor. Although a detailed understanding of the oxide cathode is not yet complete, its general features may be understood on the basis of semiconductor physics. Oxygen vacancies are produced during the·processing of the cathode material which result in the oxide becoming a strongly n-type semiconductor. The Fermi level is located near the bottom of the conduction band and does not vary its position much with temperature because of the strong n-type extrinsic semiconduction; the work function which results is of the order of 1.0 eV. Furthermore, it is believed that surface states are occupied in such a way as to tilt the energy bands down slightly at the surface, which tends to enhance emission. Because of the small work function and the presence of surface states, oxide cathodes are very efficient emitters at temperatures below 1000 deg K, compared with temperatures of the order of 2000 deg K required for tungsten.

Field emission

Fowler-Nordheim equation. If the electric field applied to a metal surface is made very large, the work function is reduced considerably by the Schottky effect and the barrier also becomes thinner as illustrated in Fig. 10. When the barrier is very thin ($\simeq 100$ Å), appreciable numbers of electrons in the metal can penetrate the barrier by quantum-mechanical tunneling. Under these conditions, emitted current densities can be

quite large because all the surface-directed electrons in the metal participate, rather than just those occupying states in the tail of the Fermi
distribution as in thermionic emission. Therefore the phenomenon is
relatively independent of temperature, and because the barrier width is
determined by the electric field, the effect is termed *field emission.*

The probability that an electron can tunnel through a potential barrier
has been shown in Chap. 6 to be an exponential function of the width
of the barrier. Consider the triangle in Fig. 10, bounded by $x = 0$, E_W,

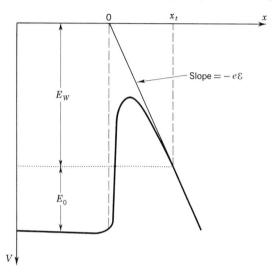

Fig. 10. Shape of the surface-potential barrier for very strong applied fields.

and the potential energy due to the field. The tunneling distance x_t at
the energy of the Fermi level can be written

$$x_t = \frac{E_W}{e\mathcal{E}} \tag{25}$$

since the slope of the field line is $-e\mathcal{E}$. Thus the field-emission current
is expected to be proportional to

$$e^{-E_W/e\mathcal{E}}. \tag{26}$$

The barrier thickness at the energy position of the Fermi level is most
important since the electron density is largest at this point. A more
detailed analysis takes into account the height of the barrier and its
detailed shape, in addition to its width. The tunneling probability is
summed over the electron-energy distribution considering only the surface-
directed electrons as in thermionic emission. The result obtained for a

slightly idealized case in which the barrier shape is assumed to be triangular is

$$J = \frac{e}{2\pi h V_B} \left(\frac{E_0}{E_W}\right)^{\frac{1}{2}} \mathcal{E}^2 e^{-4K(E_W)^{\frac{3}{2}}/3\mathcal{E}} \tag{27}$$

where
$$K^2 = \frac{8\pi^2 m}{h^2}.$$

Equation (27), known as the *Fowler-Nordheim equation* for field emission, is in good agreement with experiment, considering the experimental difficulties of realizing a smooth surface such that the externally applied field is not enhanced at minute projections. Field emission is one of the historically earliest pieces of direct evidence for quantum-mechanical tunneling by electrons.

In mks units commonly used, (27) becomes

$$J = A_F \mathcal{E}^2 e^{-B/\mathcal{E}} \qquad A/m^2 \tag{28}$$

where
$$A_F = \frac{6.2 \times 10^{-2}}{V_B} \left(\frac{E_0}{E_W}\right)^{\frac{1}{2}}$$

and
$$B = 6.8 \times 10^9 \; E_W^{\frac{3}{2}}.$$

$\mathcal{E}$ is expressed in volts per meter, and the energies are in electron volts. According to (28), detectable field-emission currents from tungsten, for example, require electric fields of the order of 10^9 V/m.

Experimentally measured field-emission currents from a single-crystal tungsten emitter are plotted in Fig. 11. The observed straight lines constitute experimental verification of the Fowler-Nordheim equation, and the work function can be obtained from the slope of the lines. By using the experimental procedures described in the next section, it is possible to determine the emission current from each crystal face separately, and hence measure the work function of that face alone.

Field-electron microscope. High fields can be obtained at the surface of an emitter if it is made into a very sharply pointed cathode. This construction is the basis of the *field-electron microscope*. A tungsten point, which can be electrolytically etched to a radius of the order of several hundred angstrom units, is sealed into one end of a glass envelope. The other end contains a fluorescent screen and a ring-shaped anode (Fig. 12). An applied potential of several thousand volts between the tungsten point and the anode is sufficient to cause high-field emission from the cathode because the electric field concentrates at the sharp point. Emitted electrons travel in straight lines to the fluorescent screen, where they produce a greatly magnified picture of the emitting surface. The magnification equals the ratio of the tube length to the cathode radius, r_2/r_1, and can easily be made to approach values of the order of 10^6. The picture observed on the screen represents the emission pattern of

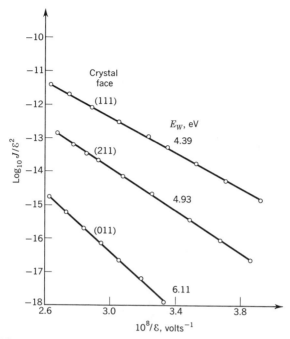

FIG. 11. Field emission as a function of applied field for different crystal faces on a tungsten surface. (*After E. W. Müller.*)

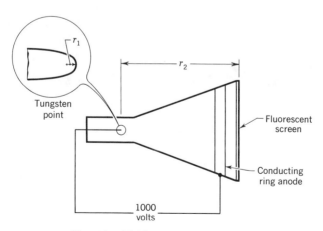

FIG. 12. Field-electron microscope.

the tungsten point produced by differences in work function over the single-crystal surface.   A typical pattern is shown in Fig. 13.   The work function depends on interatomic spacing, and the field-emission current varies rapidly with the work function, so that patterns related to the crystal structure are observed.   The symmetry of the tungsten crystal structure is clearly evident in the figure.

Field-ion microscope.   Even more useful magnifications than those of the field-electron microscope are obtained by reversing the potential

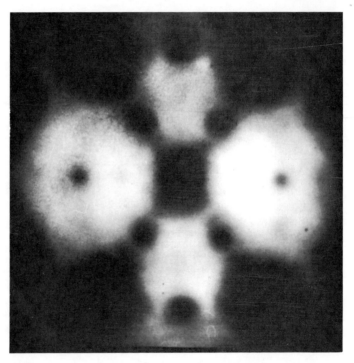

Fig. 13.   Field-electron micrograph of a tungsten tip.   (*Courtesy of E. W. Müller.*)

applied between the point and ring anode.   Residual gas atoms in the tube (usually purposely introduced) which come near to the surface of the strongly positive point are ionized by the field.   This takes place through a similar tunneling mechanism as discussed above, except that the electron tunnels from the atom to the metal under the action of the field.   The positive ions are accelerated to the screen and produce a picture of the metal surface.   The pictures observed by this *field-ion micro-scope* are capable of much greater resolution than those from the field-electron microscope because of the smaller quantum-mechanical wavelength of the heavy ions compared with that of the electron.   In the field-ion micrograph of a tungsten surface shown in Fig. 14, it is possible to distinguish spots associated with surface atoms in the structure. These patterns can be correlated with the known crystal structure.

Tunneling in insulators.   It is possible to induce field emission into any insulator which can withstand the high electric fields required.

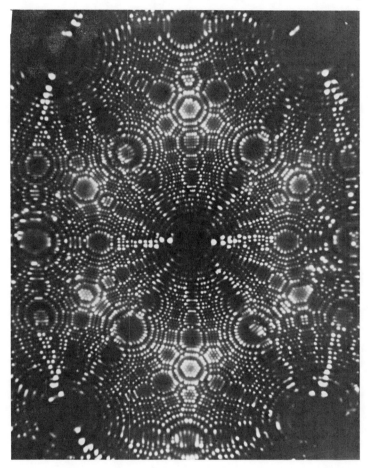

Fig. 14.   Field-ion micrograph of a tungsten tip.   (*Courtesy of E. W. Müller.*)

Consider the arrangement (Fig. 15*A*) in which a very thin insulating film is placed between two metals.   Such an arrangement can be produced by slightly oxidizing an aluminum surface to form the insulating layer and then vapor-depositing a metal electrode on top of the oxide.   Aluminum oxide is a good insulator and forms a relatively homogeneous film.   If the insulating film is sufficiently thin, $\simeq$10 to 100 Å, tunneling currents flow between the two metals in each direction.   In the absence of an applied

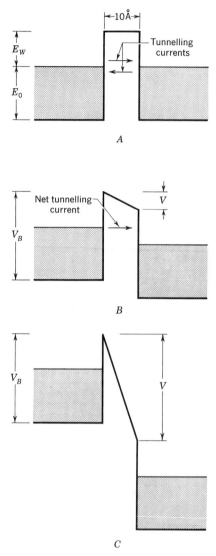

FIG. 15.   Tunneling currents through an insulating film between two metals for (*A*) no applied potential; (*B*) small applied potential; (*C*) large applied potential.

potential between the metals, their Fermi levels match, as previously discussed in explaining contact potential. If a potential is applied between the metals (Fig. 15*B*), more electrons will tunnel from the negative metal to the positive metal than in the reverse direction. The net tunneling current is proportional to the tunneling probability, the number of surface-directed electrons in the negative metal, and to the number of available states in the positive metal. As Fig. 15*B* diagrammatically indicates, only the electrons at the top of the Fermi distribution in the negative metal can find states available in the positive metal. Therefore the tunneling current is smaller than that given by (27). For small applied voltages, the tunneling current is a linear function of the applied potential, so that the film appears to have a constant resistance. The film itself does not have a conductivity, however, since it is an

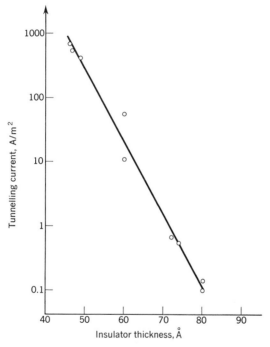

Fig. 16. Net tunneling current through aluminum oxide films as a function of thickness. (*After J. C. Fisher and I. Giaever.*)

insulator and is unaffected by the field. When the applied potential is greater than V_B (Fig. 15*C*), sufficient states are available in the positive metal so that the tunneling current is given by the Fowler-Nordheim equation. Note that in this case the barrier shape is essentially the same as in Fig. 10. This region is difficult to attain experimentally since the very high fields [$\mathcal{E} \simeq V_B/x_t = 20/(10 \times 10^{-10}) = 2 \times 10^{10}$ V/m] can cause disruptive physical changes in the insulator. The exponential dependence of the tunneling probability on the tunneling distance can be examined experimentally by measuring the tunneling current for different thicknesses of the insulating film. Data for aluminum oxide films shown in Fig. 16 clearly confirm the quantum-mechanical predictions. Note that a similar arrangement can be used to investigate electron tunneling in the superconducting state as discussed in Chap. 7.

Secondary emission

Secondary yield. Energetic electrons bombarding a solid surface can transfer energy to electrons in the solid through collisions. Such excited electrons, whose energy is raised sufficiently to surmount the surface-potential barrier, can escape (if they are surface-directed), and they are called *secondary electrons*. The number of secondary electrons produced per incident primary electron is called the secondary-emission ratio δ, or the *secondary yield*. The secondary yield is a function of the primary-electron energy since, for low values, very few electrons can be given sufficient energy to escape, while very energetic primary electrons can excite many electrons in the material. Therefore, at low primary electron energies, δ increases with energy. Very energetic primary electrons excite many electrons, but they are produced deep inside the solid because of the greater penetration of the energetic primary electrons. Secondary electrons produced in this case stand a greater chance of being reabsorbed before reaching the surface, so that δ decreases. The *secondary-yield curve*, a plot of δ as a function of primary energy, has a broad maximum of the order of 1 or 2 for primary electron energies of several hundred volts in metals. Some special composite surfaces have secondary yields as high as 10 to 50.

No completely satisfactory quantitative theory of secondary-electron emission has been developed as yet, although this is primarily due to mathematical complexities rather than a lack of understanding of the important physical principles. A useful description of the phenomena can be derived if it is assumed that the rate of production of secondary electrons is proportional to the rate of energy loss of the primary electrons and that the excited electrons are reabsorbed exponentially in traveling to the surface. Thus the contribution to the observed secondary-emission current I_s (Fig. 17), from a region dx located a distance x below the surface due to a primary current I_p, is

$$dI_s = -KI_p \frac{dE}{dx} e^{-\alpha x}\, dx \qquad (29)$$

where K is a constant
    E is kinetic energy of primary electron at x
    α is absorption coefficient of secondary electrons.
The minus sign is used because dE/dx is an inherently negative quantity since the primary-electron energy decreases as the electron penetrates

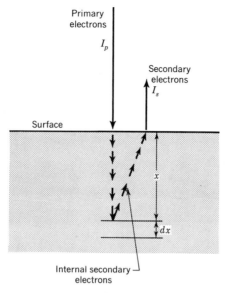

Fɪɢ. 17

into the solid. The energy of a primary electron E after penetrating a depth x is given by Whiddington's law,

$$E^2 = E_p^2 - bx \tag{30}$$

where E_p is the initial primary energy and b is a constant roughly proportional to the density of the solid. Differentiating (30),

$$\frac{dE}{dx} = \frac{-b}{2(E_p^2 - bx)^{\frac{1}{2}}}. \tag{31}$$

The maximum depth of penetration by a primary electron x_m is determined from (30) by putting $E = 0$.

$$x_m = \frac{E_p^2}{b}. \tag{32}$$

Inserting (31) into (29) and integrating from the surface to x_m, the secondary yield is

$$\delta = \frac{I_s}{I_p} = \frac{Kb}{2} \int_0^{E_p^2/b} (E_p^2 - bx)^{-\frac{1}{2}} e^{-\alpha x} \, dx. \tag{33}$$

Substituting the change of variable $y = \sqrt{(\alpha/b)(E_p^2 - bx)}$ in (33),

$$\delta = K \left(\frac{b}{\alpha}\right)^{\frac{1}{2}} e^{-(\alpha/b)E_p^2} \int_0^{\sqrt{a/bE_p}} e^{y^2} \, dy. \tag{34}$$

Finally, introducing $r = (\alpha/b)^{\frac{1}{2}}E_p$ in (34),

$$\delta = K\left(\frac{b}{\alpha}\right)^{\frac{1}{2}} e^{-r^2} \int_0^r e^{y^2}\, dy$$

$$= K\left(\frac{b}{\alpha}\right)^{\frac{1}{2}} F(r) \qquad (35)$$

where
$$F(r) = e^{-r^2} \int_0^r e^{y^2}\, dy.$$

The function $F(r)$ has a maximum value of 0.54 for a value of $r = 0.92$, so that, from (35),

$$\frac{\delta}{\delta_m} = \frac{F(r)}{F(0.92)} \qquad \text{and} \qquad \left(\frac{\alpha}{b}\right)^{\frac{1}{2}} = \frac{0.92}{(E_p)_m}. \qquad (36)$$

Finally,
$$\frac{\delta}{\delta_m} = 1.85F\left[0.92\,\frac{E_p}{(E_p)_m}\right] \qquad (37)$$

where δ_m is the maximum value of δ, which occurs when the incident primary energy is $(E_p)_m$.

Universal yield curve. Examination of Eq. (37) suggests that a *universal secondary-yield curve* exists for all materials. A plot of δ/δ_m as a

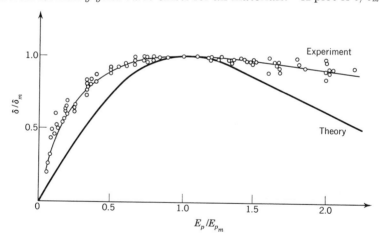

Fig. 18. Experimental and simple theoretical universal secondary-yield curves. The circles represent experimental data for aluminum, barium, beryllium, cesium, copper, gold, lithium, magnesium, molybdenum, platinum, silver, and titanium.

function of $E_p/(E_p)_m$ from (37) is compared with experimental data for several metals in Fig. 18. Agreement between (37) and experimental data is not complete because of the several simplifications introduced into the analysis. Whiddington's law is an approximation, and also it is likely that energetic secondary electrons can in turn excite other electrons.

This effect is most important at high primary energies where the deviations of (37) are most pronounced.

The magnitude of secondary-electron emission is determined by properties of the material which limit the values of $(E_p)_m$ and δ_m through the parameters b and α. Thus variations in the secondary yield are expected (Fig. 19), although it turns out that most metal surfaces are similar and that alloys, compounds, and some insulators and semiconductors tend to have high secondary yields. This difference is attributed

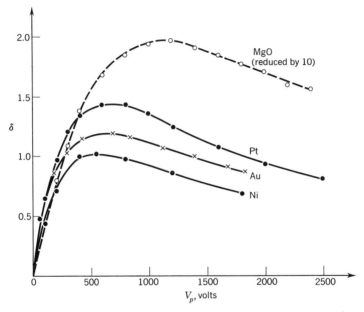

Fig. 19. Secondary-yield curves for nickel, gold, platinum, and magnesium oxide. The magnesium oxide data have been lowered by a factor of 10 in order to fit them in the figure.

to the fact that secondary electrons are emitted in a direction opposite to that of the incident primary electrons. In a collision, a primary electron tends to transfer most of its momentum in the forward direction. Therefore, in order for the secondary to be emitted, this momentum must be taken up by the crystal structure of the solid. The electrons in a metal, however, are very nearly free, and the interaction with the structure is small. The electrons in nonmetals, on the other hand, tend to interact with the ions more strongly, so that momentum transfer can be more easily effected and the secondary yield is greater.

Secondary-electron energies. The energy distribution of secondary electrons is shown in Fig. 20 for solid mercury. This distribution is typical of metals. Two groups of secondary electrons are evident.

The energy of one group equals that of the primary electrons, and the other group has an energy of a few electron volts. The first group is made up of reflected primary electrons, while the latter is the true secondary current. Some secondary electrons of intermediate energies also are observed. These may be either more energetic secondary electrons or else primary electrons that have penetrated the surface and suffered a collision which has caused them to be readmitted again. A noteworthy feature of Fig. 20 is that the true secondary electrons have energies of several electron volts. This means that the secondary current is relatively insensitive to variations in the work function; that is, the excited electrons are so energetic that their ability to escape is not

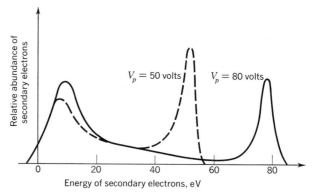

Fig. 20. Energy distribution of secondary electrons from solid mercury for two values of primary energy.

strongly controlled by the barrier at the surface. This is one reason why the secondary yields from most metals are so similar. For example, the lowering of the work function of tungsten by a monolayer of thorium, which may increase thermionic emission by a factor of 10^4, is found to increase the secondary yield by only 50 per cent.

Surface films thicker than one atomic layer can influence the secondary current, however. Most secondary electrons are produced very near the surface, so that the secondary yield is more characteristic of these layers than the bulk target material. Insulating layers can increase the proportion of reflected primary electrons because of the accumulation of surface charges, and thus increase the apparent secondary yield, particularly at primary-electron energies below $(E_p)_m$. A striking example of this behavior is the *Malter effect*, in which the surface charging is so great that a strong electric field is produced across the insulating film. This field may be sufficient to cause high-field emission from the metal.

Secondary multipliers. Secondary emission is a very useful way of amplifying weak electron currents—for example, the feeble photo-electric-emission currents discussed in the next section. A series of electrodes, each treated to have a surface of high secondary yield, are arranged as in Fig. 21. The electrodes, called *dynodes*, are maintained at successively higher potentials and shaped to direct secondary electrons from one dynode to the next. Electrons emitted from the photocathode are directed to the first dynode, where they produce many secondary electrons. These strike the second dynode, where they are again multiplied by secondary-electron emission. This process may be repeated in 10 to 12 steps, producing amplifications of the order of 10^6

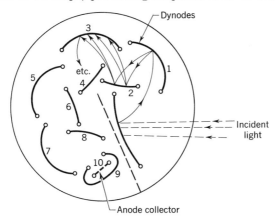

FIG. 21. Structure of a 10-stage photomultiplier which uses secondary emission for amplification.

times. Further multiplications are not generally useful because of the statistical effects (at a gain of 10^6 it is already possible to detect the emission of one photoelectron from the photocathode) and because the room-temperature thermionic emission from the photocathode is similarly multiplied.

Photoelectric emission

Einstein photoelectric equation. In many respects, photoelectric emission is analogous to secondary emission except that the primary electrons are replaced by a stream of incident photons. When a photon strikes an electron in the solid, the photon's energy $h\nu$ is converted to kinetic energy of the electron. If the excited electron has sufficient energy to surmount the surface-potential barrier, it can be emitted and detected as a *photoelectron*. The excited electron may be emitted with very little energy, or it may not be able to escape at all, depending upon its initial energy and the magnitude of $h\nu$, as illustrated in Fig. 22. The maximum energy a photoelectron can have, E_m, is given by

$$E_m = h\nu - E_W \tag{38}$$

where h is Planck's constant, and ν is the frequency of the incident radiation. This expression is known as *Einstein's photoelectric equation*. It was discovered shortly after the quantum nature of light was suggested and represents one of the earliest successes of the quantum theory.

According to (38), the maximum energy of a photoelectron is a linear function of the light frequency and is independent of light intensity. These are both difficult experimental observations to explain on a classical basis. The work function of photoemitters can be measured experimentally with (38) by determining the intercept on a plot of E_m

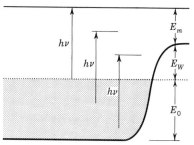

FIG. 22. Energies of electrons excited by absorption of an optical photon. E_m is the maximum energy photoelectrons can have.

versus ν. However, E_m is a rather difficult quantity to measure accurately, so that the procedure discussed in the next section is more satisfactory.

Fowler theory. Photoelectric emission is a function of temperature because, as is evident in Fig. 22, electrons occupying the highest energy states allowed by the Fermi distribution are the ones that predominate in emission. The number of these electrons is particularly influenced by temperature variations. The photoemission from a metal surface can be computed for frequencies near the *threshold* frequency ν_0 determined from (38) by setting $E_m = 0$. The least energetic photon which will cause photoelectric emission from a surface having a work function E_W has an energy $h\nu_0$. Assume that the photocurrent is proportional to the density of electrons in the metal which have surface-directed velocities sufficiently great to surmount the surface barrier after they have absorbed a photon. By analogy to (13), electrons can escape if their velocity component in the x direction is larger than

$$\tfrac{1}{2}mv_{0x}^2 = eV_B - h\nu. \tag{39}$$

The photocurrent is then found by integrating (14).

$$J = Ke \int_{v_x=v_{0x}}^{\infty} \int_{v_y=-\infty}^{\infty} \int_{v_z=-\infty}^{\infty} v_x N(v) \, dv \tag{40}$$

where K is a constant, and $N(v) \, dv$ is the electron-velocity distribution given by (11). The procedure previously invoked to integrate (14) for thermionic emission cannot be used here. This is so because electrons with energies close to the Fermi energy are most important and it is not permissible to drop unity in the denominator of the Fermi function. Equation (40) is evaluated by integrating over v_y and v_z and then expanding the integrand in the first integral in a series and integrating term by term. The expansion is valid for values of v near v_0. The result is an infinite series, the details of which are not of direct interest here. It turns out that the result has the form

$$J = AT^2 f\left(\frac{h\nu - E_W}{kT}\right) \tag{41}$$

an expression first derived by Fowler. Here A is a constant and the function $f[(h\nu - E_W)/kT]$ represents an infinite series which must be evaluated numerically. Note, however, that the function is the same for all metals and temperatures and is in the nature of a universal function similar to the previously discussed case of secondary emission. Equation (41) is in very good agreement with experimental measurements. Unfortunately, it is not possible to derive a satisfactory theoretical value for the constant A because of the complexities of the problem.

Fowler plots. The Fowler equation is useful in interpreting experimental photoelectric-emission data. Equation (41) can be put in the form

$$\ln \frac{J}{T^2} = B + F(x)$$

where $F(x)$ is the logarithm of $f(x)$, and $x = (h\nu - E_W)/kT$. $F(x)$ can be evaluated numerically and plotted and compared with experimental data which are similarly prepared by plotting J/T^2 as a function of $h\nu/kT$. This is known as a *Fowler plot*. Experimental Fowler plots have the same shape as $F(x)$. A vertical shift of the data to match $F(x)$ permits the estimation of the constant B, while a horizontal shift determines the value of E_W/kT. Because of the very satisfactory agreement of the experimental data with the Fowler theory (Fig. 23), this is the most satisfactory method for determining work functions of metals. Reasonably good agreement between work functions determined photoelectrically and from thermionic emission is usually observed, as indicated

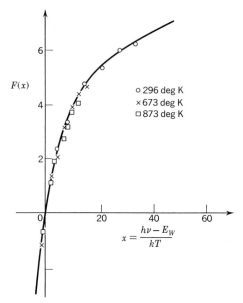

FIG. 23. Fowler plot of the photoelectric emission from silver at three temperatures.

in Table 2. Some discrepancies can be expected because of the uncertainties in measuring E_W thermionically, as discussed previously, and the fact that the thermionic-emission values are obtained at high temperatures where appreciable emission occurs. The Fowler plot satisfactorily accounts for the temperature effect in photoelectric emission, so that the work-function values derived from it may be interpreted as the true values at the absolute zero of temperature.

Table 2
Comparison of work functions determined photoelectrically
and thermionically

Metal	E_W, eV	
	Photoelectric	Thermionic
Ag	4.74	4.08
Au	4.90	4.42
Cs	1.90	1.81
Fe	4.48	4.72
Ni	4.01	4.61
Pt	6.30	5.32
Rh	4.57	4.58
Ta	4.05	4.19
W	4.58	4.52

Emission from insulators. The Fowler theory does not apply to photoelectric emission from insulators and semiconductors because the energy distribution of electrons in these materials is, in general, not degenerate. The photoelectric efficiency of certain especially prepared semiconductor surfaces is two to three orders of magnitude larger than that of metal surfaces, which is of the order of 10^{-4} photoelectrons per incident photon. One reason for this is the high reflectivity of metals, and another is the difference in binding of the electrons, as previously discussed for secondary emission. The reasons for the high efficiency of most practical photoemitter surfaces are largely unknown, however, so that the useful chemical combinations must be derived empirically, and the resulting emission curves are very complicated.

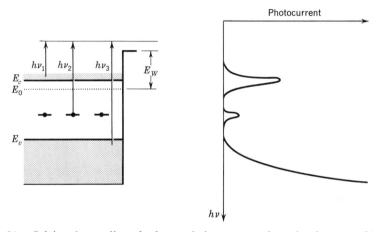

FIG. 24. Origin of complicated photoemission spectra from insulators and semiconductors.

As an example of the type of behavior that can be observed in semiconductors, consider the idealized situation in Fig. 24, where an n-type semiconductor is illuminated with monochromatic photons of several wavelengths. The semiconductor has occupied discrete levels in the forbidden-energy gap, as well as donor levels (not shown), which produce carriers in the conduction band. No photoemission is observed for very low-energy photons because insufficient energy is available for any electrons to escape. Photons with energies of the order of the energy difference between the height of the surface-potential barrier and the bottom of the conduction band may excite some conduction electrons to escape. Thus a photoemission threshold is observed at photon energies smaller than those corresponding to the work function. At somewhat larger photon energies, emission from the occupied states in the forbidden-energy gap is possible. Finally, at still larger photon energies, photoemission from the valence band can occur.   Consequently, complicated photoelectric-emission spectra can be expected from semiconductors and insulators.   The emission from occupied states other than those in the valence band is difficult to detect because of the smaller density of electrons in these levels, but it is possible to measure the effect in favorable cases.   Since the wavelength response of photoemission from semi-conductor surfaces depends on the energy-band structure, it is possible to prepare practical emitters designed for specific applications.   One surface of interest has a photoresponse similar to that of the human eye, so that the photocurrent measures the brightness of illumination as it appears to the eye.   Others can be prepared to respond only to infrared radiation, which is, of course, invisible to the eye.   A commercially

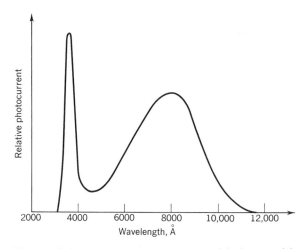

FIG. 25.   Photoemission spectrum from a commercial photosensitive surface.

important photoresponse is illustrated in Fig. 25.   Variations similar to Fig. 24 are apparent, but the origin in terms of the band structure of the solid is not well known.   Another useful application is a photocell which responds only to far-ultraviolet photons.   A *p*-type semiconductor having a wide forbidden gap has a threshold at much larger photon energies than that possible in any metal; that is, the energy difference between the valence band and the top of the surface-potential barrier can be made much greater than the work function of metals, and such a material exhibits photoemission only in the far ultraviolet.

Suggestions for supplementary reading

J. Millman and S. Seely, *Electronics*, 2d ed. (McGraw-Hill Book Company, Inc., New York, 1951).

Robert L. Sproull, *Modern physics* (John Wiley & Sons, Inc., New York, 1956).
D. A. Wright, *Semiconductors* (John Wiley & Sons, Inc., New York, 1950).

Suggestions for further reading

H. Bruining, *Physics and applications of secondary electron emission* (McGraw-Hill Book Company, Inc., New York, 1954).
K. R. Spangenberg, *Vacuum tubes* (McGraw-Hill Book Company, New York, 1948).
P. S. Wagener, *The oxide-coated cathode* (Chapman & Hall, Ltd., London, 1951).

Exercises

1. Using the data of Fig. 4, calculate the work functions of the alkali metals assuming one free electron per atom.

2. Compute the maximum Fermi energy E_0 for the idealized model used in connection with Fig. 3, assuming one free electron per atom. Explain the discrepancy. HINT: Consider the crystal structure of the alkali metals.

3. Derive Eq. (6) for the contact-potential difference between two metals by computing the thermionic-emission currents between the two metals and determining the condition for zero net current.

4. Determine the contact-potential difference between cesium and iron joined by tantalum and platinum (cesium-tantalum-platinum-iron). Can you generalize this result? (ANSWER: 2.68 eV.)

5. Calculate the thermionic emission from cesium and from tungsten at temperatures of 500, 1000, 1500, 2000, and 2500 deg K (ignore the melting point of cesium). Write down an expression for the ratio of emission from cesium to tungsten, and plot this over the same temperature range. Can you suggest a physical reason why the ratio goes this way with temperature?

6. Compute the thermionic-emission current density from germanium using data from Fig. 7 of Chap. 8 ($N_D = 10^{21}$ m^{-3}), assuming $A_0 = 5 \times 10^5$ A/m^2 deg K^2 and the energy difference between the bottom of the conduction band and infinity is 3 eV. Prepare a Richardson plot [ln (J/T^2) versus $1/T$] to see if there is any temperature region over which the work function can be determined.

7. Determine the point x_0 [Eq. (21)] for fields of 10^3, 10^4, and 10^5 V/m. Compare with interatomic distances. (ANSWER: $x_0 = 5.9 \times 10^3$, 1.9×10^3, 5.9×10^2 Å.)

8. What is the field strength required to make the barrier thickness 100 Å for a tungsten emitter? (ANSWER: $x_t = 4.5 \times 10^8$ V/m.)

9. Plot the penetration depth of primary electrons into nickel as a function of primary energy up to 1,000 V, assuming $b = 7 \times 10^{13}$ V^2/m.

10. Derive a universal secondary-yield curve, assuming that the rate of energy loss of primary electrons is constant ($dE/dx = -A$). Plot and compare with (37).

11. From the secondary-yield curve for gold in Fig. 19, determine a value for α, given that $b = 1.5 \times 10^{14}$ V^2/m. (ANSWER: 3.5×10^8 m^{-1}.)

12. Determine the threshold wavelengths for the first five metals listed in Table 2.

13. Calculate the maximum energies in electron volts of emitted photoelectrons from Ag, Cs, Ni, and Pt, when the surfaces are illuminated with 2,500 Å radiation. (ANSWER: 0.22, 3.06, 0.95 eV; for Pt there is no emission.)

14. By scaling Fig. 23, plot the relative photoemission from a tungsten surface at room temperature as a function of light wavelength. Repeat for a platinum surface.

15. Repeat Exercise 14 with the surfaces at 1000 deg K.

12. dielectric processes

So far in this book only the electrical properties of metals and semi-conductors have been considered in detail. As already shown, the conductivity in these materials results from the motion of nearly free electrons. There is a large class of materials, however, in which even the valence electrons are so tightly bound to atomic nuclei that electronic conductivity is virtually not possible. The energy-band model of an insulator is not unlike that of a typical semiconductor, except that the forbidden-energy region separating the conduction band from the valence band is much larger, of the order of several electron volts. Although insulator crystals may contain impurities, the prospective donor or acceptor states have energies that are sufficiently far removed from the band edges so that their contribution to conductivity at ordinary temperatures is negligibly small. As discussed elsewhere in this chapter, conductivity can occur by means of the diffusion of atoms, or more correctly ions, through the crystal. Ions have much lower mobilities than electrons, so that the Hall effect, which is proportional to the charge-carrier mobility, is very much smaller for such crystals than for most semiconductors.

Another way of describing an insulator is to note that the electrons are so tightly bound to the atoms that at ordinary temperatures they cannot be dislodged either by thermal vibrations or with ordinary electric fields. The negative and positive charges in each part of the crystal can be considered to be centered at the same point, and since no conductivity is possible, the localized charges remain that way essentially forever. When an electric field is applied to the crystal, the centers of the positive

338

charges are slightly displaced in the direction of the applied field and the centers of the negative charges are slightly displaced in the opposite direction. This produces local dipoles throughout the crystal, and the process of inducing such dipoles in the crystal is called polarization. The ratio of the induced dipole moment to the effective field is called the polarizability of the atom, and the dipole moment induced in a unit volume of a polarized insulator can be considered as the average of the dipole moments of all the atoms in that unit volume. It is possible that certain groups of atoms (complex ions or molecules) already possess permanent dipole moments. In crystals containing such atomic groups, an external field tends to orient the dipoles parallel to the field direction. In the absence of an external field, the dipoles are randomly oriented because of their thermal motion, so that the crystal has a zero net moment. The polarization of such *polar crystals* is strongly temperature-dependent since, even in the presence of an applied field, thermal motion tends to randomize the dipole orientations. On the other hand, the polarization of nonpolar crystals is independent of temperature since, in the absence of an external field, no dipoles exist in the crystal. The temperature dependence of polarization, therefore, can be used to distinguish polar insulators from nonpolar insulators.

Fundamental concepts

Electrostatic relations. The intensity of an electrostatic field $\mathcal{E}$ is most conveniently defined as the force per unit positive charge exerted on a test body placed in the field. Accordingly, an electrical force F acting on a test charge q results in a field intensity

$$\mathcal{E} = \frac{F}{q}.\tag{1}$$

In the mks system of units, F is measured in newtons (N) and q in coulombs (C), so that $\mathcal{E}$ has the units of newtons per coulomb (or volts per meter). According to Coulomb's law, the force between two point charges is directly proportional to the product of their charges q_1 and q_2 and inversely proportional to the square of their separation r, so that in empty space

$$F = \frac{q_1 q_2}{4\pi\epsilon_0 r^2}\tag{2}$$

where the constant ϵ_0 is called the *permittivity of empty space* and has the value of 8.85×10^{-12} F/m (farad/m or $C^2/N\text{-}m^2$). According to (1), the electric-field intensity in free space due to a single point charge q is then

$$\mathcal{E} = \frac{q}{4\pi\epsilon_0 r^2}\tag{3}$$

where r now denotes the distance from the location of the point charge to the point where $\mathcal{E}$ is measured.

When an electric field passes through a polarizable medium, the resulting polarization alters the effect of the field and it is convenient to define a new quantity, called the *dielectric displacement*, such that

$$D = \epsilon\mathcal{E} \tag{4}$$

where ϵ is called the *permittivity of the material*. In an isotropic homogeneous medium ϵ is a constant that is characteristic of the material. Corresponding to (2), the force on a charge q_1 is

$$F = \frac{q_1 q_2}{4\pi\epsilon r^2} \tag{5}$$

so that one effect of the medium is to reduce the force on the charge by the ratio ϵ_0/ϵ.

When a voltage V is applied to two parallel metal plates, a charge Q whose magnitude is proportional to the voltage develops on each plate according to

$$Q = CV. \tag{6}$$

Here C is called the capacitance of the parallel-plate capacitor and is determined by the area of the plates A and their separation l.

$$C = \frac{\epsilon A}{l} \tag{7}$$

for two plates separated by a material whose permittivity is ϵ. When the two plates are separated by vacuum (permittivity ϵ_0), then the capacitance becomes $\epsilon_0 A/l$. Insertion of a dielectric material for which $\epsilon > \epsilon_0$, therefore, increases the capacitance and decreases the potential difference required to maintain the same charge Q, according to (6). When all space between the two plates is filled by the dielectric material, the increase in the capacitance is measured by the *dielectric constant* of the material κ_e. This factor measures the ratio of the permittivity of the material to that of empty space, so that

$$\kappa_e = \frac{\epsilon}{\epsilon_0}. \tag{8}$$

Polarization density. Since the observed increase in the capacitance of a parallel-plate capacitor when a dielectric material is inserted between the plates is due to a decrease in the electric field inside the dielectric, this must be caused by the electric field of the dipoles which are induced inside the material. It is convenient, therefore, to define a *polarization density* P as the total dipole moment induced in a unit volume of the material. If the capacitance of the vacuum capacitor is denoted C_{vac}, then the new capacitance is $C = \kappa_e C_{\text{vac}}$ and the decreased field (per unit area) in the presence of a dielectric is $\mathcal{E} = (1/\kappa_e)\mathcal{E}_{\text{vac}}$. The polarization density is a measure of the change in the capacitance (or field), so that

$$P = (C - C_{\text{vac}})V = \left(\frac{\epsilon}{l} - \frac{\epsilon_0}{l}\right)V$$
$$= (\epsilon - \epsilon_0)\mathcal{E}$$
$$= (\kappa_e - 1)\epsilon_0\mathcal{E} \tag{9}$$

or, by rearranging the terms in (9),

$$\kappa_e = 1 + \frac{P}{\epsilon_0\mathcal{E}}$$
$$= 1 + \chi_e \tag{10}$$

where $\chi_e = P/\epsilon_0\mathcal{E}$ is called the electric *susceptibility*.

The polarization density is determined by three factors:

1. The *electronic polarizability* α_e produced by opposite displacements of negative electrons and positive nuclei within the same atoms

2. The *ionic polarizability* α_i produced by opposite displacements of positive and negative ions in the material

3. Contributions from the *permanent dipole moments* μ of complex ions or molecules whenever such permanent dipoles are present in the material; as shown in a later section, this contribution is $\mu^2/3kT$ per dipole.

It is possible to combine these factors analytically. Let the number of atoms or molecules per unit volume be N; then

$$P = N\left(\alpha_e + \alpha_i + \frac{\mu^2}{3kT}\right)\mathcal{E}. \tag{11}$$

The electric susceptibility is then obtained by substituting (11) in (10):

$$\chi_e = \kappa_e - 1 = \frac{P}{\epsilon_0\mathcal{E}} = \frac{N}{\epsilon_0}\left(\alpha_e + \alpha_i + \frac{\mu^2}{3kT}\right). \tag{12}$$

Note that the contribution to P from the permanent dipoles present in the material is temperature-dependent. This makes it possible to measure the dipole moments by observing the temperature dependence of the electric susceptibility. On the other hand, if no temperature dependence is observed, then it can be safely assumed that there are no permanent dipoles present.

So far the discussion has been quite general in that no restrictions have been placed on the nature of the dielectric material, that is, whether it is a gas, a liquid, or a crystalline substance. If attention is limited to crystals, it is necessary to take into account the influence, on an atom, of the internal field produced by the dipoles surrounding the atom, as well as the influence of the externally applied field. Consider the parallel-plate capacitor shown in Fig. 1. A simple way to determine the value of the electric field inside the dielectric is to insert a test charge into an imaginary cavity in the dielectric and to calculate the force acting on such a charge. When the cavity is needle-shaped and oriented with its axis parallel to the field direction and the ends are very small relative to its length (cavity a in Fig. 1), the test charge placed inside such a cavity is acted on by the average field $\mathcal{E}$, so that the force on a charge q is $q\mathcal{E}$. If the cavity has a large surface normal to the field direction, say, the disk-shaped cavity pictured in cross section by b in Fig. 1, then bound charges of density P appear on the two surfaces and the force acting on a test charge q inside such a cavity is due to the external

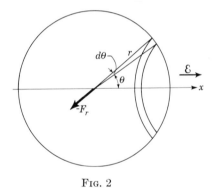

FIG. 1   FIG. 2

field and the polarization charges. The effect of these bound charges can be calculated similarly to (9), so that

$$q\mathcal{E}_{\text{internal}} = q\,\frac{D}{\epsilon_0} = q\mathcal{E} + q\,\frac{P}{\epsilon_0} \tag{13}$$

and after rearranging terms and dividing through by q,

$$D = \epsilon_0\mathcal{E} + P. \tag{14}$$

Equation (13) expresses the fundamental relation between the three quantities D, $\mathcal{E}$, and P, in any material.

The effect of the internal field on an atom is determined by considering the atom to be located in a spherical cavity. The force on a test charge inside such a spherical cavity (cavity c in Fig. 1) can be determined as follows: An enlarged view of the cavity is shown in Fig. 2. The charge density on a surface element dA of the sphere is equal to the normal component of the polarization times the surface element, that is, $P \cos \theta \, dA$. According to Coulomb's law, this charge element produces a force dF_r acting on the test charge q at the center of the sphere in the direction of r (Fig. 2).

$$dF_r = \frac{q_1 q_2}{4\pi\epsilon_0 r^2} = \frac{qP \cos \theta \, dA}{4\pi\epsilon_0 r^2}. \tag{15}$$

The effect of the charge on the whole surface of the sphere can be calculated by considering the spherical ring shown in Fig. 2, whose area

$$dA = 2\pi r \sin \theta \, r d\theta = 2\pi r^2 \sin \theta \, d\theta$$

and integrating over the whole sphere by integrating with respect to θ from 0 to π. Since it is desired to calculate the force in the direction of the electric field, $F_x = +F_r \cos \theta$, it can be calculated, with the aid of (15),

$$dF_x = + q \frac{P \cos \theta}{4\pi\epsilon_0 r^2} 2\pi r^2 \sin \theta \, d\theta \cos \theta$$

$$F_x = + \frac{qP}{2\epsilon_0} \int_0^\pi \cos^2 \theta \sin \theta \, d\theta. \tag{16}$$

This integral can be evaluated directly by making the substitution

$$z = \cos \theta \quad \text{and} \quad dz = -\sin \theta \, d\theta$$

so that

$$F_x = - \frac{qP}{2\epsilon_0} \int_{+1}^{-1} z^2 \, dz$$

$$= \frac{P}{3\epsilon_0} q. \tag{17}$$

This force is called the Lorentz force because Lorentz was the first to show how to determine the local field $\mathcal{E}_{loc}$ acting on an atom in a dielectric material. This local field is the sum of four fields,

$$\mathcal{E}_{loc} = \mathcal{E}_1 + \mathcal{E}_2 + \mathcal{E}_3 + \mathcal{E}_4. \tag{18}$$

In this equation:

$\mathcal{E}_1$ is the field intensity due to the charge density on the plates,

$$\mathcal{E}_1 = \frac{D}{\epsilon_0} = \mathcal{E} + \frac{P}{\epsilon_0}.$$

$\mathcal{E}_2$ is the field intensity due to the charge density induced on the two sides of the dielectric opposite the plates. Since this contribution to the field is opposite in direction to that of the applied field (Fig. 1), it is equal to $-P/\epsilon_0$.

$\mathcal{E}_3$ is the field intensity at the center of a spherical cavity whose radius is large compared with the size of an atom but small compared with the size of the dielectric. According to (17),

$$\mathcal{E}_3 = \frac{F_x}{q} = \frac{P}{3\epsilon_0}.$$

$\mathcal{E}_4$ is the field intensity at the center of a spherical cavity due to the dipoles of the atoms contained in that cavity. If the symmetry at the center of the sphere is cubic (as distinct from the crystal's symmetry), it can be shown that this term is equal to zero.

In the case of *dielectric isotropy*, therefore, substituting the above values in (18),

$$\mathcal{E}_{loc} = \left(\mathcal{E} + \frac{P}{\epsilon_0}\right) + \left(-\frac{P}{\epsilon_0}\right) + \frac{P}{3\epsilon_0} + 0$$
$$= \mathcal{E} + \frac{P}{3\epsilon_0} \tag{19}$$

and is called the *Lorentz field*. It has been assumed in the derivation of (19) that the region outside the sphere is a continuum having a dielectric constant κ_e. Note that the field intensity at the atom, that is, the Lorentz field, is larger than the applied field by an amount that is directly proportional to the polarization density.

Atomic polarizability. It is now possible to relate the dielectric constant of an insulator to the polarizability of the atoms comprising it. The dipole moment of a single atom p is proportional to the local field; that is,

$$p = \alpha \mathcal{E}_{loc} \tag{20}$$

where α is the *electrical polarizability* of the atom. The total polarization of an insulator containing N kinds of atoms is

$$P = \sum_{i=1}^{N} n_i \alpha_i {}^i\mathcal{E}_{loc} \tag{21}$$

where n_i is the number of i atoms having polarizabilities α_i and acted on by local field ${}^i\mathcal{E}_{loc}$.

In a dielectrically isotropic insulator, the local field inside the crystal is everywhere the same, so that it can be taken outside the summation sign in (21). Substituting the Lorentz field (19) into (21) then gives

$$P = \left(\mathcal{E} + \frac{P}{3\epsilon_0}\right) \sum_i n_i \alpha_i$$

or, after rearranging terms,

$$\frac{P}{\epsilon_0 \mathcal{E}} = \frac{\sum\limits_i n_i \alpha_i}{\epsilon_0 - \frac{1}{3}\sum\limits_i n_i \alpha_i}. \tag{22}$$

By substituting the left side of (12) for $P/\epsilon_0\mathcal{E}$ and rearranging terms, one obtains the *Clausius-Mosotti equation*

$$\frac{\kappa_e - 1}{\kappa_e + 2} = \frac{1}{3\epsilon_0}\sum_i n_i \alpha_i. \tag{23}$$

Equation (23) can be used to determine the electrical polarizabilities of the atoms if the dielectric constant is known. Conversely, the dielectric constants of new materials can be predicted from a knowledge of the individual polarizabilities of the atoms since, according to (23), the polarizabilities are additive. This relation is also valid for electronic polarizabilities α_e.

Dielectric constant

Static dielectric constant. The Clausius-Mosotti equation (23) for the dielectric constant of a material placed in a static electric field can be written as a sum of the contributions from a total of N_e electronic polarizabilities, N_i ionic polarizabilities, and N_d dipolar polarizabilities.

$$\frac{\kappa_e - 1}{\kappa_e + 2} = \frac{1}{3\epsilon_0}(N_e \alpha_e + N_i \alpha_i + N_d \alpha_d). \tag{24}$$

Consider first a material such as diamond, in which there are no ionic or dipolar contributions because all the carbon atoms are alike. The dielectric polarization then must be due to the electronic displacements within each atom. Provided that the Lorentz field in such a material is correctly given by (19),

$$\frac{\kappa_e - 1}{\kappa_e + 2} = \frac{1}{3\epsilon_0} N \alpha_e \tag{25}$$

and, for a constant α_e, the dielectric constant is a function of the number of atoms per unit volume N only. Consequently, it is not possible to test the validity of (25) in solids, and its correctness has to be verified by measurements using gases whose density can be varied more readily. It turns out that α_e does not vary significantly with frequency when a material is placed in an alternating electric field. This is true even for frequencies in the visible spectrum ($\nu \simeq 10^{15}$ sec^{-1}), in which case the dielectric constant equals the square of the refractive index n of the material. For this reason, it is sometimes called the high-frequency dielectric constant $\kappa_f = n^2$. It should be noted that the refractive index of a material actually does vary slightly with frequency and is not a constant throughout the visible spectrum.

In an ionic solid, say, an alkali halide, the positive and negative ions are attracted toward opposite ends of the electric field. Thus the ionic polarizability in (23) must also be included. In fact, it turns out that the ionic contribution is about three times larger than the electronic one in such materials. This is illustrated by a few experimental values:

	LiF	NaF	KF	RbF	LiCl	NaCl	KCl	RbCl	
κ_e	9.27	6.0	6.05	5.91	11.05	5.62	4.68	5.0	(26)
κ_f	1.92	1.74	1.85	1.93	2.75	2.25	2.13	2.19	

Temperature dependence. The dipolar polarization is a factor in a material only when it contains complex ions having permanent dipole moments. The effect of the electric field then is to orient these dipoles along the field direction since their interaction energy with the field is $-\mu\mathcal{E}$. Such alignment is opposed by thermal agitations which tend to randomize the dipole array. The dipoles are not free to rotate in a solid, but they are free to do so in a gas. A dipole of moment μ making an angle θ with the field direction contributes to the polarization a component $\mu \cos \theta$. The solid angle between θ and $\theta + d\theta$ formed by a dipole and the field direction is $2\pi \sin \theta \, d\theta$, so that the probability that a dipole lies at this angle, according to Boltzmann statistics, is

$$2\pi \sin \theta \, d\theta \, e^{-\mu\mathcal{E} \cos \theta / kT}. \tag{27}$$

The average moment contribution per dipole $\bar{\mu}$ is then determined by integrating over all angles from parallel alignment, when $\theta = 0$, to antiparallel alignment, $\theta = \pi$, so that

$$\bar{\mu} = \frac{\int_0^\pi \mu \cos \theta \, 2\pi \sin \theta \, e^{-\mu\mathcal{E} \cos \theta / kT} \, d\theta}{\int_0^\pi 2\pi \sin \theta \, e^{-\mu\mathcal{E} \cos \theta / kT} \, d\theta}. \tag{28}$$

Dividing top and bottom by 2π and letting

$$a = \frac{\mu\mathcal{E}}{kT} \qquad x = a \cos \theta \qquad dx = -a \sin \theta \, d\theta \tag{29}$$

the integrals in (28) can be written in the form

$$\bar{\mu} = \frac{\mu \int_a^{-a} x e^{-x} \, dx}{a \int_a^{-a} e^{-x} \, dx} \tag{30}$$

after multiplying both integrals by a defined in (29).

The ratio of the average moment to the true moment for a perfect dipole assembly

$$\frac{\bar{\mu}}{\mu} = \frac{e^a + e^{-a}}{e^a - e^{-a}} - \frac{1}{a} \tag{31}$$

is called the Langevin function,

$$L(a) = \coth a - \frac{1}{a} \tag{32}$$

because Langevin first derived this formula in 1905 for a paramagnetic gas. In 1912, Debye applied this relation to an electric dipole gas.

The meaning of (32) can be seen by plotting $L(a)$ as a function of $a \propto 1/T$ (Fig. 3). When a becomes large, that is, at very high field

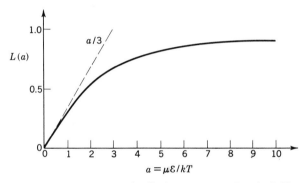

FIG. 3. Response of a perfect electric dipole gas to an electric field as determined by the Langevin function. Note that the slope of the curve is one-third when $a < 1$.

intensities or very low temperatures, the function reaches a saturation corresponding to a maximum alignment of the dipoles along the field direction. When a is less than 1, the curve in Fig. 3 has a slope of $\frac{1}{3}$, so that:

When $a \ll 1$, $$L(a) = \frac{a}{3}. \tag{33}$$

At room temperature, field intensities of about 10^9 V/m are required to approach $a = 1$. Thus, at fields that are not too large and temperatures that are not too low,

$$\bar{\mu} = \frac{\mu^2}{3kT} \varepsilon \tag{34}$$

as can be seen by substituting (33) and (29) in (31). The relative magnitudes of the competing energies determining the value of a are considered in Exercise 5. As shown there, the magnitude of kT at room temperature is many times larger than the potential-energy gain resulting from the alignment of the dipoles, so that the assumption in (33) is fully justified. It should be noted that the dipolar polarizability for permanent dipoles of moment μ is given by $\mu^2/3kT$, if the dipoles are completely free to align themselves with the field, and by $2\mu^2/3\varphi$, if an *activation energy* must be overcome before a dipole can line up. This latter case is called *hindered rotation* of the dipoles.

Frequency dependence. So far in this chapter attention has been focused on the response of a dielectric material to a static electric field. Assuming that the induced displacement, that is, the polarizability in the material, is such that the restoring force is proportional to the displacement (Hooke's law), then the behavior of the material in an alternating electric field can be treated like the problem of a harmonic oscillator. The classical differential equation determining the motion of a particle of charge e and mass m in an alternating-field intensity $\mathcal{E}_0 e^{i\omega t}$ is

$$\frac{d^2x}{dt^2} + \gamma \frac{dx}{dt} + \omega_0^2 x = \frac{e}{m} \mathcal{E}_0 e^{i\omega t} \tag{35}$$

where ω_0 is the natural angular frequency of the vibrating particle, and $\gamma\,(dx/dt)$ expresses a damping term resulting from the emission of radiation by a vibrating particle according to classical mechanics. The solution of (35) has the form

$$x(t) = \frac{e\mathcal{E}_0 e^{i\omega t}/m}{\omega_0^2 - \omega^2 + i\gamma\omega}. \tag{36}$$

Note that in a static electric field, $\omega = 0$ and the displacement (36) becomes $x = e\mathcal{E}/m\omega_0^2$, so that it is possible to define a static polarizability by $\alpha_s = ex$ and a static susceptibility

$$\chi_s = \frac{ex}{\epsilon_0 \mathcal{E}} = \frac{e^2}{\epsilon_0 m \omega_0^2}. \tag{37}$$

If α_s represents an electronic polarizability having a typical magnitude of about 10^{-30} C/m^3, the natural frequency $\nu_0 = \omega_0/2\pi$ turns out to be of the order of 10^{15} sec^{-1}. This is the reason why the electronic polarizability can be considered to be essentially independent of frequency up to frequencies corresponding to the visible spectrum. By comparison, if e and m in (37) take on appropriate values for an ion, $\nu_0 \simeq 10^{13}$ sec^{-1}, showing that the ionic polarizability is strongly frequency-dependent in the infrared part of the spectrum.

Multiplying both sides of (36) by e gives the polarizability due to an alternating field, so that, after dividing both sides by the field intensity and ϵ_0, the electrical susceptibility of a dipole gas χ_e^* containing N uncoupled dipoles (electric oscillators)

$$\chi_e^* = \frac{Ne^2/m\epsilon_0}{\omega_0^2 - \omega^2 + i\gamma\omega} \tag{38}$$

where the asterisk denotes that the susceptibility is a complex quantity. Multiplying the numerator and denominator in (38) by $(\omega_0^2 - \omega^2) - i\gamma\omega$, it is possible to separate the real and imaginary parts,

$$\chi_e^* = \frac{Ne^2}{\epsilon_0 m}\left[\frac{\omega_0^2 - \omega^2}{(\omega_0^2 - \omega^2)^2 + \gamma^2\omega^2} - i\frac{\gamma\omega}{(\omega_0^2 - \omega^2)^2 + \gamma^2\omega^2}\right]$$
$$= \chi_e' - i\chi_e''. \tag{39}$$

The meaning of the real and imaginary parts in (39) can be understood by plotting them as a function of ω close to the natural frequency ω_0, as in Fig. 4. The real part in (39) is plotted in Fig. 4A, and the imaginary part in Fig. 4B. Note that $\chi_e' = 0$ when $\omega = \omega_0$, whereas the imaginary part has a maximum at that point. The meaning of this maximum in the imaginary component is that the material absorbs energy at the

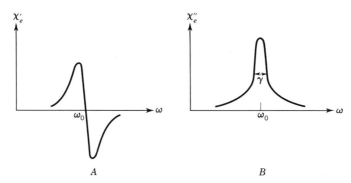

FIG. 4. Dependence of the complex electrical susceptibility on frequency near $\omega = \omega_0$. (A) The real part of the susceptibility; (B) the imaginary part of the susceptibility.

natural frequency, and this phenomenon is commonly called *resonance absorption*. As can be seen in Fig. 4A, the real part of the susceptibility is strongly frequency-dependent in this region and undergoes a change in sign. This phenomenon is called *anomalous dispersion*.

The resonance absorption is caused by the damping factor γ since, if there were no damping, that is, if $\gamma = 0$, then there would be no absorption. The damping factor has the dimensions of frequency, and its magnitude is determined by the width of the absorption curve in Fig. 4B at half maximum. The damping factor is usually expressed as an inverse average lifetime of an excited state τ; that is,

$$\gamma \equiv \frac{1}{\tau}. \tag{40}$$

By analogy with electrical LRC circuits, it is sometimes more convenient to speak of a quality factor Q, expressing the ratio of the energy stored to that which is dissipated in a half-cycle,

$$Q = \frac{\omega_0}{\gamma} = \omega_0 \tau. \tag{41}$$

It is important to realize that the field intensity in (35) is that of the field acting on the dipoles inside the material. Except in rarefied gases, this is not equal to the external field intensity. Thus, for solids the Lorentz field must be used to derive the electrical susceptibility (Exercise 9). This gives

$$\chi_e^* = \frac{e^2 N / m\epsilon_0}{\omega_0^2 - \omega^2 - Ne^2/3m\epsilon_0 + i\gamma\omega} \tag{42}$$

which can be made analogous to the susceptibility of a gas (38) by defining a new frequency $(\omega')^2 = \omega_0^2 - Ne^2/3m\epsilon_0$. From this it follows that the absorption frequency of a solid is displaced to lower frequencies compared with a gas.

According to quantum mechanics, energy changes are possible only when a transition between stationary states takes place. This means that the resonance frequency ω_0 must be interpreted as a transition frequency ω_{ij} between two states i and j. Such transitions can be produced, for example, by the interaction of an electromagnetic field with the dipole moments that it induces in passing through a material. Accordingly, the number of dipoles per unit volume N must be replaced by the difference between the densities of occupied states of type i and j; that is,

$$N \to N_i - N_j. \tag{43}$$

As discussed in Chap. 2, a transition from a higher state to a lower state is accompanied by the emission of a quantum of energy, for example, a photon, whereas the reverse transition requires the absorption of a photon. Suppose that the subscript i denotes states of higher energy than those denoted by j and the density of available states is the same for i and j states but the number of occupied states per unit volume $N_j > N_i$. In this case, the incident radiation is absorbed by transitions to the unoccupied higher states. Equilibrium requires, however, that the reverse transitions $i \to j$ also occur, so that the absorption of light causes a *forced emission* of coherent light; that is, the emitted light has the same frequency and phase as the absorbed light. The existence of this forced emission becomes evident by a small decrease in the absorption. If the relative energies of the occupied states are reversed by a field within a time that is shorter than that required for the transitions to occur, $N_j < N_i$ and the forced emission becomes dominant. It is thus possible that an incident light signal can be amplified by inducing a forced emission of coherent light in a material, and this phenomenon is called

light amplification by stimulated emission of radiation, abbreviated *laser.*
A further discussion of lasers is postponed to Chap. 14.

Electrical processes

Piezoelectricity. The behavior of polar-gas molecules in an electric
field is described by the Langevin function (32). When such a gas is
condensed to the liquid or solid state, a coupling between neighboring
dipoles ensues and hinders their rotation. In the solid state one of three
things may happen:

1. The dipoles may align themselves in some ordered array, which then
produces a net polar-axis moment in the crystal.

2. The dipoles may align themselves in an ordered array in such a way
that their dipole moments cancel each other, so that there is no net
moment.

3. The polar molecules may dissociate in the crystal into individual
positive and negative ions in such a way that there are no dipoles present.

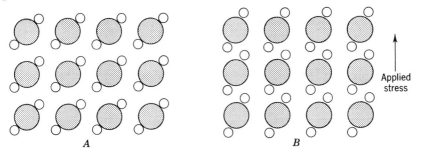

A   B   Applied stress

FIG. 5

In order to see the effect that these three types of crystal structure have
on the electrical properties of a crystal, consider first a centrosymmetric
structure composed of positively and negatively charged ions schematically represented in Fig. 5A. As discussed in Chap. 1, such a crystal
must belong to one of the 11 centrosymmetric point groups, or crystal
classes. When a mechanical stress is applied to the crystal, the atoms
are slightly displaced as shown in Fig. 5B. Since the ionic displacements are symmetrical about the symmetry centers, the charge distribution inside the crystal is not appreciably altered by the applied
stress. Next, consider the acentric crystal structure shown in Fig. 6A.
It is clearly seen there that the ions are arranged in pairs forming dipoles.
When such a crystal is deformed by an applied stress as shown in Fig. 6B,
the ions are displaced from each other in an asymmetric way, so that the
original balance of moments in the crystal is altered. This is called the
piezoelectric effect, and it is observed in crystals belonging to 20 of the 21

noncentrosymmetric crystal classes. Conversely, when an electric field is applied to a piezoelectric crystal, a dipolar realignment with the field direction is induced in the crystal, and the resulting small atomic displacements produce a mechanical strain. This is called the *inverse piezoelectric effect*.

Returning now to the classification of the three possible structure types discussed earlier in this section, it is clear that the first group is always piezoelectric, the second group is piezoelectric provided that the crystal structure lacks a center of symmetry, and the third group corresponds to centrosymmetric ionic structures which are not piezoelectric. It should be realized that the application of an electric field even to such a centro-symmetric structure causes small shifts in the relative positions of the positive and negative charges of the ions. This causes a mechanical distortion called *electrostriction* and induces dipole moments in the crystal. The interaction of the field with these induced moments

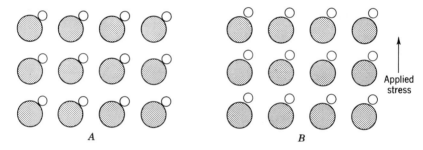

A   B   Applied
                                                                         stress

Fig. 6

produces a distortion that is proportional to the square of the field intensity. An inverse effect is not possible in such crystals because, as shown in Fig. 5*B*, the mechanical stress acts to displace the entire ions and does not create dipoles inside the material. By comparison, a piezoelectric crystal acts like a mechanical ⇌ electrical *transducer* whose response varies linearly with the electric-field intensity or the applied stress, provided that neither the field nor the stress is excessively large.

A polar structure like the one schematically represented in Fig. 6*A* has another interesting property. When a crystal having such a structure is heated, the interatomic distances are increased in an asymmetric way. Accompanying such a change in the polar axis is a change in the polarization of the crystal, so that a difference in potential is created within the crystal. This is called the *pyroelectric effect* and, obviously, can be produced either by heating or cooling the crystal. As an example of the role that crystal structure plays in determining such effects, consider the two polymorphous structures of ZnS described in Chap. 1. The cubic modification (sphalerite) has four symmetry-related polar axes along

[[111]], normal to the closest-packed planes. Similarly, the polar axis in the hexagonal wurtzite structure is [0001]. Suppose a sphalerite crystal is compressed along [111]. The Zn-S bonds parallel to the polar axis are compressed, whereas the other three bonds are flattened; that is, the originally tetrahedral angle is made smaller. This distortion, therefore, upsets the balance of the dipole moments, and a piezoelectric effect is observed. In the case of wurtzite, a compression along the c axis produces a similar effect. When a wurtzite crystal is heated, moreover, it undergoes an anisotropic thermal expansion. The resulting change in the c/a ratio thus accounts for the fact that wurtzite is pyroelectric as well as being piezoelectric. Conversely, when sphalerite is heated, the four crystallographically identical [[111]] directions expand uniformly, so that sphalerite is piezoelectric but not pyroelectric.

The application of an alternating electric field to a piezoelectric crystal causes the electric displacements to vary periodically. Generally, such displacements lag behind the alternations of the applied-field direction, so that the two are not in phase. The amount of the phase difference obviously depends on the frequency of the alternating field. For each crystal there must exist a vibrational frequency, called the resonance frequency of the crystal, at which the two are exactly in phase. As

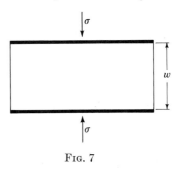

Fig. 7

already discussed in the preceding section, the crystals can absorb energy from the field at this frequency, so that this effect can be utilized to determine whether a crystal is piezoelectric or not. Such measurements can be used, for example, to determine whether a crystal structure has a center of symmetry. Alternatively, because the frequency range over which resonance occurs in most piezoelectric crystals is very sharp, such crystals (for example, quartz) can be used for frequency controls in radio transmitters and have also been used in electronic clocks.

In general, any mechanical stress can produce an electric polarization in a piezoelectric crystal; that is, it does not matter whether the stress is applied in compression, dilation, or shear. For example, consider a piezoelectric crystal placed between two metal plates, as shown in Fig. 7. If the crystal is compressed by an applied mechanical stress σ, a mechanical strain ϵ_m is produced in the crystal,

$$\epsilon_m = \frac{\sigma}{E} = \frac{\Delta w}{w} \tag{44}$$

where E is Young's modulus. The applied stress also produces a polarization density P in the crystal which is proportional to the magnitude of the applied stress:

$$P = \eta\sigma \tag{45}$$

where η is called the *piezoelectric constant*. If the two plates are connected, that is, short-circuited, $\varepsilon = 0$ and, according to (14), the electric displacement

$$D = 0 + P$$
$$= \eta\sigma. \tag{46}$$

On the other hand, if an electric field is applied to the crystal in Fig. 7 in the absence of an applied stress, then the induced strain is proportional to the applied field:

$$\epsilon_m = \eta\varepsilon. \tag{47}$$

Note that, according to (4), the electric displacement D is also proportional to the applied field intensity. When both an external field and a stress are applied to the crystal, therefore, the electric displacement becomes

$$D = \epsilon\varepsilon + \eta\sigma \tag{48}$$

and the internal strain

$$\epsilon_m = \eta\varepsilon + \frac{\sigma}{E}. \tag{49}$$

The above relations can be used in deciding the suitability of piezoelectric crystals for use as transducers in different applications (Exercise 10).

Ferroelectricity. The structures of some piezoelectric crystals are such that the dipoles are permanently lined up in the structure; that is, they are spontaneously polarized in the absence of an external field. The electric field produced by the dipoles inside the crystal, however, is often masked by charges accumulated on the crystal's surface or by twinning inside the crystal. (The consequence of twinning is that the dipole orientation in adjacent twins is related by some twin-symmetry operation in such a way that they cancel each other's moments.) As already noted, such crystals are also pyroelectric. If it is possible to reverse permanently the polarization direction of a pyroelectric crystal by applying a sufficiently intense external field, then the crystal is said to be *ferroelectric* and the phenomenon of reversing the polarization direction is called the *ferroelectric effect*. It should be noted that both piezoelectricity and pyroelectricity are inherent properties of a crystal, requiring only that the crystal structure contain a polar axis along which atomic displacements can take place asymmetrically. Ferroelectricity, on the other hand, is an effect produced in a pyroelectric crystal when it contains two juxtaposed positions along the polar axis into which the displacement can take place.

As an example of the dipolar arrays that can occur in pyroelectric crystals, consider the schematic crystal structure shown in Fig. 8. In order for a crystal to be ferroelectric, it must be able to adopt either of two crystal structures, which are identical except for the direction in which the dipoles point. One way that this can happen is by a displacement of the cation to opposite ends of its coordination polyhedron as

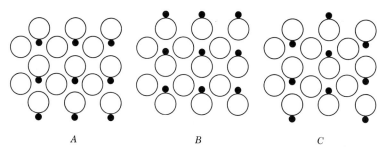

Fig. 8. Schematic representation of atomic arrangements in pyroelectric crystals. (*A*, *B*) Two equally possible structures of a ferroelectric crystal having inverted polarities; (*C*) antiferroelectric crystal.

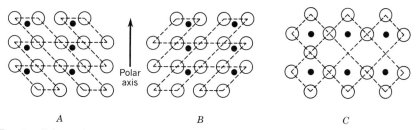

Fig. 9. Schematic representation of ferroelectric and paraelectric structures. (*A*, *B*) Two equally possible structures of a ferroelectric crystal having dipole-moment components in opposite directions parallel to the polar axis of the crystal; (*C*) high-temperature symmetric structure does not have a polar axis and is therefore paraelectric.

shown in Fig. 8*A* and *B*. Suppose the larger open circles in Fig. 8 represent anions and the smaller black circles represent cations. Then the dipoles in the structure shown in Fig. 8*A* are oriented so that their negative poles point upward whereas those in the array shown in Fig. 8*B* point downward. If the dipoles in alternate polyhedra point up and down (Fig. 8*C*), then there is no net moment in the crystal and it is said to be *antiferroelectric*. Another way that a dipole array in a crystal can be altered is illustrated in Fig. 9. The two structures shown in Fig. 9*A* and *B* differ in that the dipole moments have oppositely directed components along the polar axis.   This is so because the cations in these two structures are displaced to two geometrically different but crystallographically equivalent corners of the polyhedra.   This pair can be distinguished from the pair in Fig. 8 by the fact that the ionic displacements have taken place to adjacent rather than opposite corners.

As already noted in a preceding section, the contribution to the polarization from permanent dipoles in a crystal is temperature-dependent. This is so because the thermal motion of the atoms increases with temperature.   If both structures in Fig. 9A and B have equal potential energies, it follows that, above some transformation temperature, the crystal will adopt an "average" structure as the one depicted in Fig. 9C. The symmetrical atomic array in such a structure destroys the dipoles, and the crystal is no longer ferroelectric.   This is called the *para* modification of the structure, and the crystal is said to be *paraelectric*, quite similar to the ferromagnetic $\rightleftharpoons$ paramagnetic transitions of crystals discussed in the next chapter.   The transition temperature is usually called the *Curie* point, in honor of Pierre Curie, who first discovered the piezoelectric effect with his brother in 1880 and later studied the temperature dependence of magnetic materials.   The dielectric constant of ferroelectric crystals increases anomalously at the Curie point.   The temperature dependence of the dielectric constant in such materials is similar to the magnetic-susceptibility dependence on temperature in ferromagnetic materials.

$$\kappa_e = \frac{C}{T - T_C} + \kappa_f \tag{50}$$

where C is called the Curie constant

$\quad$ T_C is the transition temperature

$\quad$ κ_f is the high-frequency dielectric constant representing electronic contributions to the polarization.

When $T \simeq T_C$, $\kappa_e \gg \kappa_f$ and the electronic contribution can be neglected.

Below the transition temperature, ferroelectric crystals usually consist of multiple twins.   Each twin individual is spontaneously polarized in a specific direction; however, the directions of polarization of neighboring twins are not parallel.   Consequently, *ferroelectric domains* exist in the crystals as illustrated in Fig. 10, which shows an electron micrograph of domains formed in a $BaTiO_3$ crystal.   The uncompensated electric charges present on the surfaces of a uniformly polarized insulator produce a depolarizing field which makes such a uniformly polarized crystal unstable. The presence of adjacent domains having opposing directions of spontaneous polarization, on the other hand, serves to reduce the depolarizing field and stabilizes the crystal.   This energy is stored in the domain walls, and an equilibrium condition is reached between the number and size of domains present and the thickness of the walls that separate them. There is thus a similarity between ferroelectric domains and the analogous domains formed in ferromagnetic crystals (Chap. 13). On an atomic scale, however, the two kinds of domains are different. As shown in the next chapter, the spontaneous magnetic moments in ferromagnetic crystals are produced by a coupling between spin moments of electrons on

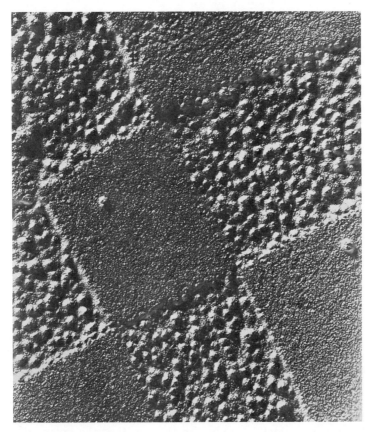

FIG. 10. Electron micrograph of the surface of a single crystal of $BaTiO_3$ showing the individual ferroelectric domains magnified about 30,000 times. (*Courtesy of D. P. Cameron.*)

adjacent atoms. The electric dipole moments in ferroelectric crystals, on the other hand, are produced by a change in their crystal structure below the Curie point along a direction determined by the structure rather than by the applied-field direction. Consequently, the strains produced by such a distortion (electrostriction) are much greater in ferroelectric crystals than the analogous magnetostriction in ferromagnets. In fact, as discussed in the next chapter, the magnitude of magnetostriction depends on the relative magnetic-field direction, which can be arbitrarily chosen.

As might be expected from the analogy to ferromagnetism, the presence of domains in a ferroelectric crystal produces a hysteresis in the polarization when an alternating field is applied. A typical hysteresis curve for a ferroelectric crystal is shown in Fig. 11. To start, suppose that the crystal has equal numbers of domains with oppositely directed polarizations. When an electric field is applied parallel to a crystallographic direction, the domains whose polarization is more nearly parallel to the field direction have a lower energy, so that they grow in size at the expense

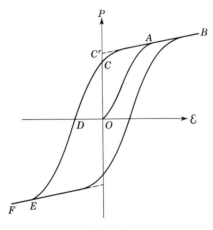

Fig. 11

of the antiparallel domains. As the field is increased, therefore, the total polarization of the crystal increases rapidly (curve OA in Fig. 11) until a saturation value is reached (AB), at which point the crystal is a single domain. This is usually accompanied by a distortion of the crystal in the form of an elongation along the polarization direction. When the field later decreases to zero, a number of domains retain their orientation parallel to this crystallographic direction, so that the polarization does not return to zero. This polarization is called the *remnant polarization* and is shown at C in Fig. 11. If the line AB is extrapolated backward to C', then the polarization value of a single domain at zero applied field is obtained. This is so because the entire crystal acts as a single domain in the AB region. Thus C' represents the magnitude of the *spontaneous polarization* of a single domain. When there are equal numbers of oppositely directed, spontaneously polarized domains in the crystal, its polarization again becomes zero. It is necessary, therefore, to apply a reverse field to the crystal in order to reverse the polarization directions of a sufficient number of domains.   The magnitude of the field required to remove the remnant polarization is called the *coercive field* and is indicated by D in Fig. 11.   As the magnitude of the reverse field is increased further, the saturation value of polarization in the reverse direction is reached (EF).   Reversing the field again then traces out the curve FEB, and so forth.   It follows from this discussion that an insulator crystal having a suitable crystal structure can exhibit the ferroelectric effect only if the coercive field required to reverse its polarization direction is not so large as to cause electric breakdown of the crystal.   By comparison, ferroelectric crystals having relatively low spontaneous polarizabilities are frequently used in computers as memory cells, the direction

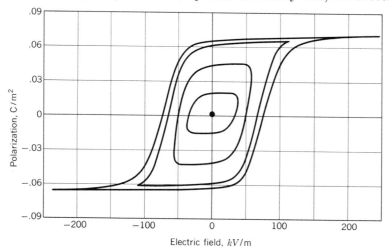

FIG. 12.   Hysteresis loops of barium titanate for several values of the maximum applied electric field.   (*Courtesy of Marvin E. Anderson.*)

of polarization indicating, say, a plus or a minus sign.   The effect that the applied field has on the hysteresis curves is illustrated in Fig. 12.

There are several groups of crystals that are known to be ferroelectrics. The crystals typifying each group are rochelle salt, $NaK(C_4H_4O_6)\cdot4H_2O$; potassium dihydrogen phosphate, KH_2PO_4; barium titanate, $BaTiO_3$; and guanidine compounds such as $C(NH_2)_3Al(SO_4)_2\cdot6H_2O$.   It should be borne in mind that each of these groups includes a number of isomorphous and isostructural crystals which also are ferroelectrics.   As already indicated, the property of ferroelectricity is related to crystal structure.   So far it has not been possible, therefore, to devise a single theory to explain this effect in different crystals, although several phenomenological theories have been proposed for rochelle salt, potassium dihydrogen phosphate, and barium titanate.

$BaTiO_3$ has the perovskite structure shown in Fig. 13.   The Ba and

O atoms can be seen to form closest-packed planes in Fig. 13 normal to the [[111]] directions of the cube, so that the resulting structure can be thought of as a cubic closest packing of such sheets containing a titanium atom in the octahedral void at the center of each unit cube. Another view of this structure is shown in Fig. 14; in this case the origin of the unit cell has been shifted to the titanium-atom position. This is the structure of BaTiO$_3$ above 393 deg K, and as a consequence of its symmetry it represents the paraelectric structure of barium titanate. Below

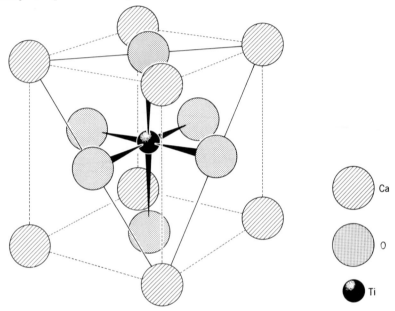

Ca

O

Ti

F$_{\text{IG}}$. 13.   Perovskite (CaTiO$_3$) structure.   The atoms are shown reduced in size for clarity.

393 deg K, the structure becomes distorted by an elongation in the polarization direction along one of the cube axes and a slight contraction at right angles to it.   The resulting structure is tetragonal, with $c/a = 1.04$ at room temperature.   Since there are three possible polar axes in a cube, such a polarization can instantaneously occur in six different directions, so that it is not surprising that a single cubic crystal becomes twinned upon transition to the tetragonal phase.

The actual atomic displacements responsible for this transformation are illustrated in Fig. 15.   The actual magnitudes of the shifts in BaTiO$_3$ are shown in the projection of its structure on (100) in Fig. 15A.   The titanium atom moves toward one of the oxygen atoms in its coordination octahedron, while the oxygen atoms move in the opposite direction by a like amount.   By comparison, a much more drastic displacement of the ions takes place in the isomorphous PbTiO$_3$, as shown in the projection

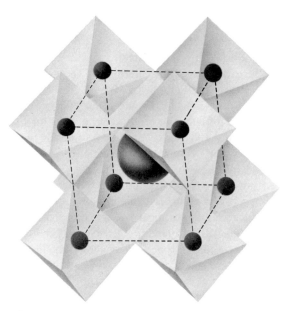

FIG. 14. Crystal structure of BaTiO₃. The oxygen octahedra coordinating Ti⁴⁺ are shown at the corners of the unit cell. Note that the large Ba²⁺ ion is coordinated by 12 oxygens since it forms part of the cubic closest-packed array of Ba and O atoms. (Compare with Fig. 13.)

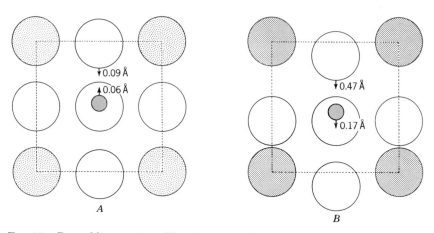

FIG. 15. Perovskite structure (Fig. 13) projected on (100). The open circles represent oxygen atoms, the small dark circles titanium, and the large shaded circles the other cation. (A) Tetragonal distortion in BaTiO₃; (B) tetragonal distortion in PbTiO₃.

of its structure on (100) in Fig. 15*B*.   In lead titanate, both the oxygen
and titanium ions are displaced in the same direction relative to the lead
atoms and by larger amounts, so that $c/a = 1.06$ in PbTiO$_3$.

Below room temperature ($\simeq$278 deg K) another transformation takes
place, resulting in a discontinuous change in the polarization direction
parallel to [110] in the original cubic structure.   The resulting orthorhombic structure is best described by taking *a* parallel to [110], *b* parallel to
[1$\bar{1}$0], and *c* parallel to [001] in the original cube.   A third transformation
occurs at approximately 193 deg K, at which point the polarization direction becomes [111] of the original perovskite cube and the structure has
rhombohedral (hexagonal) symmetry.   The reasons for the stability of
these various structures can be explained by applying known principles
of thermodynamics.   The result is a phenomenological theory which postulates that the local field produced by the polarization increases at a
faster rate than the elastic restoring forces binding the ions in the crystal.
The relative displacements of the titanium ions inside the oxygen octahedra coordinating them are then stabilized at different positions in the
three different structures.   The resulting buildup in the local field's intensity is sometimes called the *polarization catastrophy*.

It is of interest to observe what happens to the dielectric constant
near the transformation temperature.   Assuming dielectric isotropy, (23)
can be written after rearrangement of terms,†

$$\kappa_e = \frac{1 + \dfrac{2}{3\epsilon_0} \sum_i n_i \alpha_i}{1 - \dfrac{1}{3\epsilon_0} \sum_i n_i \alpha_i} \tag{51}$$

where n_i is the number of i atoms having polarizabilities α_i.   When
$\sum_i n_i \alpha_i = 3\epsilon_0$, the dielectric constant becomes infinite while P has some
finite value and $\mathcal{E} = 0$, according to (22).   This is the condition for the
so-called polarization catastrophy described above.

For small deviations of the sum in (51) from its critical value, it is
possible to substitute for $\frac{1}{3}\epsilon_0 \Sigma n_i \alpha_i$ the quantity $1 - \delta$, where $\delta \ll 1$.
Substituting in (51) then gives $\kappa_e \simeq 3/\delta$, and assuming that the devia-
tion δ is temperature-dependent according to

$$\delta \propto \frac{C}{3}(T - T_C) \tag{52}$$

---

† This assumption is strictly valid only for the metal atoms in BaTiO$_3$.   The
environment of oxygen atoms is not cubic; however, the effect of the dipoles in the
spherical cavity used in calculating the local field in (17) can be neglected for present
purposes.

the Curie-Weiss law (50) near the transition temperature

$$\kappa_e \propto \frac{C}{T - T_C} \tag{53}$$

is obtained. The plot in Fig. 16 shows the dependence of the dielectric constant on temperature in a ferroelectric crystal. An identical curve is obtained in a plot of the specific heat as a function of temperature near

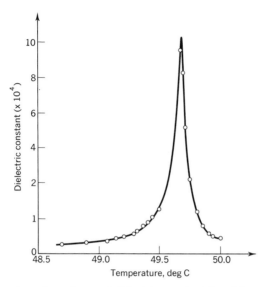

FIG. 16. Temperature dependence of dielectric constant in triglycene sulfate near the Curie point.

the Curie point. Such curves are called lambda curves, because of their shape, and are characteristic of transitions of this type.

Ionic conductivity. As mentioned in the introduction to this chapter, when an electric field is applied to an ionic insulator, an electric current may be produced. Since the number of electrons occupying quantum states in the conduction band of an insulator is negligibly small, it is necessary to postulate that the current is carried by positively or negatively charged ions or both. Now, in a crystal in which all the ions occupy their appropriate equipoints, the motion of ions is not possible, since there is no way for an ion to move without leaving its proper site. It is necessary to conclude, therefore, that the current carriers are either interstitial ions or else vacant ionic sites. The mechanism whereby interstitial ions or vacant sites are created is purely statistical in nature and has already been described in Chap. 5. It will be recalled from that chapter that ion vacancies are called Schottky defects and interstitial ions plus compensating ion vacancies are called Frenkel defects.

The generation of Schottky defects proceeds by the migration of, say, a positive ion to the surface of the crystal, thereby leaving a positive vacancy behind in the crystal. In order to preserve local charge neutrality, it is necessary that a negative ion also migrate to the crystal's surface, creating a negative vacancy thereby. The result of the generation of such pairs of vacancies is that the volume of the crystal increases while its density decreases. On the other hand, a Frenkel defect is produced when an ion migrates into an interstitial position in the structure which is sufficiently far removed from the resulting vacancy so that direct recombination can be ruled out. This process does not disturb the macroscopic charge neutrality, nor does it appreciably change the volume or the density of the crystal. It is thus possible to determine which of these two types of imperfections is present in a crystal by careful density measurements.

The process whereby ions or vacancies move through the crystal is called *ionic diffusion*. Suppose that a layer of radioactive sodium (Na^{23}) is deposited on one side of a sodium chloride crystal, and the crystal is maintained at some elevated temperature for a finite period of time. It can be shown by subsequent sectioning of the crystal and measuring the radioactivity of successive sections that the radioactive sodium diffuses through the crystal in a regular way; that is, the concentration of Na^{23} atoms varies regularly with depth in the crystal. The diffusion process can be expressed analytically by *Fick's law*, which states that the number of atoms crossing a unit area per unit time, that is, the flux of atoms J, is proportional to the gradient of their concentration N. Considering diffusion in only one direction,

$$J = -D \frac{dN}{dx} \tag{54}$$

where D is the diffusion coefficient and the minus sign indicates that the flux is in a direction opposite to the direction of the gradient; that is, if the gradient increases to the right, diffusion proceeds to the left.

One usually distinguishes the diffusion of atoms normally present in the crystal by calling this process *self-diffusion*. The atomic mechanisms whereby self-diffusion can occur are pictured in Fig. 17. Four distinct processes are possible:

A. Direct interchange between two atoms. Because of the need to displace neighboring atoms and, in the case of ionic compounds, the large repulsive forces between atoms of like charge, this mechanism requires a relatively large amount of energy.

B. *Migration of an interstitial atom* subsequent to formation of a Frenkel defect.

C. *Atomic displacement into a neighboring vacancy* subsequent to formation of a Schottky defect. This is usually described as a *diffusion of vacancies* in the opposite direction.

D. *Diffusion of pairs of vacancies* by means of atomic movements into either of two adjacent vacancies. Since the pair moves by the motion of either vacancy, provided that the pair is not split up in the process, the energy required to move the pair is actually less than that required to move an isolated vacancy.

The energy required for any of the above processes to operate, that is, the sum of the energies of defect formation and subsequent migration, is called the activation energy of that process. Without distinguishing

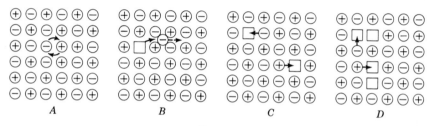

A   B   C   D

FIG. 17

between the different processes, it has been observed that the temperature dependence of the coefficient of self-diffusion is related to the activation energy φ by

$$D = D_0 e^{-\varphi/kT} \tag{55}$$

where D_0 is a constant for the crystal.

The ionic conductivity due to monovalent ions of one sign, that is, positive or negative ions but not both, is

$$\sigma_{\text{ionic}} = eN\mu_{\text{ionic}} \tag{56}$$

where N is the number of ionic sites of one sign, and the mobility of these ions, according to Einstein, is

$$\mu_{\text{ionic}} = \frac{eD}{kT}. \tag{57}$$

Combining (56) and (57) and substituting (55) for D, the ionic conductivity is given by

$$\sigma_{\text{ionic}} = \frac{e^2 N}{kT} D$$

$$= \frac{e^2 N}{kT} D_0 e^{-\varphi/kT}. \tag{58}$$

Figure 18 shows an idealized plot of ln σ as a function of reciprocal temperature for an alkali halide crystal. According to (58), the slope of the straight line in such a plot can be used to determine the activation energy. As shown in Fig. 18, the curve consists of two straight-line portions; hence there are two activation energies for the two different temperature ranges. For sodium chloride, $\varphi_1 = 1.80$ eV in the high-temperature region, and $\varphi_2 = 0.77$ eV in the lower-temperature region. The reason for the two slopes can be explained as follows: At high temperatures, the thermal energy is sufficiently great to create vacancies, and the activation energy represents a sum of the energies required for vacancy generation and the motion of ions into the vacancies. At lower temperature, the thermal energy is only large enough to allow the migration of atoms into vacancies already present in the crystal. Recent

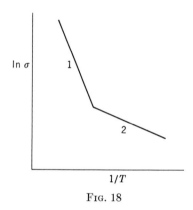

FIG. 18

experimental evidence has shown that these vacancies are primarily due to the inclusion of divalent metal atoms in substitution for the monovalent metal atoms in the crystal's structure. For each divalent metal thus incorporated in the crystal, a positive vacancy must be present nearby in order to maintain charge neutrality. Similar to the case of semiconductors, the low-temperature conductivity is said to take place in the extrinsic region, because the presence of divalent impurity atoms is required for conductivity to occur in this temperature region. On the other hand, the high-temperature conductivity is characteristic of the crystal and is called the intrinsic conductivity for this reason.

In view of the relatively small mobilities of ions, the Hall effect in these crystals is usually too small to be measured accurately, and it is not possible to determine the sign of the charge carriers by such means. It is possible to devise a simple alternative experiment for this purpose, however. Suppose that two identical crystals of silver chloride (AgCl) are placed between two silver electrodes and an electric field is applied, as shown in Fig. 19. Then the following situations are possible. If the electric current is carried by Ag^{1+} ions, they will move toward the negative electrode, so that, after a while, the negative electrode is increased in size at the expense of the positive electrode. In this case, the salt acts merely as a conductor of positive ions and is not otherwise affected. If, on the other hand, the current is due to negative-ion migration, then the anions will collect on the positively charged electrode, where they will combine with the silver to form new AgCl. Thus, after a while, the positive electrode will appear to have lost silver to the salt crystal adjacent to it, which, concurrently, must increase in size. If both conduction processes are operative, an intermediate result should be obtained. It turns out that, in AgCl and in most alkali halides, the bulk of the current is carried by the metal ions. In barium and lead halides, conversely, the majority carriers are the negative halide ions.

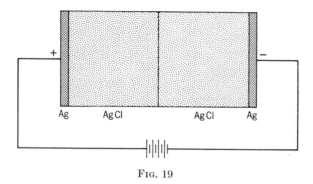

Fig. 19

It is interesting to note, in passing, that the self-diffusion in metals also proceeds predominantly by means of vacancies rather than through the migration of interstitials or the direct exchange between pairs of atoms. This is so because the voids in metal closest packings, just like the empty tetrahedral voids in halogen closest packings, are too small to accommodate the metal atoms without distorting the structure. Suppose that a piece of brass (CuZn) is surrounded on two sides by copper and that wires of some relatively inert metal such as molybdenum are imbedded at the interfaces in the Cu-brass-Cu *diffusion couple* (Fig. 20). Upon subsequent heating of the sample to allow diffusion to occur, it is observed that zinc diffuses out of the brass and into the copper and the molybdenum wires on opposite sides of the brass core move toward each other. This is called the *Kirkendall effect*, and it is explained by postulating that the diffusion coefficient of zinc is greater than that of copper, so that zinc diffuses out of the brass faster than copper can diffuse into the brass. The net flow of mass out of the brass, or in other words, the net flow of vacancies into the brass, causes it to decrease in size as evidenced by the motion of the wires.   The Kirkendall effect has also been observed in Cu-Sn, Cu-Au, and other metal systems.

Electric breakdown.   When an insulator is placed between two metal plates and a very large electric field is applied, a relatively large electric

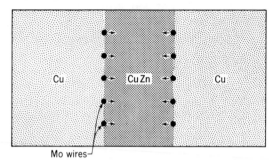

Fig. 20.   Cu-CuZn-Cu diffusion couple.   As the zinc diffuses out of the brass into the adjacent copper, the Mo wires move toward each other.

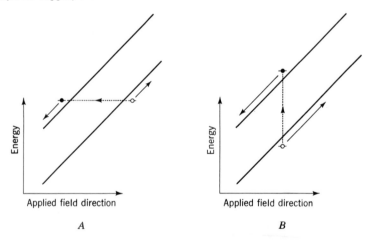

Fig. 21.   Electric breakdown in semiconductors.   (*A*) Zener breakdown;   (*B*) avalanche breakdown.

current may flow between the plates, if the electric-field intensity exceeds a critical value.   This process is called *electric breakdown* in an insulator. When the crystal is actually a semiconductor, the breakdown can occur by two distinct processes.   As described in Chap. 8, the application of an electric field skews the energy-band model, as shown in Fig. 21.   If the applied field is sufficiently large, there is a possibility that some of the electrons occupying states in the valence band can move into quantum states of like energy in the conduction band, as shown in Fig. 21*A*.

This mechanism was first proposed by C. Zener and is called *Zener breakdown* (see also page 279). Another possible breakdown mechanism is shown in Fig. 21*B*. If a free electron gains more kinetic energy from the applied field than it loses in electron-atom collisons, then it is possible for it to lose this excess energy in a collision with a valence electron. This process results, therefore, in the production of an electron-hole pair, and now the two electrons can, in turn, produce four more conduction electrons, and so forth.† This rapid increase in the number of conduction electrons is called the *avalanche breakdown* and requires relatively high fields ($\simeq 10^7$ V/m in Ge).

The disruptive breakdown processes in true insulator crystals are not as well understood. Nevertheless, it is possible to distinguish five distinct processes:

1. *Thermal breakdown* is produced in insulators when heat is generated by the ionic currents faster than the crystal can dissipate it. Since the conductivity of heat in insulators is small, it is possible that temperatures in excess of the crystal's melting point can be attained in parts of the crystal. The resulting local melting of the crystal increases the ionic mobilities, and electric breakdown ensues.

2. *Electrolytic breakdown* can occur whenever conducting paths are present in the crystal. These paths form with the aid of imperfections in the crystal, such as dislocations, or along the dendritic or lineage structure interfaces sometimes present in crystals.

3. *Dipole breakdown* may be caused in insulators either by polarizable atoms (molecules) or by permament dipoles already present. When such dipoles surround a stressed region, they can produce local impurity or imperfection states lying in the forbidden-energy gap of the crystal. The much lower ionization potentials of electrons attached to dipoles then facilitate breakdown.

4. *Collision breakdown* is similar to the avalanche breakdown in semiconductors. Because of impurities in the crystal, or because some electrons have been injected from the contacts, there are some electrons available for conduction. When the energies of these electrons become sufficiently great, they collide with other electrons, producing ever-increasing numbers of electrons and holes.

5. *Gas-discharge breakdown* can occur whenever the insulator contains occluded gas bubbles. For example, the silicate layers in a micaceous crystal are frequently separated by thin adsorbed layers of air or other gases. Since the electric field required to ionize the gas ($\simeq 10^6$ V/m) is much less than that required for electric breakdown in an insulator ($\simeq 10^8$ V/m), the gas ionizes first, and the gas ions bombard the internal

---

† Actually, the holes also contribute to this process.

crystal surfaces, causing them to deteriorate, until complete electric breakdown of the insulator occurs.

Suggestions for supplementary reading

Walter P. Cady, *Piezoelectricity* (McGraw-Hill Book Company, Inc., New York, 1946).
Helen D. Megaw, *Ferroelectricity in crystals* (Methuen & Co., Ltd., London, 1957).
G. Shirane, F. Jona, and R. Pepinsky, Some aspects of ferroelectricity, *Proc. Inst. Radio Engrs.*, vol. 43 (1955), pp. 1738–1793.
Charles P. Smyth, *Dielectric behavior and structure* (McGraw-Hill Book Company, Inc., New York, 1955).
Arthur R. von Hippel, *Dielectrics and waves* (John Wiley & Sons, Inc., New York, 1954).
Arthur R. von Hippel (ed.), *Molecular science and molecular engineering* (The Technology Press of The Massachusetts Institute of Technology and John Wiley & Sons, Inc., New York, 1959).

Suggestions for further reading

Peter Debye, *Polar molecules* (Dover Publications, Inc., New York, 1945).
Adrianus J. Dekker, *Solid state physics* (Prentice-Hall, Inc., Englewood Cliffs, N.J., 1957).
Charles Kittel, *Introduction to solid state physics*, 2d ed. (John Wiley & Sons., Inc. New York, 1956).

Exercises

1. Two parallel plates, 0.10×0.25 m^2 in area, are separated by sodium nitrate, $\kappa_e = 5.2$, and are permanently connected to a 250-V battery.
  (a) What is the capacitance of this capacitor for $l = 0.05$m?
  (b) What is the charge on the plates?
  (c) What is the induced dipole moment per unit volume in the dielectric?
  (d) What is the electric-field intensity in the dielectric?
2. A capacitor is constructed from two concentric metal spheres whose radii are 3 and 4 cm, respectively. The space separating the two spheres is completely filled with sulfur, $\kappa_e = 4.0$. What is the capacity of this spherical capacitor? (ANSWER: 5.33×10^{-11}F.)
3. Consider the parallel-plate capacitor in Fig. 1. Why must the ends of the needle-shaped cavity be small? Assuming that the ends have an area of 4 mm^2, what is the force acting on a test charge placed into such a cavity?
4. What is the dipole moment of a NaCl molecule in a vapor? Assume that the molecule consists of Na^{1+} and Cl^{1-} ions separated by 2.5 Å.
5. Consider an ideal dipole gas placed in a field intensity of 3×10^5 V/m. The dipole moment of a gas molecule is of the order three *debye* $= 10^{-29}$ C-m. Calculate the potential energy $\mu\varepsilon$ of a dipole in this field in mks units and show that kT at room temperature, expressed in joules, is about 1,000 times larger.
6. When a dielectric is placed in an alternating field $\varepsilon = \varepsilon_0 e^{i\omega t}$, the displacement D usually lags behind. This can be expressed by

$$D = D_0 \cos(\omega t - \delta) = D_0 (\cos \delta \cos \omega t + \sin \delta \sin \omega t)$$
$$= D_1 \cos \omega t + D_2 \sin \omega t.$$

Expressing the complex dielectric constant $\kappa_e^* = \kappa_e' - i\kappa_e''$, the relation between D and $\mathcal{E}$ becomes $D = \kappa_e^* \epsilon_0 \mathcal{E}_0 e^{i\omega t}$. Show that this leads to a relation for the so-called *loss factor* $\tan \delta = \kappa_e''/\kappa_e'$.

7. The current density in a capacitor of unit area is

$$J = \frac{dD}{dt}$$

and the energy dissipated per second in a unit volume of the dielectric is given by

$$\frac{\omega}{2\pi} \int_0^{2\pi/\omega} I\mathcal{E}\, dt.$$

Show that the energy losses in a dielectric placed in an alternating field of intensity $\mathcal{E} = \mathcal{E}_0 \cos \omega t$ are proportional to $\sin \delta$ by making use of the results of Exercise 6. Note that the *loss factor* $\tan \delta = \sin \delta$ only when δ is small.

8. Derive an expression for the frequency-dependent complex dielectric constant $\kappa_e^* = \kappa_e' - i\kappa_e''$ by analogy with the derivation of (39) in the text. HINT: The polarization density of an ideal gas $P^* = \epsilon_0(\kappa_e^* - 1)\mathcal{E}_0 e^{i\omega t} = 4\pi\epsilon_0 N\alpha_e^* \mathcal{E}_0 e^{i\omega t}$.

9. Derive Eq. (42) in the text by assuming that the Lorentz field correctly gives the internal field in a solid dielectric. HINT: The polarization density now is $P^* = \epsilon_0(\kappa_e^* - 1)\mathcal{E}_0 e^{i\omega t} = 4\pi\epsilon_0 N\alpha_e^* \frac{1}{3}(\kappa_e^* + 2)\mathcal{E}_0 e^{i\omega t}$.

10. The resonance frequency of a piezoelectric crystal is given by

$$\nu_0 = \frac{v}{2w} = \frac{1}{2w}\sqrt{\frac{E}{\rho}}$$

where v is velocity of sound in crystal
$\quad w$ is width of crystal
$\quad E$ is Young's modulus
$\quad \rho$ is density of crystal.

(a) What is the resonance frequency of a quartz crystal 2.5 mm wide?

(b) What must be the size of a crystal to be useful in the frequency control of oscillators in the kilocycle range? In the megacycle range?

(c) Is quartz suitable for use in both ranges? (Look up the values of density and Young's modulus in a handbook.)

11. Consider the piezoelectric crystal shown in Fig. 7.

(a) When the electrodes are not connected to each other, what is the open-circuit voltage when a stress σ is applied?

(b) What is the mechanical strain in this case?

(c) What is the electrical polarization in this case?

(d) How do these quantities compare with the case of short-circuited electrodes?

12. Assume that only Na^{1+} contributes to the ionic conductivity in NaCl. If the measured conductivity of NaCl at 600 deg K is $10^{-4}\ \Omega^{-1}\text{-m}^{-1}$, what is the value of the diffusion coefficient per unit area for sodium in NaCl ($a_{\text{NaCl}} = 5.63\ \text{Å}$)?

13. magnetic processes

When a material is placed in an inhomogeneous magnetic field, it is either attracted by the strong part of the field or repelled by it. This means that a magnetic field is induced in the material and interacts with the external field. When the induced field opposes the external magnetic field the phenomenon is called *diamagnetism*. When the induced field aids the external one it is called *paramagnetism*. The total magnetization induced in the material is proportional to the applied-field strength. Defining this magnetization M as the total magnetic moment per unit volume, it can be related to the applied magnetic intensity H by

$$M = \chi H \tag{1}$$

where χ is called the *magnetic susceptibility* of the material.† If the material contains N magnetizable particles per unit volume, each of which has a magnetizability α_m, the magnetic susceptibility per unit volume of the material is obviously

$$\chi = N\alpha_m. \tag{2}$$

Note that the magnetizability is the magnetic counterpart of electrostatic polarizability discussed in Chap. 12. The origin of the magnetiza-

† In addition to the magnetic-field strength H, another quantity is commonly used in the discussion of magnetism, namely, the *magnetic flux density* B (analogous to the electric displacement D discussed in the preceding chapter), which is defined

$$B = \mu_P H$$

where μ_P is the *magnetic permittivity* of a material. The magnetic permittivity of empty space, μ_0, is $4\pi \times 10^{-7}$ H/m (henry per meter).

bility α_m in (2) can be explained by considering the structure of the atoms comprising a material.

According to Ampère's law, a moving current I encircling an area A produces a magnetic field at right angles to the plane of the current loop. Provided that the radius of the loop is small in comparison with the distance from the loop at which this magnetic field is measured, the field produced is identical with that of a magnetic dipole whose moment is

$$\mathbf{\mu} = IA\mathbf{n} \tag{3}$$

where $\mathbf{n}$ is a unit vector normal to the plane of the moving current. Consider an electron of charge $-e$ describing a circular orbit of radius r (Fig. 1). If the angular velocity of the electron is ω, the current flowing in the loop is $-e\omega/2\pi$ and the magnetic dipole moment associated with this electron orbit then has the magnitude

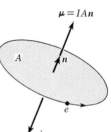

$\mathbf{\mu} = IA\mathbf{n}$

$$\mu = -\frac{e\omega}{2\pi} \times \pi r^2 = -\frac{e\omega r^2}{2} \tag{4}$$

The angular momentum of the electron $m' = m\omega r^2$, where m is the mass of the electron, so that

$$\mu = -\frac{e\omega r^2}{2} \times \frac{m}{m} = -\frac{e}{2m} m'. \tag{5}$$

FIG. 1

The minus sign in (5) means that the angular-momentum vector points in the opposite direction from the dipole-moment vector as shown in Fig. 1.

It is clear from the above discussion that a moving electron in an atom has associated with it a magnetic field. Actually, two kinds of motion must be distinguished. One is the orbital motion, determined by the azimuthal quantum number l, whose angular momentum can assume $m = 2l + 1$ specific orientations in space relative to some external-field direction. The other is the moment associated with the electron spin, which can have either of two opposite directions. The orbital angular momentum can be combined vectorially with the spin to give the total angular momentum represented by the quantum number $j = l + m_s$. In discussing atoms containing several electrons, it is convenient to combine all the l vectors to form a resultant L, all the spins m_s to form a resultant S, and finally the total angular momentum J formed by vectorial combination of L and S. Before describing the relative importance of each type of motion, it is necessary to consider the interactions between the magnetic moments in an atom and an external field.

Atomic considerations

Zeeman effect. The spherical charge distribution in an *s* orbital can have a density variation that depends only on the radial distance from the nucleus. There is no preferred directional motion for the electron, so that an *s* orbital does not have a net angular momentum and, according to (5), no magnetic moment. All other orbitals have electron densities that vary as a function of three space coordinates, so that they have nonspherical charge distributions and a net angular momentum. In an isolated atom the moments of the various lobes of a particular orbital average to zero because their relative orientations in space cannot be

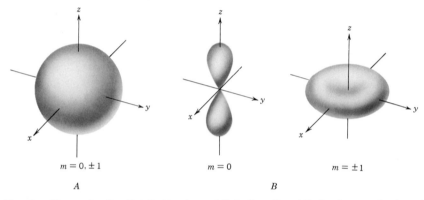

$$m = 0, \pm 1 \qquad\qquad m = 0 \qquad\qquad m = \pm 1$$

$$A \qquad\qquad\qquad\qquad B$$

Fig. 2.   Charge density distribution in *p* orbitals (*l* = 1).   (*A*) In absence of external field the spatial distributions cannot be resolved; (*B*) in presence of a unidirectional field parallel to *z*, two different orientations for *m* = 0 and *m* = ±1 can be resolved.

specified uniquely. The charge density then appears to be distributed as shown in Fig. 2*A*. In the presence of an external field, however, the permanent magnetic dipole moments associated with each type of orbital motion can be distinguished. Consider the three possible *p* orbitals in the presence of a magnetic field directed along the *z* direction. The coupling between the magnetic moments of these orbitals and the external field leads to a splitting of the energy of the otherwise degenerate *p* level into three distinct energy levels as shown in Fig. 3. This leads to the well-known Zeeman effect and is a consequence of the interaction energy, which is determined by the inclination of the dipole to the field direction. The possible magnetic-moment components along the *z* direction in Fig. 3 are determined by (5) to be

$$-\frac{e}{2m} \times \frac{h}{2\pi}, \qquad 0, \qquad +\frac{e}{2m} \times \frac{h}{2\pi}. \tag{6}$$

In view of (6) it is convenient to define a quantity called the *Bohr magneton:*

$$\mu_B = \frac{eh}{4\pi m} = 9.27 \times 10^{-24} \text{ J/T}$$

where J/T is joule per tesla (joule per weber per square meter).

Assuming a homogeneous magnetic field, that is, if the intensity of the field at right angles to z is uniform, one of the p orbitals ($m = 0$) aligns itself parallel to z while the other two ($m = \pm 1$) lie in the xy plane

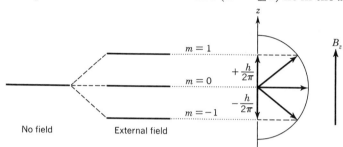

FIG. 3. Splitting of energy levels corresponding to p orbitals (shown to the left) due to space quantization of the orbital magnetic moments (shown to the right). The components of the angular momentum along the applied-field direction differ by $h/2\pi$; the magnetic moment components differ by $h/2\pi \times e/2m$.

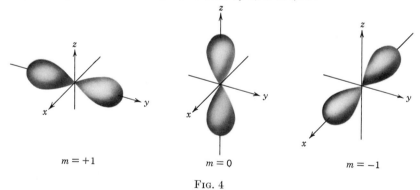

FIG. 4

(Fig. 2B). Their exact orientation relative to x and y is not fixed by the field, so that the charge density of these two orbitals appears as shown to the right in Fig. 2B. Note that the uncertainty principle is not disturbed by this orientation of the orbitals since the direction of the angular-momentum vector is not fixed about the z axis. If the external magnetic field is not unidirectional, however, all three orbitals become localized in space. This is the case, for example, when the atom is in a crystal, that is, in a field produced by the surrounding atoms. The spatial orientation of the three p orbitals in a cubic field is shown in Fig. 4. Analogous distributions of the five d orbitals in a cubic field are shown on page 73. In the presence of a unidirectional field, these five orbitals can have only three different distributions corresponding to $m = 0$, $m = \pm 1$, and $m = \pm 2$, obtained by rotating the localized orbitals about the z direction.

When the interaction of the spin moments with an external field is considered, it appears by analogy with (6) that the magnetic-moment component along the field should be

$$\frac{e}{2m} \times \frac{\frac{1}{2}h}{2\pi} = \frac{1}{2}\mu_B. \tag{7}$$

Actually, this is only approximately correct because (5) does not apply to electron spins. It turns out that the magnetic-moment component along the field direction is equal to g times $\frac{1}{2}$ Bohr magneton, where $g = 2.0023$.

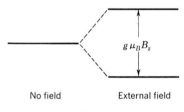

No field   External field

Fig. 5

This produces a splitting of the energy level into two levels as shown in Fig. 5. The difference in energy between the two levels is

$$\Delta E = g\mu_B B_z \tag{8}$$

so that g can be said to measure the amount by which the level is split up and is usually called the *spectroscopic splitting factor* for that reason.

In an atom, the positively charged nucleus is also spinning, so that it has a magnetic moment, usually expressed in *nuclear magnetons*. By analogy with the Bohr magneton, a nuclear magneton is defined

$$\mu_n = \frac{eh}{4\pi m_p} = 5.05 \times 10^{-27} \text{ J/T} \tag{9}$$

where m_p is the mass of a proton. Note that the nuclear magneton is about a thousand times smaller than a Bohr magneton, so that the nuclear contribution to the magnetizability of an atom is usually neglected. As shown later in this chapter, it is nevertheless possible to make use of the weak nuclear moments to study molecular structure by observing their interaction with external magnetic fields.

Magnetic gyroscopes. The angular momenta of the orbitals shown in Fig. 2B precess about the applied-field direction so that the magnetic moments behave like quantized gyroscopes. The magnetic moment μ is related to the quantized mechanical angular momentum m' by the *gyromagnetic ratio*

$$\gamma = \frac{\mu}{m'} = \frac{e}{2m} \tag{10}$$

according to (5). A magnetic field acting in the z direction (Fig. 6)

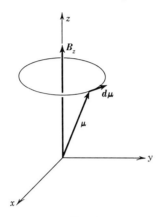

FIG. 6

produces a torque on the dipole whose magnitude is

$$\frac{\text{Torque}}{\text{Volume}} = \mathbf{\mu} \times \mathbf{B}. \tag{11}$$

The torque causes a time rate of change of the angular momentum

$$\frac{d\mathbf{m'}}{dt} = \mathbf{\mu} \times \mathbf{B} \tag{12}$$

and multiplying both sides of (12) by γ,

$$\frac{d\mathbf{\mu}}{dt} = \gamma(\mathbf{\mu} \times \mathbf{B}). \tag{13}$$

The three components of $d\mu/dt$ along x, y, and z are

$$\begin{aligned}
\frac{d\mu_x}{dt} &= \gamma(\mu_y B_z - \mu_z B_y) \\
\frac{d\mu_y}{dt} &= \gamma(\mu_z B_x - \mu_x B_z) \\
\frac{d\mu_z}{dt} &= \gamma(\mu_x B_y - \mu_y B_x).
\end{aligned} \tag{14}$$

If the field acts only in the z direction, that is, if $B_x = B_y = 0$, then $d\mu_z/dt = 0$ and the relations in (14) become

$$\frac{d\mu_x}{dt} = \gamma\mu_y B_z$$
$$\frac{d\mu_y}{dt} = -\gamma\mu_x B_z. \tag{15}$$

and

In order to solve these two simultaneous equations, differentiate both sides of each equation in (15) with respect to t.

$$\frac{d^2\mu_x}{dt^2} = -\gamma^2 B_z^2 \mu_x$$
$$\frac{d^2\mu_y}{dt^2} = -\gamma^2 B_z^2 \mu_y \tag{16}$$

and

after substituting (15) for $d\mu_x/dt$ and $d\mu_y/dt$ on the right side of (16). It is clear from (16) that $d\mathbf{\mu}/dt$ lies in the xy plane at right angles to the applied-field direction.

The two components in (16) can be compared with a linear oscillator whose frequency is

$$\omega_L = \gamma B_z = \frac{e}{2m} B_z \tag{17}$$

and is called the *Larmor frequency*. Note that one component of the torque in (15) is positive while the other is negative, so that they are 90 deg out of phase with each other in space and time. This is the reason for the precession of the dipole moment about z. The Larmor frequency is proportional to the applied-field strength and the gyromagnetic ratio, but not to the angular position, so that it is not quantized.

Magnetic susceptibility

Diamagnetism. The total magnetic moment per unit volume induced in a material by an external field is determined by the magnetic susceptibility of the material according to (1). From the above discussion it is clear that the plane of the electronic orbital precesses about the applied-field direction and produces an induced magnetic moment. According to Lenz's law, the magnetic field produced by an induced current opposes the change in the magnetic field which produces it. The reason for this can be seen with the aid of (5). The angular momentum of the precessing orbital is $m\omega_L r_{xy}^2$, where m is the mass of an electron and $r_{xy} = (x^2 + y^2)^{\frac{1}{2}}$ is the projection of the orbital radius on a plane normal to the applied-field direction. The induced magnetic moment then is

$$\mu_{\text{induced}} = -\frac{e}{2m} \times m\omega_L r_{xy}^2$$

$$= -\frac{e}{2m} \times \frac{emB_z}{2m} r_{xy}^2$$

$$= -\frac{e^2 B_z}{4m} r_{xy}^2 \tag{18}$$

after substituting (17) for ω_L. The minus sign in (18) shows that the induced moment is opposed to the applied field B_z in agreement with Lenz's law.

When a spherical charge distribution in an atom is considered, the mean square distance of the electrons from the nucleus is

$$\bar{r}^2 = \bar{x}^2 + \bar{y}^2 + \bar{z}^2 = \tfrac{3}{2}\bar{r}_{xy}^2. \tag{19}$$

The total magnetic moment per kilogram molecule can be determined next by substituting (19) in (18) and multiplying it by Avogadro's number N_0. The diamagnetic susceptibility χ_D then becomes

$$\chi_D = \frac{M}{H} = \frac{\mu_0 N_0 e^2}{6m} \sum \bar{r}^2 \tag{20}$$

and is known as *Langevin's equation*, in which the mean square orbital radius is summed over all the orbitals in the atom. In solids containing N atoms per unit volume, N_0 should be replaced by NZ, where Z is the number of electrons each atom contains.

Langevin's equation remains valid when quantum mechanics is applied provided that the electron distribution is symmetrical and $\bar{r}^2$ is calculated from electronic wave functions. Since $\chi_D \propto \bar{r}^2$, the outer electrons make the largest contribution to the diamagnetic susceptibility. Assuming $\bar{r}^2 \simeq 10^{-20}$ m², the magnitude of the diamagnetic susceptibility is approximately equal to 10^{-5} J/T^2. A very nearly spherical charge distribution obtains when the outer shell of an atom is closed. This is the case in inert gases and in ions, particularly cations. The diamagnetic susceptibilities of a few atoms are compared as follows:

	H_2	He	Li^{1+}	Na^{1+}	F^{1-}	Cl^{1-}	
χ_D	-2.01	-1.91	-0.7	-6.1	-9.4	-24.2×10^{-5} J/T²	(21)

The values in (21) are approximate because the susceptibility is affected by the atomic environment, or crystal structure. It is clearly evident, however, that χ_D increases as the atomic number increases, as predicted by the theory described above. Similarly, when the diamagnetic susceptibilities of ions having the same number of electrons but a different nuclear charge are compared, the susceptibility should be inversely proportional to the nuclear charge because the higher charge draws the outer electrons in closer, decreasing $\bar{r}$.   That this is so is illustrated as follows:

$$
\begin{array}{cccccc}
 & \text{S}^{2-} & \text{Cl}^{1-} & \text{A} & \text{K}^{1+} & \text{Ca}^{2+} & \\
\chi_D & -27.6 & -24.2 & -19.1 & -14.6 & -10.7 \times 10^{-5}\,\text{J/T}^2 & \quad(22)
\end{array}
$$

The influence of the atomic environment on the susceptibility was first studied by Pascal, who found that the molecular susceptilibity of organic molecules is equal to the sum of the atomic susceptibilities plus additional terms due to the nature of the intramolecular bonds.   It also has been found that crystals having anisotropic structures exhibit marked anisotropies in their diamagnetic susceptibilities.   Elements like antimony, bismuth, and graphite have susceptibilities that can be as large as 10^{-3}. Nevertheless, the diamagnetic susceptibility of solids is clearly very small, so that the diamagnetism produced by an external field is small according to (1).

The agreement between measured and calculated values of the diamagnetic susceptibility is remarkably good for many materials.   In fact, it is possible to test the correctness of the wave functions used to calculate $\bar{r}^2$ by this means.   Because the diamagnetic susceptibility is very small, however, its effect on the magnetic properties of most materials has little practical significance.   The force F with which a diamagnetic material of volume $\mho$ is repelled when placed in a magnetic field of strength B and gradient dB/dz is determined by

$$ F = \chi_D \mho B \frac{dB}{dz}. \tag{23} $$

Since χ_D is negative, the force acts in a direction opposite to that of increasing field gradient dB/dz and the material is repelled by the strong part of the field.

The electric currents which produce the magnetic dipoles discussed above are limited to atomic or, at best, to molecular orbitals and are believed to circulate without encountering any resistance.   An analogous situation arises in a metal, in which currents formed by free electrons can be similarly induced and are called *eddy currents*.   The analogy breaks down, however, because eddy currents are opposed by the electrical resistance of the metal, so that they do not build up a significant magnetic moment.   By comparison, the eddy currents in a superconductor are unopposed and extend throughout the entire conductor, so that a magnetic moment proportional to the field is produced and remains even after the external field is removed.   This ability of superconductors to retain the currents induced by an external field for long periods makes them very useful for devices requiring a "memory," such as computers, for example.

Paramagnetism. As shown above, when an atom has a closed outer shell, the spin moments cancel each other and a weak diamagnetism due to the orbital motion of the electrons results. The spin moments of unpaired electrons, however, can align themselves either parallel or antiparallel to the field. Electrons that have parallel spin components have a lower energy (Fig. 5), so that such alignment is preferred. This leads to an induced field that aids the applied field, so that M/H is positive. (The magnetic moments due to the orbital angular momenta of unpaired electrons similarly can have components parallel and antiparallel to the external field.) This effect is called *paramagnetism* and was first studied systematically by P. Curie late in the nineteenth century. Curie found that the paramagnetic susceptibility of a material χ_P is inversely proportional to the absolute temperature.

$$\chi_P = \frac{C}{T} \tag{24}$$

where C is called the Curie constant and is a positive quantity characteristic of the material. Note that a positive susceptibility means that a paramagnetic material is attracted toward the strong end of a magnetic field according to (23).

Langevin later applied classical statistical mechanics to paramagnetism in gases and found that the Curie constant could be expressed theoretically.

$$C = \frac{\mu_0 N_0 \bar{\mu}^2}{3k} \tag{25}$$

where N_0 is Avogadro's number

    k is Boltzmann's constant

    $\bar{\mu}$ is average magnetic moment per molecule.

It is a fairly simple matter to calculate the paramagnetic susceptibility of a solid material when interactions between the electrons are neglected. Suppose a material placed in a magnetic field of strength H contains unpaired electrons contributing a total of n moments per unit volume of which n_- have spins that are parallel to the field and n_+ have antiparallel spins. The potential energy for the parallel magnetic moment is

$$-V = -\mu_B B.$$

The number of moments per unit volume of each type can then be determined from the Boltzmann relation, so that

$$n_- = Kne^{+V/kT}$$

and $$n_+ = Kne^{-V/kT} \tag{26}$$

where K is a proportionality constant, and the net magnetization per unit volume

$$M = \mu_B(n_- - n_+) = \mu_B Kn(e^{V/kT} - e^{-V/kT}). \tag{27}$$

The statistically averaged moment per dipole is equal to the net magnetization divided by the total number of moments per unit volume $n = n_- + n_+$, so that, utilizing (26) and (27),

$$\bar{\mu} = \frac{M}{n} = \mu_B \frac{Kn(e^{V/kT} - e^{-V/kT})}{Kn(e^{V/kT} + e^{-V/kT})}$$

$$= \mu_B \tanh \frac{V}{kt}. \tag{28}$$

As already shown, $V \ll kT$, so that $\tanh(V/kT) \simeq (V/kT)$ and

$$\bar{\mu} = \frac{\mu_B^2 B}{kT}. \tag{29}$$

The magnetic susceptibility defined in (1) then becomes

$$\chi_P = \frac{M}{H} = \frac{n\bar{\mu}}{H} = \frac{\mu_0 n \mu_B^2}{kT}. \tag{30}$$

The paramagnetic susceptibility in (30) is inversely proportional to the absolute temperature T, a result similar to others obtained by applying statistical mechanics to solids. It also explains why the induced magnetization M disappears immediately after the external field H is removed, since the alignment of the magnetic moments depends on the presence of an external field. Once the field is removed, the energy difference between the two kinds of moments disappears and thermal motion of the electrons randomizes their orientations, so that, on the average, the net magnetization $M = 0$.

The Curie constant corresponding to (30) is

$$C = \chi_P T = \frac{\mu_0 n \mu_B^2}{k} \tag{31}$$

and differs slightly from the Langevin relation (25). This difference arises from the fact that the atoms in a gas enjoy three degrees of freedom, so that the number of magnetic dipoles per kilogram-mole is given by $N_0/3$. [It is left to Exercises 8 and 9 to show that (31) reduces to (25) when the quantum-mechanical procedure for calculating spin angular momenta is utilized.] Relation (30) correctly describes the magnitude and temperature dependence of susceptibility for all paramagnetic materials except metals. According to (30), the paramagnetic susceptibility of a metal containing about 10^{28} dipoles/m^3 is 10^{-3} J/T^2 at room temperature, whereas the experimentally determined value is about 10^{-5} J/T². This hundredfold disparity comes about because classical mechanics assumes that all the free electrons in a metal can gain energy from the magnetic field. (See also the discussion of electronic specific heat in Chap. 6.) Similarly, the temperature dependence of the paramagnetic susceptibility predicted by the Curie relation (24) is not observed in metals as it is in other materials.

The observed weak and temperature-independent paramagnetism of metals can be explained by applying Fermi-Dirac statistics to this problem. The distribution of electrons in a metal is shown in Fig. 7 by

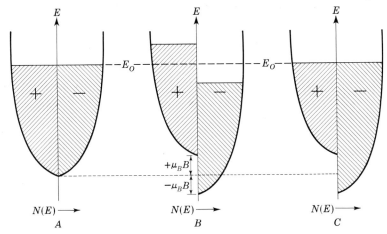

FIG. 7. Density-of-states curves for a metal. The density of states is shown twice, the states having plus spins to the left, and those having minus spins to the right, of the energy axis. (*A*) Distribution of states in absence of external field; (*B*) instantaneous shift of energies when field is applied; (*C*) equilibrium occupation of states while external field is present.

plotting the parabola-shaped function representing the density of states as a function of energy. In the absence of an external magnetic field, the electrons occupy all the available states having energies less than the Fermi energy E_0. They are shown divided into two groups in Fig. 7*A*, accordingly as their spins are plus or minus. When an external field is applied, the magnetic moments due to the spins line up either parallel or antiparallel to the field. Assuming that a plus denotes electrons which have **antiparallel** moments, these electrons undergo a shift in energy $+\mu_B B$, as shown in Fig. 7*B*. This situation is not stable, and at equilibrium some of the electrons in the antiparallel states undergo transitions to the lower-energy parallel states. This leads to the situation shown in Fig. 7*C*. The magnitude of the energy shift, even due to a fairly strong field, $B = 10$ T, is very small, $\mu_B B \simeq 10^{-3}$ eV, so that at room temperature the thermal energy, $kT \simeq 0.03$ eV, is sufficiently greater to keep the number of electrons having parallel and antiparallel moments very nearly equal. Not only does this explain the very small value of the observed susceptibility, but also why the susceptibility is virtually independent of reasonable changes in temperature.

In addition to metals, other materials also can be paramagnetic. The basic requirement for paramagnetism is that the atoms must have electrons in partially filled shells since the magnetic moments due to orbital angular momenta and spins compensate to zero in a closed shell. This means that ionic crystals such as the alkali halides or covalent crystals in which each orbital contains two electrons with opposite spin cannot be paramagnetic. On the other hand, the atoms in the transition series and the rare-earth elements have incomplete shells, so that they can form paramagnetic compounds. For example, the $4f$ electrons in rare-earth elements lie "inside" the atoms and are effectively screened from adjacent atoms by partly filled $5s$ and $5p$ shells. Thus their magnetic properties in solids are quite similar to those of isolated atoms. Since there are $14\,f$ states and only $10\,d$ states, the paramagnetism of the rare earths is generally greater than that of the transition elements. Quantum-mechanical calculations of the expected paramagnetic susceptibility make use of Hund's rule requiring unpaired electrons to have parallel spins and are fairly complicated. As shown by J. H. Van Vleck, when the higher allowed energy levels are considered, the calculated and experimentally determined susceptibilities agree fairly well. An interesting consequence of such calculations is that the effective number of electrons per atom contributing to the magnetic moment turns out to be nonintegral. This is further discussed in connection with the magnetic properties of the iron group in the next section.

Ferromagnetism

In the discussion of magnetism so far, the possibility of an interaction between the magnetic moments of adjacent atoms has been neglected. This is a reasonable procedure to follow when the atoms have closed outer shells, in which case the induced diamagnetism is so weak that such interactions can be safely neglected. Similarly, the unfilled inner $4f$ shells of the rare earths are sufficiently screened by the outer electrons so that the paramagnetism is essentially due to the magnetization of isolated atoms. When the transition metals are considered, however, the outermost shells contain unpaired electrons, and it is reasonable to expect that they can interact with similarly unpaired electrons in neighboring atoms. Thus it is not particularly surprising to find that iron retains a permanent magnetization even after the external magnetic field has been removed.

This permanent magnetization was first explained by P. J. Weiss, in 1907, who postulated that an internal molecular field was present in such ferromagnetic materials which favored the parallel alignment of the magnetic moments of adjacent atoms. The Weiss field wM is frequently called an exchange field because it presumably comes about by means of a quantum-mechanical exchange interaction between the electrons of adjacent atoms. It is proportional to the magnetization M, so that the effective field acting inside the material can be expressed

$$H_{\text{effective}} = H + wM. \tag{32}$$

By analogy to a paramagnetic material, the net magnetization per unit volume is obtained by combining (1), (24), and (32), so that

$$M = \frac{C}{T}(H + wM). \tag{33}$$

Recalling that $\chi = M/H$, divide both sides of (33) by H, so that

$$\chi = \frac{C}{T} + w\frac{C}{T}\chi$$

and

$$\chi\left(1 - w\frac{C}{T}\right) = \frac{C}{T}$$

and finally,

$$\chi = \frac{C}{T - wC} = \frac{C}{T - \Theta} \tag{34}$$

where $\Theta = wC$ is a constant characteristic of the material, normally called the *Curie temperature*. In order to see the difference that the Weiss field makes in the paramagnetic susceptibility dependence on temperature, the Curie relation (24) is shown plotted in Fig. 8A and the Curie-Weiss relation (34) in Fig. 8B. As long as $T \gg \Theta$, both curves are similar. As $T \to \Theta$, the susceptibility tends to infinity, and for temperatures less than the Curie temperature, it has no meaning; that is, there exists a spontaneous magnetization in the material without the presence of an external field. This is the phenomenon of *ferromagnetism* and is analogous to the polarization catastrophy in ferroelectrics. Equation (34) is known as the Curie-Weiss law.

The magnetic moments due to orbital motion of nearly free electrons in a metal is negligible in comparison with the spin moments. Whereas the gyromagnetic ratio for orbital moments is $e/2m$, for ferromagnetic materials it is about e/m, the same value that it has for electron spins. Assuming, therefore, that only the spin system need be considered, it is possible to determine the magnetic behavior of a ferromagnetic material below the Curie temperature.   In the absence of an external field,

$$H_{\text{effective}} = wM_s$$

where $M_s = n\bar{\mu}$ is the spontaneous magnetization at some temperature 0 deg K $< T < \Theta$.   According to (28),

$$\frac{\bar{\mu}}{\mu_B} = \tanh \frac{\mathbf{u}_0 \mu_B w M_s}{kT}. \tag{35}$$

Multiplying the numerator and denominator on the left of (35) by the number of moments per unit volume n,

$$\frac{n\bar{\mu}}{n\mu_B} = \frac{M_s}{M_0} \tag{36}$$

where M_0 is the spontaneous magnetization at $T = 0$ deg K when all the available spins are lined up parallel to each other.

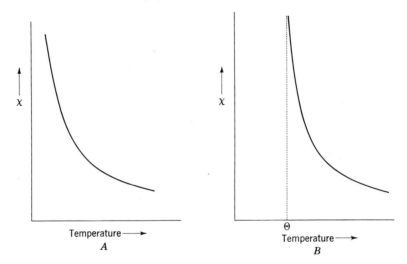

Fig. 8.   Paramagnetic susceptibility dependence on temperature (A) according to the Curie relation; (B) according to the Curie-Weiss law.

Combining (36) with (35) and substituting Θ/C for w,

$$\frac{M_s}{M_0} = \tanh \left(\frac{\Theta}{T} \frac{M_s}{M_0} \right). \tag{37}$$

Equation (37) expresses the relation between the normalized magnetization M_s/M_0 and the normalized temperature T/Θ.   It is called a thermomagnetic equation of state for the ferromagnetic phase of a material and is shown plotted in Fig. 9. At zero degrees, the magnetization is at a maximum and $M_s/M_0 = 1.0$. As the temperature increases, thermal motion randomizes successively more of the parallel alignments, until, just below the Curie temperature, the alignment starts to disappear very rapidly and, above the Curie temperature, the spontaneous magnetization is gone. This behavior is quite similar to order-disorder transitions in binary allows in which an ordered array of atoms at low temperatures becomes progressively more disordered with increasing temperature until, above the transition temperature, all vestiges of an ordered arrangement are lost. The theoretical curve shown in Fig. 9 is very closely followed by the actual curves experimentally derived for Fe, Co, and Ni,

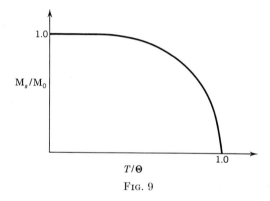

FIG. 9

although their respective spontaneous magnetization and Curie constants are quite different.

Exchange interactions. The first theoretical explanation of the Weiss field in ferromagnetic materials was proposed by Heisenberg in 1928. He used a quantum-mechanical approach similar to the Heitler-London treatment of the hydrogen molecule discussed in Chap. 3. In this treatment, an exchange interaction between electrons in different quantum states is shown to lead to a lower energy provided that the spin quantum numbers of both states are the same; that is, the spins are parallel. By analogy to the hydrogen molecule, the strength of the exchange interaction depends rather critically on orbital overlap, that is, on the interatomic separation, and it may in fact change its sign as this separation is varied. It can be shown that, as two atoms approach each other, the electron spins of unpaired electrons in each atom assume parallel orientations. As they are brought closer together, the spin moments are maintained parallel by increasing forces. As the interatomic distance is decreased still further, however, these exchange forces decrease, until finally they pass through zero and an antiparallel spin orientation is favored. This can be expressed quantitatively by a potential-energy difference V_{ij}, which is proportional to the dot product of the total spins $\mathbf{S}_i$ and $\mathbf{S}_j$, representing the angular moments (in units of $h/2\pi$) of the ith and jth states:

$$V_{ij} = -J_{ij}\mathbf{S}_i \times \mathbf{S}_j \qquad (38)$$

where J_{ij}, the exchange integral, measures the interaction between the spins and is a function of the interatomic separation. This dependence of the exchange integral is illustrated in Fig. 10. It was first shown by

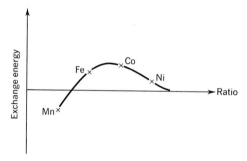

FIG. 10.   Dependence of exchange integral on interatomic separation expressed as a ratio of one-half the interatomic distance in a crystal and the average radius of the 3d shell. The corresponding values for Mn, Fe, Co, and Ni are indicated by crosses.

Bethe that, for the iron group, the conditions favoring parallel orientations occur when the ratio between one-half the interatomic distance in a crystal and the average radius of the 3d shell is greater than 1.5. Some typical values of this ratio for these metals are as follows:

	Mn	Fe	Co	Ni	
Ratio	1.47	1.63	1.82	1.98	(39)

As can be seen, the ratio for manganese is slightly less than 1.5, which explains why manganese crystals are not ferromagnetic whereas compounds containing manganese atoms spaced farther apart are. Notable examples are the Heusler alloys Cu_2MnSn and Cu_2MnAl. (Note that these ferromagnetic alloys contain neither Fe, Co, nor Ni.)

According to the zone theory, two factors determine the internal energy of a metal crystal. The first is the kinetic energy of the electrons determined by the Fermi energy, which makes a positive contribution to the total energy, and the second comes from the exchange interaction between the electrons and makes a negative contribution. A crystal, then, is ferromagnetic when the negative energy due to the exchange integral is greater than the positive-energy term due to the kinetic (Fermi) energy of the electrons, which tends to randomize the spin directions. The exchange integral used in this calculation is somewhat different from the one proposed by Heisenberg, in that it is based on the electrostatic interaction between one electron and the electrons in an identical fictitious crystal having a hole in place of this electron. It turns out that this exchange integral is independent of interatomic distance and does not vary too much from atom to atom in the metal. In order to explain ferromagnetism, therefore, it is necessary to compare the Fermi energies of different metals. Although only nickel has been examined in detail, the results can be extended to show that the Fermi energy is smaller than the exchange energy in Fe, Co, and Ni. As expected, it is somewhat greater in the case of Mn. It should be noted that a small Fermi energy denotes a narrow energy zone, that is, a large number of electrons having very closely spaced energies. Thus this calculation supports the earlier conclusions that narrow, partially filled d zones and d orbitals, which do not overlap in neighboring atoms to any appreciable extent, are necessary and sufficient for ferromagnetism to occur.

More recently, Zener has proposed a different explanation for the occurrence of ferromagnetism. The exchange integral, as calculated in the Heitler-London model for the bonding of two atoms, is negative. Heisenberg, therefore, was forced to postulate that the exchange integral was negative only when the atoms were brought very close together and became positive when the unfilled d shells did not overlap appreciably. It has never been proved, however, that such a reversal in the sign of the exchange integral actually occurs. Zener has suggested that the exchange integral always remains negative and that ferromagnetism is due to an interaction between electrons occupying quantum states in the partially filled d and s zones. The proposed interaction, which accounts for the excess of parallel spin moments of one kind, is similar to the interaction (Hund's rule) that causes unpaired electrons in different orbitals to have parallel spins. According to this model, an overall decrease in the energy is obtained when the electrons occupying states in the d zone have parallel spin moments and the s-zone electrons contribute a small moment in the same direction. Because of the approximate nature of the calculations used to support the above-stated competing theories, it is not possible to decide, at present, which one describes most correctly the interactions responsible for ferromagnetism.

On an atomic scale, the reason for the permanent magnetic moment of Fe, Co, and Ni atoms is best understood by considering their electronic structures. Each atom has two $4s$ electrons and an incomplete $3d$ shell.

If interactions favoring parallel spins are assumed, the electrons are distributed among the available states as follows:

$$\text{Fe} \quad 3d^6 \;\boxed{\uparrow\downarrow}\,\boxed{\uparrow}\;\boxed{\uparrow}\;\boxed{\uparrow}\;\boxed{\uparrow}\qquad 4s^2\;\boxed{\uparrow\;\downarrow}$$

$$\text{Co} \quad 3d^7 \;\boxed{\uparrow\downarrow}\,\boxed{\uparrow\downarrow}\,\boxed{\uparrow}\;\boxed{\uparrow}\;\boxed{\uparrow}\qquad 4s^2\;\boxed{\uparrow\;\downarrow}\tag{40}$$

$$\text{Ni} \quad 3d^8 \;\boxed{\uparrow\downarrow}\,\boxed{\uparrow\downarrow}\,\boxed{\uparrow\downarrow}\,\boxed{\uparrow}\;\boxed{\uparrow}\qquad 4s^2\;\boxed{\uparrow\;\downarrow}\;.$$

According to this scheme, the atomic moments of these metals should be, respectively, 4, 3, and 2 Bohr magnetons. The actual values turn out to be:

$$
\begin{array}{cccc}
 & \text{Fe} & \text{Co} & \text{Ni} \\
\text{Atomic moment} & 2.22 & 1.70 & 0.61 \text{ Bohr magnetons}
\end{array}\tag{41}
$$

The nonintegral number of electron spins per atom determined from saturation-magnetization measurements given above is explained by making use of the zone theory as follows: The $3d$ and $4s$ zones of these

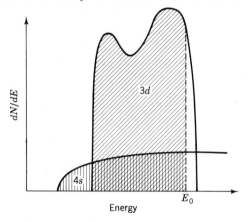

Fig. 11. Schematic representation of overlapping $3d$ and $4s$ zones of nickel. On an average, 9.4 of the 10 available $3d$ states per atom and 0.6 of the 2 available $4s$ are occupied. (*After Slater.*)

metals overlap, as shown for the case of nickel in Fig. 11. Since the relative occupation of these two zones is determined by the Fermi energy E_0, it is assumed that, on the average, 0.6 electrons per atom occupy states in the $4s$ zone and 9.4 electrons per atom occupy states in the $3d$ zone. Of these, 5 must have one kind of spin and 4.4 the other, leaving an excess of 0.6 unpaired spins per atom. The values listed in (41) have been independently confirmed by neutron-diffraction measurements. It is possible to check this model in still another way also. Suppose copper is alloyed with nickel. Since both metals have identical crystal structures and closely similar atomic radii, the copper atoms substitute directly for nickel atoms in the host structure. Copper has one more electron than nickel, and it is reasonable to expect that this electron will prefer the lower-lying $3d$ levels of nickel. Consequently, as more copper is added, the magnetic susceptibility should decrease, becoming zero when all the $3d$ states are filled at the composition 60 per cent Cu–40 per cent Ni. Saturation-magnetization measurements of a number of copper-nickel alloys have confirmed this, and in fact it has been shown that when 30 per cent zinc or 15 per cent silicon is added, the saturation magnetization similarly goes to zero.

Domain structure. The above discussion has explained why iron, cobalt, and nickel are ferromagnetic, but it has not accounted for such

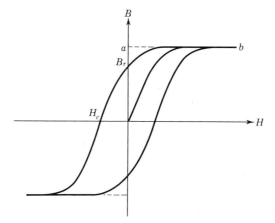

FIG. 12. Hysteresis loop of a ferromagnetic material.

physical properties of magnets as, for example, the familiar hysteresis-loop characteristic of the induced magnetization in an alternating magnetic field. It should be recalled that the magnetic induction B is related to an external-field intensity H by

$$B = \mu_P H \qquad (42)$$

where μ_P is the magnetic permittivity of the material. By analogy to the dielectric constant discussed in Chap. 12, it is sometimes convenient to employ a dimensionless constant $\kappa_m = \mu_P/\mu_0$ called the *permeability*. When the relation between B and H in ferromagnetic materials is examined (Fig. 12), it becomes apparent that the permeability is not constant and that B is not linearly dependent on H. When a magnetically neutral iron bar is placed in a magnetic field, the magnetic induction B increases with increasing H until it reaches a saturation value. Upon removal of the external field, the magnetic induction returns to a value B_r called the remnant magnetization, and a coercive field H_c must be applied in the reverse direction in order to again render the bar nonmagnetic. Increasing the external field further in the reverse direction then leads to saturation, and so forth. The dashed line *ab* denotes the hysteresis curve of a single ferromagnetic domain which is commonly called a square loop. Note that, unlike the hysteresis effect in ferroelectrics (page 358), the saturation represented by *ab* is horizontal. This is so because all the spins are parallel in the ferromagnetic crystal at this point whereas further increases in the dipole strengths of ferroelectric crystals are possible in an external field.

Even before the atomic mechanism of ferromagnetism was clearly understood, Weiss assumed that a single crystal of iron consists of small regions, or *domains*, within each of which the unpaired electrons have all their spins parallel to some direction but not to that of the spins in a neighboring domain. By using quantum-mechanical considerations it can be shown that this is caused by the relatively small, but nevertheless finite, contribution to the magnetization from the orbital angular moments of these electrons. When an external field is applied, the domains that have net moments parallel to the direction of the field have their energy reduced; whereas those domains that do not, have their energy increased. The crystal's energy, obviously, can be lowered if all the domains align themselves parallel to the field. This can be accomplished in either of two ways. Either the direction of magnetization of an entire domain changes at once or a domain that is favorably oriented grows in size at the expense of a less favorably oriented domain. When the external field is reversed in direction, all the domains must be reoriented, and owing to several causes discussed below, this requires an additional field to overcome the factors opposing such reorientation, so that a hysteresis loop results.

It turns out that magnetization is an anisotropic property in crystals. For example, it is easier to magnetize iron (body-centered cubic) along [[100]] than along any other directions, and most difficult along [[111]]. It is common practice to refer to these as *easy* and *hard directions* of magnetization, respectively. To show that this is a structure-dependent property, the easy directions in nickel are [[111]] (normal to the closest-packed layers) and the hard directions are [[100]]. Similarly, the easy direction in the hexagonal closest packing of cobalt is [0001], normal to the hexagonal closest-packed layers. The magnetization of single crystals along certain directions is invariably accompanied by changes in their physical dimensions. For example, iron crystals expand along the direction of magnetization and contract at right angles to it, so that the total volume tends to remain the same. In a nickel crystal, the dimensional changes are reversed; it contracts along the magnetization direction and expands in directions normal to it. Thus the magnetization of ferromagnetic crystals depends on the way that the structure is strained, a phenomenon called *magnetostriction*. As might be expected, magneto-striction is related to the elastic properties of the crystal and can be used to explain the formation of magnetic domains in crystals. With six [[100]] directions equally easy, it is natural that a single crystal of iron has domains containing magnetic moments aligned parallel to each of these six directions. Obviously, the domains must be separated by boundaries in which the magnetic moments undergo a gradual trans-formation from one orientation to the other. The exchange energy favoring parallel alignment of all moments prefers thick boundaries. It is opposed by the magnetic anisotropy, which favors a minimum deviation from the easy directions of magnetization. A compromise between these opposing forces is reached, and it turns out that the boundaries have thicknesses of the order of hundreds of atoms in actual crystals. The domains themselves can have various sizes, determined primarily be the free energy of the boundaries, whose contribution to the crystal's energy decreases with increasing domain size, since the ratio of boundary area to domain volume decreases. Lest it be deduced from this that a crystal consisting of a single domain has the lowest possible energy, it should be realized that such a crystal is a permanent magnet whose magnetic field contains magnetic energy. Thus the domain structure proposed by Weiss serves to lower the overall energy, thereby neutralizing the individual permanent magnets.

When a weak magnetic field is applied parallel to an easy direction of a crystal, the domains having lower energies, due to their more favorable orientation, grow in size, because of the reorientation of the moments in the boundaries. The resulting movement of boundaries can actually be observed by sprinkling a fine powder of Fe_2O_3 on a polished crystal surface. The powder particles align themselves along the domain boundaries, and a *Bitter pattern* is obtained. Several domains can be seen in the Bitter pattern of a single crystal of iron shown in Fig. 13. It has been observed that the movement of domain boundaries is impeded by imperfections and proceeds in jumps. These jumps can be made audible by surround-ing the crystal with an induction coil connected to an amplifier. This is called the *Barkhausen effect*, and is even more noticeable if the domains change their magnetization directions by rotation. The latter effect becomes more prominent when the strength of the field is increased rapidly.

Ferromagnetic materials. The manufacture of suitable magnetic devices requires the use of polycrystalline materials rather than single crystals. The magnetization of a polycrystalline material is impeded, however, by the random orientation of the grains. In a random array, only a fraction of the grains have their easy direction of magnetization parallel to the field direction. Moreover, magnetostriction in these crystallites produces strains in their neighbors, which, in turn, affect the ease of magnetization in a more complicated way. Thus several factors must be considered in selecting suitable materials, not the least of which is the relative ease of fabrication, which serves to determine the ultimate cost. When a permanent magnet is desired, say, for driving the cone in a loudspeaker, then a large area in the hysteresis loop and extreme

Fig. 13.   Magnetic domains in a single crystal of iron.   (*Courtesy of R. V. Coleman and G. G. Scott.*)

magnetostriction are assets. The large area of the loop indicates that, after saturation magnetization is attained, the external field can be removed without appreciable loss in the induced magnetization. Pronounced magnetostriction means that it is more difficult to change the direction of magnetization of a grain once the material has reached saturation for a particular direction. A typical material used for such purposes is alnico, whose composition expressed in atomic per cent is 62Fe-26Ni-12Al. Following fabrication, such a material is annealed at an elevated temperature for a prolonged time period, during which a new phase precipitates out of the grains and mechanically hardens the material. The magnetic consequence of this precipitation hardening is that the coercive force required to remove a subsequently induced magnetization is very large. The specific alloy composition is also chosen so as to permit a reasonably high saturation magnetization per unit volume, so that the desired magnetic-field strengths can be obtained without requiring excessively large magnets.

When a ferromagnetic material is placed in an alternating magnetic field, the area enclosed by the loop in Fig. 12 measures the energy lost per cycle in inverting and reinverting the direction of magnetization. In designing transformers, therefore, a ferromagnetic material having a very small hysteresis-loop area is desired. Minimizing hysteresis losses also requires that magnetostriction be minimized, so that a fairly stress-free material must be used. One way of doing this is to induce a preferred instead of a random orientation for the grains in a transformer plate. Mechanical rolling of a metal into sheet form automatically produces some preferred orientation. Upon subsequently annealing the sheet while it is in a strong magnetic field, growth of those grains that have their easy magnetization direction parallel to the field is abetted. The resulting material, for example, Permalloy (Ni_3Fe), tends to resemble a single crystal and can attain permeabilities as high as 100,000 and has very small coercive field values. It also has been pointed out that ferromagnetic alloys having integral numbers of Bohr magnetons per atom exhibit virtually no magnetostriction. Although the reason for this has not been clearly established, use can be made of this observation to predict what materials are suitable for applications requiring minimum hysteresis losses (see Exercise 11).

Other types of magnetism

Exchange interactions. When two ferromagnetic atoms are separated from each other by another atom, for example, an oxygen atom in Fe_2O_3, the two metal atoms can still interact with each other. A number of different exchange mechanisms have been proposed to explain this phenomenon. They differ primarily in the way that the two p electrons donated to the central oxygen ion are assumed to interact with the unpaired electrons in the adjacent metal ions. Four distinct mechanisms have been proposed:

1. *Superexchange.* In this model, one of the two p electrons of oxygen is transferred to the incomplete half-shell of one of the two metal ions, say, Me_1. According to Hund's rule, its spin is parallel to that of the metal electrons occupying states in the unfilled shell. The other p

electron in the same oxygen orbital must have an antiparallel spin direction in accord with the Pauli exclusion principle. Finally, the indeterminacy principle requires that such a configuration can also be formed with the other metal atom, Me_2. This is illustrated in Fig. 14A and leads to a ferromagnetic coupling since the spins of both metal atoms are maintained parallel to each other. When the two metal ions have closed half-shells (Fig. 14B), then the p electron shared with one of the metal atoms, say, Me_1, is antiparallel to those in the Me_1 subshell. A similar sharing of the other p electron with the Me_2 subshell leads to an interaction favoring antiparallel spin alignments between the two metal ions.

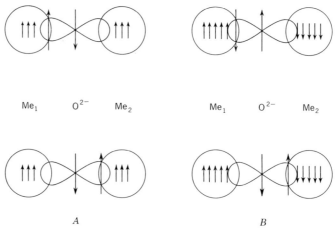

FIG. 14. Two equivalent superexchange models for (A) ferromagnetic exchange between the partly filled half-shells of the two metal atoms via the p orbital of the central oxygen ion; (B) antiferromagnetic exchange between the closed half-shells of two metal ions.

This is called *antiferromagnetic* coupling. Since the superexchange interaction depends on orbital overlap, it is strongest for collinear configuration Me-O-Me and is weakest when the two Me-O bonds form a right angle.

2. *Indirect exchange.* This mechanism is similar to superexchange except that it is argued that the potential energy of the p electrons is lowered when both electrons participate in antiferromagnetic coupling simultaneously. This results from wave-function calculations which show that the energy is lowest when both p electrons are near the metal ions, not unlike the conclusion reached in the discussion of electron-pair bonding in Chap. 3.

3. *Double exchange.* This model was first proposed by Zener and is similar to his explanation of the exchange interaction in ferromagnetic metals. It will be recalled from that discussion that Zener proposed that

the coupling between two adjacent metal atoms was actually antiferromagnetic since the exchange integral is assumed to remain negative. When an oxygen atom is interposed between the two metal ions, the antiparallel coupling between the two p electrons leads to an antiferromagnetic coupling between the adjacent metal atoms.

4. *Semicovalent exchange.* In this model it is assumed that the bonding between the atoms in metal oxides has an appreciable covalent character. Accordingly, the electrons on the metal ions are assumed to be distributed among hybridized orbitals. It is then shown that electron-pair bonds formed with the central oxygen atom lead to an antiferromagnetic coupling between the two metal atoms.

Antiferromagnetism. Although the above four mechanisms differ in detail, they lead to essentially the same result, namely, that two metal atoms separated by an oxygen atom between them can interact in such a way that the spins of their unpaired electrons are, respectively, antiparallel. If equal numbers of atoms of both kinds are present, the material has zero net magnetization and is said to be antiferromagnetic. Returning to the discussion of the effective magnetic field inside a material (32), this means that the Weiss constant w is negative. The magnetic susceptibility equivalent to (34) in an antiferromagnetic material, therefore, has the form

$$\chi = \frac{C}{T + wC} = \frac{C}{T + \Theta_A} \tag{43}$$

where Θ_A now is called the *asymptotic* Curie temperature, because the transition from paramagnetic to antiferromagnetic behavior occurs not at Θ_A, but at another temperature, T_N.

As already stated, a negative exchange interaction results from antiparallel spin moments in adjacent atoms. In a crystal, this means that alternate atoms have their spin moments oriented parallel to each other but adjacent atoms do not. It is convenient to picture such an arrangement by considering the atoms to be arranged on two interpenetrating sublattices,[†] so that the spin moments of atoms in one sublattice array have an opposite sense from the spin moments of atoms in the other sublattice array. Below the transition temperature, the spin moments of both sublattice arrays are ordered, as shown diagrammatically in

---

[†] Recently it has been shown that the 230 space groups can be extended to include *antisymmetry* operators which have the property of repeating a white motif into a black motif and the black one back into a white motif. (Actually, several such repetitions are possible before the cycle is completed.) These so-called *color space groups* include the symmetry groups of all possible arrangements of parallel and antiparallel spin directions and have proved very useful in the study of ferromagnetic and antiferromagnetic crystals.

Fig. 15, and no spontaneous magnetization is possible, since the moments exactly cancel when summed over the entire crystal. Above the transition temperature, the order disappears because of thermal motion of the electrons, and the paramagnetic behavior of the crystal can be described by considering the behavior of the atoms in the two sublattice arrays separately. Denoting the two sublattice arrays by the subscripts A and B, respectively, it follows from (33) that

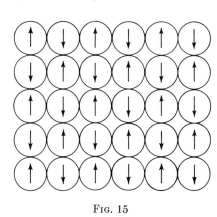

$$M_A = \frac{C'}{T}(H - \alpha M_A - \beta M_B) \quad (44)$$

and

$$M_B = \frac{C'}{T}(H - \alpha M_B - \beta M_A) \quad (45)$$

where α is the interaction parameter for two atoms having like spin moments (AA or BB), and β is the interaction parameter for two unlike atoms.

Fig. 15

The total magnetization is the sum of partial magnetizations like (44) and (45), so that

$$\begin{aligned}
M &= M_A + M_B \\
&= \frac{C'}{T}[(H - \alpha M_A - \beta M_B) + (H - \alpha M_B - \beta M_A)] \\
&= \frac{C'}{T}[2H - (\alpha + \beta)(M_A + M_B)] \\
&= \frac{2C'}{T}[H - \tfrac{1}{2}(\alpha + \beta)M].
\end{aligned} \quad (46)$$

Letting $C = 2C'$, since the numbers of A and B atoms are equal, and rearranging terms,

$$\chi = \frac{M}{H} = \frac{2C'}{T + C'(\alpha + \beta)} = \frac{C}{T + \frac{1}{2}C(\alpha + \beta)} = \frac{C}{T + \Theta_A} \quad (47)$$

which is the same as relation (43) already encountered above.

The transition, or Néel, temperature is determined by Eqs. (44) and (45), since it is the temperature for which M_A or M_B has a finite value in the absence of an external magnetic field. Setting $H = 0$ and eliminating M_A and M_B in these two simultaneous equations, a quadratic equation is obtained whose positive root gives the Néel temperature

$$T_N = (\beta - \alpha)C' = \tfrac{1}{2}(\beta - \alpha)C. \quad (48)$$

This is the temperature at which a kink in the magnetic-susceptibility curve occurs, as shown in Fig. 16. Note that the paramagnetic susceptibility follows the curve predicted by the Curie relation for $T > T_N$ and tends toward minus the Curie temperature below the Néel point. This is the reason why Θ_A in (43) is called the asymptotic Curie temperature. A similar anomalous behavior is observed in the temperature dependence of the specific heat and thermal-expansion coefficient of antiferromagnetic materials.

Ferrimagnetism and ferrites. According to the previous discussion of ferromagnetic and antiferromagnetic interactions, it appears likely that there are materials in which both types of interactions can occur simultaneously. This is the case for the iron oxide magnetite, Fe_3O_4, which has Fe^{3+} and Fe^{2+} ions distributed among the octahedral and tetrahedral voids of a cubic closest packing of oxygen ions, as shown in Fig. 17. If all the iron atoms are assumed to have parallel spin moments (ferromagnetic interactions), it can be shown that the total saturation moment is about 14 Bohr magnetons per formula weight. The measured value, however, is

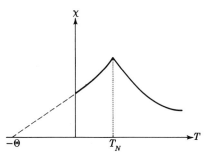

FIG. 16. Magnetic susceptibility of an antiferromagnetic material.

4.08 Bohr magnetons. This discrepancy was first explained by Néel, who showed that some of the exchange interactions were actually antiferromagnetic. The residual magnetization is thus due to an excess of spin moments of one kind.

The crystal structure of Fe_3O_4, shown in Fig. 17, is called the *spinel structure*, after the mineral spinel, $MgAl_2O_4$, in which it was first determined (see also Chap. 9). The face-centered cubic unit cell contains 32 closest-packed oxygen atoms and 16 metal atoms occupying one-half of the available octahedral voids and 8 metal atoms occupying one-eighth of the available tetrahedral voids. It thus can accommodate eight formula equivalents of the type AB_2O_4, where A and B represent any metal ions whose total charge must equal $+8$. Since it is possible to satisfy this condition by a variety of combinations of divalent, trivalent, and higher-valent ions, it is not surprising that a large number of metal oxides adopt the spinel structure. The specific properties of each oxide depend, of course, on the valence of the ions and their relative distribution among the octahedral and tetrahedral voids. In Fe_3O_4, the unit cell contains eight Fe^{3+} ions in tetrahedral voids and eight Fe^{3+} plus eight Fe^{2+} ions in octahedral voids. Néel explained the low saturation magnetization of Fe_3O_4 by assuming that the spins of the Fe^{3+} ions were antiparallel. Note, for example, in the front left corner in Fig. 17, that a tetrahedral Fe^{3+} ion is joined to an octahedral Fe^{3+} ion by a straight line passing through the cornermost oxygen ion. As already discussed, this leads to an antiferromagnetic exchange interaction according to, say, the superexchange mechanism. Since there are equal numbers of Fe^{3+} in octahedral and tetrahedral voids, their magnetic moments cancel and

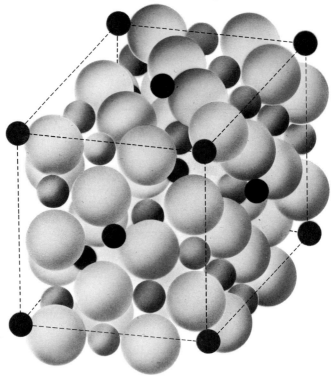

Fig. 17. Crystal structure of ferrite, Fe_3O_4. The origin of the outlined face-centered cubic unit cell has been placed in a tetrahedral void of the oxygen closest packing occupied by Fe^{3+}. One-half of the octahedral voids (larger dark spheres) are equally occupied by Fe^{3+} and Fe^{2+} ions.

the resultant moment of Fe_3O_4 must be due to the Fe^{2+} ions in octahedral voids. The relative disposition of the ions and their spin directions relative to a closest-packed oxygen layer is shown schematically in Fig. 18. The calculated moment of an Fe^{2+} ion is approximately 4 Bohr magnetons, in agreement with the saturation-magnetization measurement of 4.08 Bohr magnetons per Fe_3O_4 formula weight.

The various oxides adopting the spinel structure and containing iron are commonly called *ferrites*, and this magnetic phenomenon is called *ferrimagnetism*. They are characterized by an electrical resistivity that is as much as 10^{16} times higher than that of metals at room temperature. This means that eddy currents cannot exist in ferrites, so that they can be used as magnetic materials at frequencies extending into the microwave range. Strictly speaking, two kinds of spinel structures must be distinguished. In the normal spinel structure, for example, that of $MgAl_2O_4$, the A atoms are in the tetrahedral sites and the B atoms in the octahedral sites. In the inverse spinel structure, half of the B atoms occupy tetrahedral voids and the octahedral sites are equally shared by A and B atoms. The correct formula denoting the inverse spinel structure then is written $B[AB]O_4$, for example, $Fe^{3+}[Fe^{3+}Fe^{2+}]O_4$ for magnetite. In the analogous nickel ferrite, $Fe^{3+}[Ni^{2+}Fe^{3+}]O_4$, use can be made of this information to calculate the resultant saturation magnetization. The moments of Fe^{3+} and Ni^{2+} are 5 and 2 Bohr magnetons, respectively. The net moment per formula weight then must be 2 μ_B, since the moment of

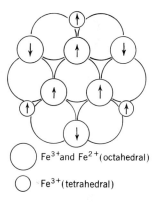

$\bigcirc$ Fe^{3+} and Fe^{2+} (octahedral)

$\bigcirc$ Fe^{3+} (tetrahedral)

$$Fe^{3+} \text{ in tetrahedral sites is } -5\,\mu_B$$
$$Fe^{3+} \text{ in octahedral sites is } +5\,\mu_B$$
$$Ni^{2+} \text{ in octahedral sites is } +2\,\mu_B$$
$$\overline{\text{Net moment} = +2\,\mu_B} \quad (49)$$

FIG. 18. Spin directions in Fe_3O_4. Fe^{3+} in tetrahedral and in octahedral sites have antiparallel spins. The net magnetization is thus due to Fe^{2+} ions in octahedral sites.

Such considerations make it possible to design a ferrite to yield any desired net moment, within limits imposed by the constituent atoms. This is so because the ferrite structure shown in Fig. 17 is stable for wide variations of x in such formulas as $Fe[Ni_xFe_{2-x}]O_4$. Further examples of such design calculations are left to Exercise 14.

Other metal oxides may also be ferromagnetic or antiferromagnetic. This is true of certain oxides crystallizing with the perovskite structure described in the preceding chapter. For example, $(La_xCa_{1-x})MnO$ is ferromagnetic when $x \simeq 0.3$ and antiferromagnetic outside this composition region. Most sulfides of iron, nickel, cobalt, and manganese are similarly either ferromagnetic or antiferromagnetic. The wide range of magnetic materials thus available has made it possible to devise a number of new devices. For example, some of the ferrites have virtually square hysteresis loops. Upon application of an external field that is larger than the coercive force, the magnetization instantaneously becomes saturated in the field direction. In the absence of eddy currents and other losses, the switching can be done, not only rapidly, but also very efficiently, so that such ferrites find application in computer memories and similar devices.

Magnetic resonance

Paramagnetic resonance. According to Eq. (8) the interaction of an electron's spin with a magnetic field results in two energy levels for the electron separated by

$$\Delta E = g\mu_B B. \tag{8}$$

Transitions can be induced between these states by the absorption of energy from an electromagnetic field of frequency ν such that

$$h\nu = g\mu_B B. \tag{50}$$

When such energy absorption occurs, the phenomenon is called *paramagnetic resonance.* According to (50), resonance can occur when either

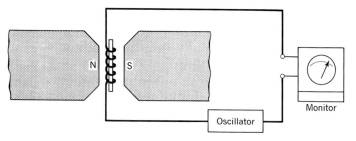

FIG 19. Paramagnetic-resonance arrangement. The sample is placed between the poles of a magnet and is surrounded by a coil producing an alternating field at right angles to the static field. When the frequency of the alternating field, controlled by the oscillator, coincides with the resonance frequency of the sample, the absorption of energy is detected in the monitor.

the frequency is varied while the field strength is held constant or the field strength is varied at a constant frequency. The transitions between the levels in (8) correspond classically to a tipping of the precessing magnetic gyroscopes, so that their magnetic-moment components along the static-field direction change from, say, $+\mu_B$ to $-\mu_B$. In a paramagnetic-resonance apparatus, therefore, the sample is placed between two poles of a magnet and surrounded by a coil carrying the alternating field at right angles to that of the static field (Fig. 19). When the condition (50) is satisfied by the alternating field, the moments can change their orientation with respect to the constant field, and the energy absorbed from the alternating field is detected in a suitable monitor. A plot of the absorption as a function of the static-field strength then shows peaks at resonance, which are expressed in terms of the half-width in magnetic-field units rather than frequency (or wavelength).

Returning to (50), it is possible to express this relation in terms of an angular frequency $\omega = 2\pi\nu$. After substituting $\mu_B = eh/4\pi m$,

$$\omega = g\,\frac{e}{2m}\,B \tag{51}$$

and it is clearly evident that the absorption frequency depends on the strength of the magnetic field. Comparing (51) with the Larmor frequency (17), that is, the classical description of the characteristic frequency of a dipole in a magnetic field, it can be seen that the Larmor frequency differs from the correct value for an electron by the constant g, called the spectroscopic splitting factor. This difference is due to the quantum-mechanical properties of electron spin.

The magnetic resonance of electrons is called *electron paramagnetic resonance* (EPR) and was first discovered by a Soviet scientist, E. K. Zavoisky, in 1944. This discovery was not immediately exploited in Russia, but rather by scientists in the United States, Great Britain, and Holland. The small mass of the electron and its relatively fast spin require the use of frequencies in the range of 10^4 megacycles, so that microwave waveguides are necessary to carry the alternating field to the sample. In general, the electrons in a material are paired up, so that the two energy levels of (8), as sketched in Fig. 5, are equally populated and no absorption of radiation takes place. In order to observe EPR, therefore, it is necessary that the material contain unpaired electron spins. Conversely, this means that EPR can be used to detect localized electrons, for example, at trapping sites in an insulator or at impurity atoms in a semiconductor.

The actual resonance parameters, namely, the field strength and half-width, are determined by the total field acting on the electron. This, in turn, is comprised of the external field and a local field determined by its environment in the crystal structure. This makes EPR very useful in exploring the inner structure of a single crystal since variations in the resonance parameters with crystallographic direction can be used to determine structural details. Conversely, when the crystal structure is accurately known, the location of localized electrons can be deduced from the symmetry properties of the EPR spectrum.

Paramagnetic resonance analogous to EPR but arising from nuclear magnetic moments is termed *nuclear magnetic resonance* (NMR) and was independently discovered at about the same time (1946) by Purcell and by Bloch. Because of the heavier mass of nuclei compared with electrons, NMR takes place at much lower frequencies, as predicted by (51). In a magnetic field of 0.5 T, for example, the resonant frequencies are of the order of 10 megacycles. The resonance width is quite sharp in liquids ($\simeq 10^{-7}$ T) and much broader in solids ($\simeq 10^{-3}$ T). This is so because the nuclear moments are affected by any local fields present inside the material, in addition to the external field. In liquids, any local fields set up by neighboring atoms are short-lived, so that, on the average, each nucleus finds itself in a similar environment and responds in the same way. The environments of different nuclei in a crystal, conversely, are not alike, so that a stronger field is required to invert some of the nuclear moments than is required for others, and resonance occurs over a wider range of field intensities. Some crystals, for example, cyclohexane, nevertheless have a sharp resonance response well below their melting points. This anomalous behavior indicates that the molecules in these crystals are free to rotate even in the solid, averaging out the local fields thereby. This is an example of the application of NMR to the study of the structural features in crystals. (Another way that this can be done is to utilize isotopes whose heavier nuclei give rise to absorption spectra at different field strengths.) An alternative way to make use of NMR is first to align the nuclear moments in a liquid and then to shut off the fixed field. The nuclei are then free to change their precession direction and may precess about any other field direction that is present. This is the principle underlying the operation of magnetometers mounted in satellites which are used to measure small variations in the relatively weak magnetic fields in outer space.

Ferromagnetic resonance. In the case of EPR and NMR, individual magnetic dipoles present in a material can respond to the incident electromagnetic field independently of each other. When a ferromagnetic material is placed in a high-frequency alternating field, however, the external field cannot act on the individual spin moments because they are internally coupled. Consequently, in *ferromagnetic resonance* it is necessary to consider the interaction of the radiation with the resultant magnetization M of the solid rather than with individual spins. The resonance frequency for this interaction is given by a relation analogous to (51),

$$\omega = g \, \frac{e}{2m} \, \sqrt{BH}. \tag{52}$$

The resonance frequencies are in the microwave range, similar to EPR, because basically both phenomena arise from electron spin interactions.

Equation (52) is useful for determining the g values of ferromagnetic materials. Some representative values are compared as follows:

	Ni	Co	Fe	Fe_3O_4	Permalloy	
g	2.20	2.22	2.15	2.2	2.1	(53)

These results are in good agreement with similar values obtained in gyromagnetic experiments.

Cyclotron resonance. A phenomenon related to paramagnetic resonance is observed when a semiconductor is placed in a steady magnetic field. The conduction electrons (or holes) in the crystal are caused to travel in spiral orbits about the magnetic-field direction. If the radius of their orbit is r, the centrifugal force acting on an electron moving with a velocity v is

$$F = \frac{m^*v^2}{r} \qquad (54)$$

where m^* is the effective mass of an electron or a hole. At the same time, the electron feels a force due to the applied field called the Lorentz force,

$$F = evB. \qquad (55)$$

At equilibrium, the two forces just balance:

$$\frac{m^*v^2}{r} = evB$$

and the angular frequency of the electron is

$$\omega = \frac{v}{r} = \frac{e}{m^*} B. \qquad (56)$$

Next, suppose that an alternating field is introduced at right angles to the steady field H. When the frequency of the alternating field is $\nu = \omega/2\pi$, then resonance absorption takes place and the effect is called *cyclotron resonance* by analogy to the spiral orbits executed by electrons accelerated in the magnetic fields of a cyclotron. Note that (56) follows directly from (51) by substituting $g = 2$ for the spectroscopic splitting factor for an electron.

Cyclotron resonance in semiconductors requires alternating fields in the radio-frequency range. It is very useful for measuring the effective mass of electrons or holes, as can be seen from Eq. (56). In fact, it is one of the most effective ways of determining the band structure of transition metals and semiconductors. This is so because it is possible to determine the effective mass as a function of direction in a crystal by changing the orientation of the crystal relative to that of the applied field.

Suggestions for supplementary reading

L. F. Bates, *Modern magnetism*, 3d ed. (Cambridge University Press, London, 1951.)
John C. Slater, *Quantum theory of matter* (McGraw-Hill Book Company, Inc., New York, 1951), especially pp. 399–434.

Arthur R. von Hippel, *Molecular science and molecular engineering* (The Technology Press of the Massachusetts Institute of Technology and John Wiley & Sons Inc., New York, 1959).

C. Zener, Impact of magnetism on metallurgy, *Trans. AIME*, vol. 203 (1955), pp. 619–630.

Suggestions for further reading

Adrianus J. Dekker, *Solid state physics* (Prentice-Hall, Inc., Englewood Cliffs, N.J., 1957).

Charles Kittel, *Introduction to solid state physics*, 2d ed. (John Wiley & Sons, Inc., New York, 1956).

N. F. Mott and H. Jones, *The theory of the properties of metals and alloys* (Clarendon Press, Oxford, 1936; reprinted by Dover Publications, Inc., New York, 1958).

E. C. Stoner, *Magnetism and matter* (Methuen & Co., Ltd., London, 1934).

Exercises

1. Consider the motion of a $1s$ electron in hydrogen in terms of the Bohr model. Assuming that its motion is arbitrarily restricted to a circular orbit lying in a plane, what is the magnetic moment associated with such an orbit? How does its magnitude compare with 1 Bohr magneton?

2. Is there any reason why the magnetic moment of the fictitious orbit described in Exercise 1 *should* equal 1 Bohr magneton? Justify your answer. (ANSWER: Yes.)

3. Calculate the Larmor frequency for the precession of an electronic spin moment in a magnetic-field strength of 5×10^{-2} and 5×10^{-1} T. Repeat for the nuclear spin moment of hydrogen.

4. The magnetic moment of a dipole is related to its angular momentum by relation (10) in the text. Make a sketch, similar to Fig. 6, showing the vectors **B**, **m**′, and $d\mathbf{m}'/dt$. With the aid of this drawing, show that Eq. (12) correctly describes the motion of a vector **m**′ precessing about **B** with the Larmor frequency (17). From your vector analysis, what is the direction of $\boldsymbol{\omega}_L$? Show it in your sketch.

5. Many crystals exhibit magnetic anisotropy in that their magnetic suceptibility has different magnitudes along two principal crystallographic directions. This difference $\Delta\chi$ can be measured by suspending a square crystal, containing one of the principal directions within the area of the square, by a fine fiber attached to the center of the square, inside a magnetic field of strength B. If the volume of the crystal is $\mathcal{V}$, the force couple exerted on it

$$G = \frac{\mathcal{V}}{2} \, \Delta\chi \, B^2 \sin 2\theta$$

where $\Delta\chi$ is the susceptibility difference per unit volume, and θ measures the rotation of the principal direction relative to the field direction. In most materials, θ is quite small.

After the external field is removed, the restoring couple per angular twist of the fiber k_0 is related to the time of the swing t_0 and moment of inertia I by $k_0 = 4\pi^2 I/t_0$. In a magnetic field, therefore, the restoring couple per angular twist, then, is $k_1 = k_0 + G/\theta$. Obtain an expression for $\Delta\chi$ if the new time of oscillation is t_1. [ANSWER: $\Delta\chi = (4\pi^2 I/\mathcal{V}B^2)(1/t_1^2 - 1/t_0^2)$.]

6. Suppose the crystal in Exercise 5 is calcite (density = 2,720 kg/m³ and horizontal area normal to the fiber, 0.5×0.5 cm).

(a) Calculate the moment of inertia per unit volume if $I = m(a^2 + b^2)/12$ for a rectangular block whose sides are a and b.

(b) Calculate $\Delta\chi$ for calcite if $t = 12.8$ sec when $H = 0$ and $t = 2.69$ sec when $B = 0.2475$ T. Express your answer in susceptibility per cubic meter and also per kilogram of calcite.

Note that this experiment has the advantage that the field gradient dB/dz in (23) need not be measured.

7. Calculate the paramagnetic susceptibility per unit volume of cesium at 300 deg K. Make use of the procedures discussed in Chap. 6, Exercise 3, to determine the density of free electrons. (ANSWER: 9.54×10^{-5} J/T^2.)

8. By analogy to the derivation of (30) in the text, show that $\chi_P = \mu_0 n(g^2/4)\mu_B^2/kT$ for electron spins, where g is the spectroscopic splitting factor. This is the expression for χ_P derived by the early quantum theory.

9. According to quantum mechanics, the total spin angular momentum for electron is $[S(S + 1)]^{\frac{1}{2}}$, rather than $S = \frac{1}{2}$ assumed in the early quantum theory. Show that when the average magnetic moment $\bar{\mu}$ is expressed in these terms, it leads to the classical equation (25) for the Curie constant.

10. The saturation magnetization of a ferromagnetic crystal measures the total number of magnetic dipoles that can be lined up in a strong field. If (41) in the text correctly expresses the effective number of dipoles per atom, what is the saturation magnetization per unit volume in body-centered cubic iron? (The density of iron is 7,860 kg/m^3.)

11. It has been shown by J. L. Snoek that alloys having an averaged integral number of Bohr magnetons per atom show very little, if any, magnetostriction. Can you suggest two simple alloys in the cobalt-nickel system for which this should be the case? What does the Snoek rule predict for alnico?

12. Zener has suggested that the body-centered cubic structure of chromium is stabilized by a double-exchange mechanism leading to antiparallel spins in adjacent atoms.

(a) Show that this is possible in a body-centered cubic structure by making a sketch of the structure indicating the spin directions in neighboring atoms.

(b) Prove that such an alignment of spins is not possible in either of the two kinds of closest packings adopted by other metals.

(c) What kind of magnetic behavior should chromium exhibit if Zener's hypothesis is correct?

13. According to (41) in the text, each atom in an iron crystal contributes 2.22 Bohr magnetons on the average. Yet the total number of Bohr magnetons per Fe$_3$O$_4$ formula in magnetite was calculated to be 14 μ_B. How can you justify this calculation? How do you account for the observed value of 4.08 μ_B per Fe$_3$O$_4$? HINT: Compare the electronic structures of Fe^{3+} and Fe^{2+} and check Fig. 18.

14. Consider nickel ferrite having the general formula Fe[Ni$_x$Fe$_{2-x}$]O$_4$. What must the value of x be to give a net moment per formula weight of 1 Bohr magneton? HINT: Remember to maintain charge neutrality.

15. Demonstrate that the resonance frequencies encountered in EPR are approximately one thousand times larger than those required for NMR. Why does ferromagnetic resonance require microwave frequencies? HINT: Make use of the Larmor frequencies calculated in Exercise 3.

14. optical processes

In discussing the interaction of materials with electromagnetic radiation such as visible light, for example, it is necessary to keep in mind the dual nature of such radiation. As already shown in the discussion of photoconductivity in Chap. 10, it is sometimes convenient to think of light as composed of photons having an energy $h\nu$ which can be absorbed by electrons in a material via transitions to excited states. Alternatively, when the radiation is thought of as an electromagnetic wave, it can be represented by an alternating electric field of frequency $\nu = \omega/2\pi$ propagating through the material. It is common practice in optics to denote the ratio of the velocity of light in vacuum c to its velocity in a material medium v by an *index of refraction*,

$$n = \frac{c}{v}. \tag{1}$$

The velocity of light in a material medium is also determined by its dielectric constant κ_e, so that

$$v = \frac{c}{\sqrt{\kappa_e}} \tag{2}$$

and

$$n = \sqrt{\kappa_e}. \tag{3}$$

The relation between n^2 and κ_e predicted by (3) is not always obeyed because the dielectric constant of a material depends on frequency, as discussed in Chap. 12. According to Maxwell's equations, this relation can be correctly expressed by

$$\kappa_e = n^2 - \varkappa^2 \quad \text{and} \quad n\varkappa = \frac{\chi_e''}{\nu} \tag{4}$$

408

where $\varkappa$ is an extinction coefficient and χ_e'' is the imaginary part of the complex electrical susceptibility (Chap. 12) $\chi_e^* = \chi_e' - i\chi_e''$. Here

$$\chi_e' = \frac{Ne^2}{\epsilon_0 m} \left[\frac{\omega_0^2 - \omega^2}{(\omega_0^2 - \omega^2)^2 + \gamma^2\omega^2} \right]$$

and

$$\chi_e'' = \frac{Ne^2}{\epsilon_0 m} \left[\frac{\gamma\omega}{(\omega_0^2 - \omega^2)^2 + \gamma^2\omega^2} \right]. \tag{5}$$

Relations (4) and (5) can be used to discuss the optical properties of a material. The imaginary part of the susceptibility χ_e'' has an appreciable value only when $\omega \simeq \omega_0$, in which case resonance absorption is said to take place. The real part χ_e', therefore, is of greater practical importance

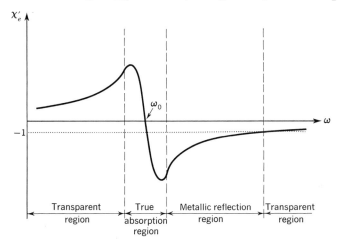

FIG. 1. Optical properties of materials at different frequencies of the incident beam.

because it has a nonzero value whenever $\omega \neq \omega_0$. According to the discussion in Chap. 12, the electrical susceptibility of a material is a measure of its polarizability per unit electric field and

$$\kappa_e = 1 + \chi_e' = 1 + \frac{Ne^2}{\epsilon_0 m} \left[\frac{\omega_0^2 - \omega^2}{(\omega_0^2 - \omega^2)^2 + \gamma^2\omega^2} \right] \tag{6}$$

for frequencies differing from the natural frequency ω_0.

The dependence of χ_e' on frequency is shown in Fig. 1. On the long-wavelength (low-frequency) side of ω_0, the magnitude of $\omega_0^2 - \omega^2$ is considerably larger than $\gamma\omega$, so that κ_e is positive and larger than unity, according to (6). Furthermore, χ_e'' is zero, so that $\varkappa = 0$ according to (4) and $\kappa_e = n^2$ as predicted by (3). This is the case when a material is completely transparent, and is characteristic of most ionic and molecular materials in the visible portion of the electromagnetic spectrum (Exercise

1).   The magnitude of χ'_e increases as $\omega \to \omega_0$, but does not go to infinity because of the damping term in the denominator.   It passes through zero when $\omega = \omega_0$ and again reaches a maximum on the other side of the natural frequency.   This corresponds to the absorption region for the material, as already discussed in Chap. 12.   As the frequency increases further, $\omega_0^2 - \omega^2$ becomes negative.   At sufficiently large frequencies, κ_e is positive, according to (6), but less than unity; that is, the material is optically less dense than vacuum.   At frequencies not very much greater than the natural frequency, it is possible for κ_e to become negative. When this happens, n tends to zero, according to (4), and $\varkappa$ is finite (Exercise 2).   This leads to total reflection of the incident radiation and is the reason why this region is labeled the metallic-reflection region in Fig. 1. Note that when χ'_e is less negative than -1, the refractive index again becomes positive although it is less than unity.

The optical properties of materials have been discussed so far in terms of the classical theory.   As shown, it is possible to explain the transmission, absorption, and reflection of light thereby.   In this connection it should be noted that the reflection coefficient for normal incidence is given according to the classical theory by

$$R = \frac{(n - 1)^2 + \varkappa^2}{(n + 1)^2 + \varkappa^2}. \tag{7}$$

The classical theory considers an assembly of independent oscillators, however, and is not capable, therefore, of explaining the observed behavior of specific materials.   Thus Drude had first to postulate the existence of free electrons in metals before the above-described total-reflection region could be used to explain the characteristic luster of metals.   Quantum mechanics, or more appropriately, the band theory of solids, can be used to explain the optical behavior of specific materials as shown next.

Absorption of light

Metals.   Assuming that the free-electron model is valid for metals, the absorption of light by a metal should be similar to that of a collection of free oscillators, so that it follows the predictions of the classical theory described above.   The free electrons actually absorb the incident radiation by transitions to higher allowed quantum states.   According to quantum mechanics, such transitions are allowed only between states lying in different zones, or bands.   In terms of the zone model, this means that transitions can occur between states in two zones provided that the wave vector k in both zones has the same direction, so that momentum is conserved in the transition.   Suppose the Fermi surface in the first zone is as shown in Fig. 2. The states corresponding to this surface in the second zone then lie along the dashed arcs. Similar transitions are possible also between other occupied states in the first zone and corresponding states in the second zone. The allowed transitions are not unlike the photoelectric effect discussed in Chap. 11, except that the electrons do not actually leave the crystal. This process is sometimes called the *internal photoelectric effect*.

The smallest energy difference results when the transition takes place between states at the Fermi surface and the dashed arcs in Fig. 2. The energy of a state is related to the wave vector k, describing it by $E = h^2k^2/8\pi^2m$, according to the discussion in Chap. 7. Assuming that the energy gap between the two zones in Fig. 2 just vanishes, that is,

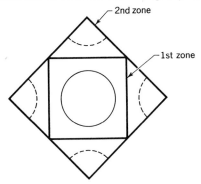

FIG. 2. First and second Brillouin zones of a cubic crystal having one monovalent atom per lattice point of its primitive cubic lattice. The Fermi surface lying in the first zone is shown by the circle. The dashed arcs in the second zone mark the positions of states having corresponding wave numbers.

that the energies of the states on both sides of the boundary are continuous according to the free-electron model, it is possible to determine the minimum energy required for a transition to take place. This leads to a minimum absorption frequency

$$h\nu_{\min} = \frac{[(2k_{hkl} - k_0)^2 - k_0^2]h^2}{8\pi^2m} \qquad (8)$$

where k_0 is the wave number of a state at the Fermi surface, and k_{hkl} is the wave number of a state at the first Brillouin-zone boundary as defined in Chap. 7. The wavelength corresponding to this frequency then can be calculated by setting $\lambda = c/\nu_{\min}$ as shown in Exercise 3. Some typical wavelength values thus deduced for monovalent metals are as follows:

	Na	K	Rb	Cs	Cu	Ag	
$\lambda_{\min}$	6,200	9,100	11,000	12,300	3,700	4,700 Å	(9)

The wavelengths calculated in (9) should be considered as order-of-magnitude estimates only, but they do suggest that alkali metals should absorb in the infrared region while the noble metals should absorb in the visible or ultraviolet region. These findings are corroborated by experimental evidence. The alkali metals are generally transparent to visible light, but not to infrared radiation, and appear to agree with the predictions of the classical theory. In copper and silver, the internal photoelectric absorption starts at 5,750 and 3,100 Å, respectively. The differences between these numbers and (9) are due to the simplified assumptions used in calculating the latter. It is possible to arrive at a qualitatively correct explanation of the long-wavelength limit, as well as some of the structure observed in the absorption band itself, by considering the detailed distribution of the density of occupied states in these metals. It should be noted in this connection that the linear-absorption coefficient in the absorption region is quite large ($\simeq 10^7$ m^{-1}), so that very thin foils must be used in the measurements. An important complication arises from the surface irregularities in such films, so that the experimental results obtained for different surface treatments show pronounced variations. The color of metals is also a consequence of their absorption spectra. Thus copper is reddish, because the reflection coefficient decreases as the wavelength of the incident light decreases, and silver is blue-white, because absorption does not begin until the ultraviolet part of the spectrum.

Semiconductors. The application of the band theory to the elucidation of observed optical properties of solids is most easily demonstrated in the case of semiconductors. Most semiconductors, including germanium, silicon, and indium antimonide, are transparent to infrared radiation. The linear-absorption coefficients of two typical germanium crystals are shown in Fig. 3. Starting with the longer wavelengths and moving toward the shorter wavelengths (moving from right to left in Fig. 3), it is observed that the absorption coefficient decreases, indicating that the crystals become more transparent to shorter-wavelength radiation. This is consistent with the greater energy of the photons of shorter-wavelength radiation ($E = hc/\lambda$). At a wavelength of 2 μ (1 $\mu = 10^{-6}$ m), however, the absorption coefficient begins to increase abruptly, until at approximately 1.7 μ it becomes extremely high. The reason for this abrupt increase in absorption becomes evident when the band model of germanium is considered. The energy of the radiation corresponding to a wavelength of 1.7 μ is 1.15×10^{-19} J, or 0.72 eV, which is the width of the energy gap in germanium. Thus a photon having this energy is capable of exciting an electron occupying a quantum state at the top of the valence band to a quantum state lying at the bottom of the conduction band. In the process, the energy of the photon is absorbed and the crystal appears to be opaque to radiation of this energy. Still shorter wavelength radiation also can excite electrons from lower-energy quantum states to higher states in the conduction band. This results in an ever-increasing number of electrons occupying states in the conduction band. These nearly free electrons behave quite similarly to the free electrons in a metal in that they can absorb additional energy by transition to still higher energy states in the conduction band and reradiate it by transitions to unoccupied lower-energy states. Thus most semiconductors exhibit a characteristic metallic luster when viewed with visible light. This is borne out by the two curves in Fig. 3. The

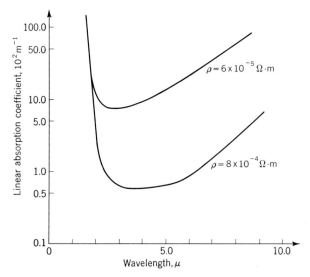

FIG. 3. Absorption coefficients of two *n*-type germanium crystals at room temperature.

lower-resistivity germanium crystal has a larger density of conduction electrons and a greater absorption coefficient at all wavelengths. It should be noted that the actual transitions in the conduction band take place through electron-phonon interactions because direct transitions are not allowed. Thus the temperature of a semiconductor has an important influence on its absorption characteristics. Temperature has another effect also. As the temperature of a semiconductor increases, the Fermi level shifts, as described in Chap. 8, and the density of conduction electrons increases.

The transmission spectrum of a silicon crystal is shown in Fig. 4. The energy gap in silicon is 1.09 eV, so that the transmission drops to zero at approximately 1.1 μ. That the absorption process in silicon actually generates conduction electrons can be verified by noting that the crystal becomes photoconductive, as discussed in Chap. 10. At longer wavelengths the transmissivity drops again, although not to zero. Since the incident photons have lower energies in this range, the increased absorption must take place via electronic transitions from states lying in the forbidden-energy region to the conduction band or else from quantum states lying in the valence band to the localized states. It follows from this discussion that absorption measurements can be used to determine the energy-band model of a semiconductor. The main absorption edge occurs at an energy corresponding to the forbidden-energy gap, while the other absorption maxima (transmission minima) determine the energies of the localized states. In practice, such measurements must be carried out at low temperatures so that thermal excitation

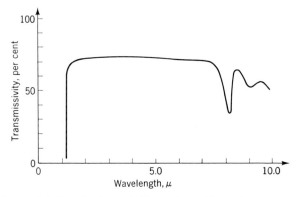

Fɪɢ. 4.   Transmissivity of polycrystalline *p*-type silicon at room temperature.

of the electrons does not mask the transitions to be studied. This is particularly true if transitions to or from acceptor or donor states are to be detected.

Insulators. The optical properties of semiconductors described above resemble those of metals in the visible region and those of insulators in the infrared region of the electromagnetic spectrum. True insulators, by comparison, have large forbidden-energy regions, so that they are transparent to visible light. At sufficiently high energies, however, they should become absorbent according to the classical theory discussed in the introduction to this chapter. The absorption spectra of several alkali halides are shown in Fig. 5. As can be seen, absorption occurs in the ultraviolet region but the absorption peaks show a structure not predicted by relations (5). This means that the absorption process is not a simple polarization effect, and a different explanation for the observed maxima must be sought.

A mechanism for the absorption of photons in alkali halides was first proposed by Frenkel in 1931. He assumed that each absorbed photon produces an electron-hole pair. The electron does not undergo a transition to a quantum state in the conduction band, since photoconductivity does not result, but remains localized in the vicinity of the hole. Such an electron-hole pair is called an *exciton*, and it is held together by the Coulomb attraction of two opposite charges. The Coulomb potential

$$V = -\frac{e^2}{4\pi\epsilon r} = -\frac{1}{\kappa_e}\frac{e^2}{4\pi\epsilon_0 r} \tag{10}$$

is modified by the dielectric constant κ_e of the crystal and is inversely proportional to the separation r between the electron and the hole. Thus the exciton energies are not unlike those of hydrogen and must be less than that of the energy gap so that exciton levels lie in the forbidden-energy region of the crystal (Fig. 5). The binding energy of such an

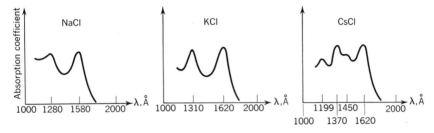

FIG. 5. Representative absorption spectra of alkali halides. (*After E. G. Schneider and H. M. O'Brian, and R. Hilsch and R. Pohl.*)

electron-hole pair can be calculated from a slightly modified hydrogen formula (Chap. 3),

$$E_n = -\frac{m_0^* e^4}{\kappa_f^2 \epsilon_0^2 8 n^2 h^2} \tag{11}$$

where all the terms have their usual meaning, except that the effective mass of an exciton $m_0^* = m_e^* m_h^*/(m_e^* + m_h^*)$ represents an averaging of the effective masses of an electron m_e^* and a hole m_h^*. The high-frequency dielectric constant κ_f is used in (11) because only electronic polarizabilities need be considered at these frequencies. The exciton levels for the first few values of the principal quantum number n (not to be confused with the refractive index also denoted by n) are shown relative to the bottom of the conduction band in Fig. 6. The positions of the first absorption maximum for several alkali halides are as follows:

	LiCl	NaCl	KCl	RbCl	CsCl	
Position of first peak	1,430	1,580	1,620	1,660	1,620 Å	(12)
Exciton level	8.6	7.8	7.6	7.4	7.6 eV	

Returning to the absorption spectra shown in Fig. 5, the tail at longer wavelengths depends on imperfections present in the crystal (discussed in the next section). The peaks correspond to transitions to the different exciton levels shown in Fig. 6, which, according to (11), should become more closely spaced as the dielectric constant increases (Exercise 4). The absorption peaks have a finite breadth, even at temperatures near the absolute zero. This broadening is due to thermal vibration of the atoms, since a sharpening of the peaks does occur when the crystal is cooled. This means that the exciton levels are not sharp, but constitute narrow bands. The exciton, or electron-hole pair, can move

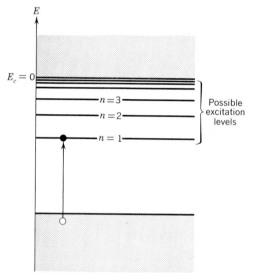

FIG. 6. Possible exciton levels in an insulator. The electron (black dot) in the first level and the hole (open circle) in the valence band constitute an exciton.

through the crystal a distance of about 1 μ before recombination takes place (Exercise 5). Because the exciton is electrically neutral, such motion does not constitute an electric current and the mobility is difficult to measure. There is, nevertheless, both theoretical and experimental evidence to support this model.

When the energy of an incident photon is sufficient to remove completely the valence electron from the vicinity of its parent atom, the situation can be described on the energy-band model in Fig. 6 as a transition to the continuum of states in the conduction band. Exactly at what point in the ultraviolet spectrum of alkali halides photoconductivity ensues has not been clearly established as yet. The relatively less pronounced structure in the low-energy portion of the absorption spectrum is attributed to overlapping bands in this continuum, by analogy to similar structure observed in the absorption spectra of noble metals. By comparison, the absorption maxima of molecular crystals are much sharper and resemble more nearly the spectra of isolated molecules. This is a consequence of the strong intramolecular forces in such crystals, and the hydrogenic model discussed above cannot be applied to molecular crystals. It turns out, moreover, that the absorption spectra of other polar crystals also differ from those of alkali halides. The absorption spectra of AgCl, AgBr, TlBr, and CdI_2 have very pronounced tails on the long-wavelength side of the first absorption maximum, extending even into the visible spectrum, and the peaks are not as pronounced. On the other hand, several very distinct maxima are observed in Cu_2O, and their relative positions correspond very closely to the exciton energies predicted by (11).

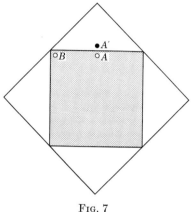

Fɪɢ. 7

The reason for the existence of excitons in insulators can be illustrated quite simply in terms of the zone model. Consider the first and second Brillouin zones of a fictitious crystal having a primitive cubic structure (Fig. 7). If the crystal is an insulator, the first zone is completely filled and the second zone is completely empty. Now the highest energy level in the first zone lies at the corner of that zone, whereas the lowest energy level in the second zone lies at the center of the boundary between the zones. This means that less energy is required for a transition from a point B to A' than from A to A'. According to quantum-mechanical selection rules, a transition $B \rightarrow A'$ is not allowed because their momentum, or **k** vectors, are different. Thus the transition having the lowest allowed energy is $A \rightarrow A'$ and represents a transition to the conduction band. If an electron at A' can pair up with a hole at B, however, the electron-hole pair can be generated by a lesser energy than that required for a transition $A \rightarrow A'$, and the resulting exciton level must lie in the forbidden gap. It was shown by Frenkel that it is possible to describe the momentum of an exciton by a wave vector **k**, having allowed values between $-\pi/a$ and $+\pi/a$, and that the exciton bands exist throughout the whole crystal and are not like the localized levels produced by impurity atoms in semiconductors. Although the exciton bands are continuous, they are not conduction bands, because the electron-hole pairs are electrically neutral.

Color centers

F **centers.** Although insulators having large energy gaps are transparent to visible light, they sometimes appear colored. The color results from a selective absorption of part of the visible spectrum and is believed to be caused by imperfections present in the crystal. These imperfections are called *color centers* and serve to introduce localized states in the forbidden-energy region of an insulator similar to the effect of impurities in semiconductors The presence of such levels in the forbidden gap permits the absorption of longer-wavelength radiation, that is, of photons having energies smaller than the width of the energy gap. Several kinds of imperfections can give rise to such levels; for example, interstitial zinc atoms in ZnO color the otherwise transparent crystal yellow. The most studied color centers are those produced by an excess of metal atoms present in nonstoichiometric alkali halides and are called *F* centers, after the German word for color, *Farbe*. The colors produced depend on the crystal; for example, excess lithium in LiF renders it pink, excess potassium turns KCl blue, while excess sodium in NaCl makes the crystals yellow. Although the color centers described below have been studied in artifically colored crystals, the profusion of elements abounding in nature is responsible for the coloration of naturally occurring minerals, including gem stones.

Suppose an alkali halide crystal is heated in an atmosphere containing an excess of the alkali metal vapor. When an alkali metal atom deposits on the crystal's surface, it dissociates into a cation and an electron. An anion from the crystal may then move next to the cation and start a new layer growing on its surface, while the anion vacancy diffuses into the interior. The generation of vacancies by this process is clearly demonstrated by a decrease in the crystal's density determined with the aid of x-ray diffraction measurements. The free electron similarly can diffuse into the crystal, and when it encounters a vacancy, it becomes attracted by its positive charge. The trapping of electrons at anion vacancies is required in order to maintain local charge neutrality. The trapped electron has a ground-state energy determined by the environment of the vacancy, which is usually expressed by an effective dielectric constant. It is possible to calculate hydrogenic energy levels for the electron by an equation like (11) provided that the effective dielectric constant is known. These energy levels lie in the forbidden-energy region of the crystal and, as in hydrogen, progress from relatively widely spaced levels to an almost continuous set of levels just below the bottom of the conduction band. When the crystal is irradiated with white light, some component of this light has the appropriate energy to excite the trapped electron to one of the higher-lying states, so that it becomes absorbed in the process and the absorption spectrum of a crystal containing F centers exhibits a characteristic absorption peak close to the visible portion of the spectrum. It does not change when an excess of another metal is introduced in the crystal provided that the foreign atoms can substitute for the metal atoms normally present in the host crystal. This corroborates the

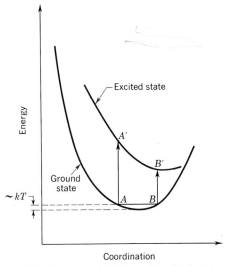

FIG. 8. Relative energies of a ground state and excited state of a color center.

assumption that the absorption peak results from transitions to excited states determined by the trapped electron.

The above-mentioned dependence of the energy levels of an F center on the atomic environment of a vacancy suggests that the absorption maximum should shift to shorter wavelengths (higher energies) as the interatomic distances in the crystal decrease. Such a shift in the peak position is actually observed when a crystal is cooled or heated. It turns out that the product of the frequency of the absorbed light times the square of the interatomic distance remains constant in alkali halide crystals, whereas the absorption maximum has a finite breadth, even at very low temperatures, and becomes broader with increasing temperature. The reason for this can be seen by considering the temperature dependence of the energy of a color center. Figure 8 shows the energy variations of the ground state and first excited state as a function of the coordination of a vacancy. At some finite temperature, the ground-state energy is not the minimum of the lower curve, but lies approximately kT above it, because the thermal vibrations of the coordinating ions alternate between A and B. This means that the energy of the absorbed radiation can range between that of a transition like $A \rightarrow A'$ or $B \rightarrow B'$. The width of the absorption peak obviously depends on the difference between the two energies corresponding to A' and B'. As the temperature increases, AB becomes displaced to higher energies and the width of the absorption peak increases. Because of this width, it usually is called a band, namely, *F band*. The tail appearing on the short-wavelength side of the peak is similarly called the *K band*.

The upper curve in Fig. 8 is determined by the change in the environment of a vacancy when the trapped electron is in an excited state. This is usually expressed by a change of the effective dielectric constant in the vicinity of such a vacancy. Whereas an alkali halide crystal is normally diamagnetic because the ions have closed outer shells (Chap. 13), the presence of trapped electrons introduces some paramagnetic character. This makes it possible to study the environment of an F center by electron-paramagnetic-resonance measurements. As might be expected from the above described model for F centers, the total number of photons absorbed should be directly proportional to the number of vacancies present in the crystal. The results of careful experiments show that the total absorption is indeed proportional to the number of excess metal atoms incorporated, that is, the number of F centers produced.

Other kinds of centers. Because the excitation levels to which an electron is excited from an F center lie very near the bottom of the conduction band, it is possible for such electrons to become thermally excited to states lying in the conduction band of the crystal. This leads to photoconductivity, which is actually observed under suitable conditions. It is, of course, necessary that electrons flowing to the positive electrode be replaced by other electrons injected into the crystal by the negative electrode, so that space charges do not build up inside the crystal. In the absence of any external field, such a conduction electron may wander into the vicinity of an F center and become captured by it. This produces a new kind of center having a higher energy level because the two electrons are not as tightly bound to a single vacancy. Such color centers are called F' centers. To test this model, suppose that a crystal containing F centers is irradiated by light having sufficient energy to excite the trapped electrons to the excited states. If this process takes place at room temperature, the absorption in the F band is gradually decreased following such irradiation, and this process is called *bleaching*. Following the bleaching, a new band, called the F' *band*, is found to lie on the long-wavelength side of the original F band (Fig. 9A). The energy-band model of the process responsible for its formation is shown in Fig. 9B. The two trapped electrons are first shown in their respective F centers by the black dots. After one of them has migrated to the other, via transitions to and from the conduction band, both are shown occupying the F' level next to an empty F level. When a crystal containing such newly formed F' centers is, in turn, irradiated with suitable longer-wavelength light, the F' centers are broken up and the F centers restored. This reversibility is consistent with the energy-band model proposed in Fig. 9B and serves to confirm it. In carrying out such measurements, it

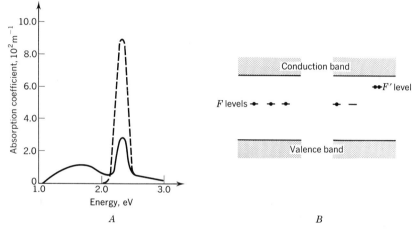

FIG. 9. Comparison of an F and F' band. (A) The absorption spectrum following bleaching is shown by the solid curve. The initial F-band absorption is shown by the dashed curve. (B) Energy-band model for F centers and for an empty F center and an F' center.

is very important to control the temperature precisely, so that it is not too high to mask these transitions and not too low to prevent their occurrence in sufficient quantity for ready observation.

It is possible to produce color centers in other ways also. For example, when an alkali halide crystal contains excess halogen ions, the accompanying cation vacancies can trap holes quite similarly to the way that anion vacancies trap electrons. The color centers thus produced are called V centers, and crystals containing them exhibit several absorption maxima, usually designated V_1 bands, V_2 bands, and so forth. Alternatively, color centers can be produced in crystals by irradiating them with very high energy radiations, say, x-rays or γ-rays. The process in this case first consists of the production of recoil electrons by the incident photons, followed by the interaction of these high-energy electrons with valence electrons in the crystal to produce electron-hole pairs. The "freed" electrons and holes can migrate through the crystal until they encounter vacancies normally present in all such crystals (Chap. 5) and become trapped. Consequently, such irradiated crystals usually contain both F and V centers. Unlike stoichiometrically deficient crystals, however, the crystals that are colored by radiation can be easily bleached by reirradiating them with visible light or by annealing them. This is so because the excited electrons and holes ultimately can recombine with each other. By comparison, nonstoichiometric crystals contain a "built-in" excess of electrons or holes, so that their color cannot be permanently removed without changing them chemically. Still other kinds of color centers have been observed also; however, their nature is not as well understood, so that they are not considered further here.

Photographic process. When silver halides are irradiated with visible light, F-center formation is not observed. Instead, free silver precipitates out of the crystal structure and the precipitates quickly attain colloidal dimensions. These regions become opaque to visible light and form the basis of the familiar photographic process. An explanation of this process was first proposed by Gurney and Mott in 1938 and appears to be correct, except for some details that have not yet been satisfactorily worked out. An incident photon striking a crystal of, say, AgBr produces an electron-hole pair. Since AgBr is photoconductive in the visible region, the electron is excited to a quantum state lying in the conduction band. The electron migrates to the surface, where it becomes trapped by some impurity, possibly Ag_2S present in small amounts. An interstitial Ag^{1+} ion present in the crystal as a Frenkel defect then is attracted by the charge of the trapped electron, and the two combine to form a neutral silver atom. Suppose a second photon is absorbed and produces another electron-hole pair. The electron can migrate through the crystal until it combines with the neutral silver atom to form an Ag^{1-} ion. (It is, of course, possible that the electron is trapped by the same impurity that catalyzed the formation of the first neutral silver atom.) This negative charge again attracts an interstitial Ag^{1+} and combines with it to form Ag^0. The reasonableness of this process is supported by the relatively high mobility of Ag^{1+} ions in AgBr as compared with the relative immobility of the anions. In order to preserve charge neutrality, however, it is necessary to assume that the positive holes recombine with bromine ions on the surface and permit the neutral bromine atoms to escape from the crystal.

As the above process continues, the dimensions of the free-silver precipitate increase in size. The process must take place at the surface (or at an internal crack), so that there is adequate room for its growth. There is some evidence from electron microscope observations in support of the formation of filamentary growths of this type. When the precipitate reaches colloidal dimensions of submicroscopic size, it is said that a *latent image* has formed in the grain. By suitable chemical development, it is possible to abet its growth until most of the silver in the grain is deposited on this nucleus. This is the normal process used in rendering opaque the exposed AgBr grains in a photographic emulsion. A fixer is then used to make the unexposed grains insensitive to visible light. It is possible, however, to increase the size of the metal colloids without chemical agents. This can be done by continuing to expose the sensitized grain to visible light. The latent image continues to grow in size by the same process that initially formed it until enough Ag^{1+} ions are converted to free silver to coat the surface. The crystal then appears black when viewed with visible light, and this process is called the *print-out effect*. Absorption of light by such a crystal does not show the temperature dependence that is observed for crystals containing F centers. Although the formation of a latent image presupposes the existence of impurities on the surface, once a nucleus is formed, the print-out effect proceeds at a rate that appears to be independent of small amounts of impurities present. By comparison, the removal of silver or silver sulfide from the surface of grains in a photographic emulsion greatly decreases the speed of the film response. Conversely, the addition of dyes to the surface of such grains, called dye sensitization, can alter the spectral region in which latent images are formed.

Emission of light

Luminescence. There are several ways to excite a valence electron to a higher-energy state in the crystal. This can be done by the absorption of light or other radiation, as discussed above, or, for example, by a strong electric field. After a time interval ranging from microseconds to minutes, the electron returns to its ground state unless a photochemical reaction takes place, for example, the formation of a latent image. The transition to the ground state must be accompanied by the dissipation of the extra energy. Usually part or all of the extra energy is dissipated as heat by interactions with the vibrating atoms in a crystal. Alternatively, the electron can lose its energy by the emission of a photon of light, according to the relation

$$h\nu = E_{\text{excited}} - E_{\text{ground}} \qquad (13)$$

and the general process is called *luminescence*. If the crystal emits light simultaneously with its excitation, the process is called *fluorescence*; if it takes a microsecond or longer to emit its light photons, then the process is called *phosphorescence* and the crystal is said to be a *phosphor*.

Most crystals do not luminesce in their pure state. In fact, there is a great amount of evidence showing that luminescence occurs only when the crystal contains certain impurity atoms called *activators*. Suppose an incident photon excites a valence electron from an activator atom to a higher-energy state. The energy of the subsequently emitted photon is always less than the original excitation energy. The reason for this can be seen with the aid of Fig. 10. The ground-state energy as a function of the coordination of the activator atom is represented by the lower curve. When a photon is absorbed, it must expend an energy sufficient for the transition $B \to B'$ to occur. The configuration is changed by the excited electron since its orbital is now enlarged and alters the potential distribution in its vicinity. The energy dependence of the new situation is

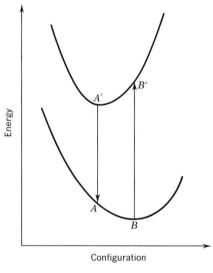

Fig. 10

depicted by the upper curve in Fig. 10. After reaching thermal equilibrium, the excited state has the energy corresponding to A', so that, when the electron returns to its ground state, it undergoes a transition $A' \to A$. The extra energies $E_{B'} - E_{A'}$ and $E_A - E_B$ are expended as heat. It is clear from this discussion that the light emitted depends on the relative energies of the excited and ground states (Fig. 10), so that it is characteristic of the activator atom. This process is called *characteristic luminescence*, therefore, and is discussed further in the next section. Because the emitted light has a longer wavelength, there is virtually no self-absorption of this light.

The different kinds of luminescence possible in crystals also depend on the means employed to excite the electrons. Several categories can be distinguished, including:

1. *Photoluminescence,* in which the excitation is accomplished by the absorption of photons. The radiation source may be infrared, visible, ultraviolet, or x-radiation; for example, this is the way that incident radiation is made visible in phosphorescent coatings used in x-ray fluoroscopy and in infrared-image converters. One of the earliest applications of photoluminescence is in fluorescence lamps, in which the insides of the glass envelopes are coated with a phosphor. Ultraviolet radiation produced by a mercury-glow discharge inside the lamp is then converted to visible light without appreciably heating the lamp.

2. *Cathodoluminescence* results when a phosphor is bombarded by high-energy electrons or cathode rays. This is the process that is operative in the screens of cathode-ray tubes and television tubes.

3. *Electroluminescence* occurs when a phosphor is placed in an insulating medium having a very high dielectric constant and an alternating electric field is applied across the crystal. The very large electric-field strength built up inside the phosphor is sufficient to "empty" an activator quantum state. When the excited electron subsequently returns to its ground state, visible light is emitted. The light given off is a function of the voltage and frequency of the applied field. Luminescent panels operating at 400 volts and 3,000 cps have been used to illuminate entire rooms. It is similarly possible that future television screens will use electroluminescence rather than cathodoluminescence in television sets that will hang like pictures on a wall.

4. *Thermoluminescence* is actually a misnomer, because the electrons are not initially excited by thermal means. Instead, the electrons are excited by other means at very low temperatures and become trapped in states lying in the forbidden-energy region. The low temperature retards the emptying of these states, so that luminescence does not occur until the temperature of the phosphor is subsequently increased. Thermoluminescence, therefore, is particularly useful in studying the energy levels of impurity quantum states because the amount of thermal energy absorbed by the electrons during heating can be correlated to the energy of the emitted radiation.

5. *Triboluminescence* is produced when two fairly hard insulators, for example, two quartzite rocks, are briskly rubbed together. Although its present application is limited to entertaining guests in a darkened parlor, the electronic excitations taking place at the rubbed surfaces are probably similar to the more carefully studied processes described in this chapter.

Phosphors. The role that an activator ion plays in a phosphor depends on its position in the crystal structure. If it occupies a substitutional position, that is, when it is in a correct metal site, it behaves differently than when it is incorporated interstitially. In either case, charge neutrality must be preserved and sometimes requires the addition of other kinds of atoms called *coactivators*. For example, when Ag^{1+} is added to ZnS, the amount of silver that can be included is limited by the need to form S^{2-} vacancies. If the sulfur ions are simultaneously replaced by Cl^{1-} ions, however, the amount of substitutional silver can be increased. The inclusion of coactivators, therefore, leads to a charge compensation in the crystal (Chap. 9). Another way that such compensation can be accomplished is by adding equal amounts of monovalent and trivalent cations to the crystals during their growth.

The emission curves of zinc sulfide phosphors activated with copper, silver, manganese, and excess zinc are compared in Fig. 11. The divalent

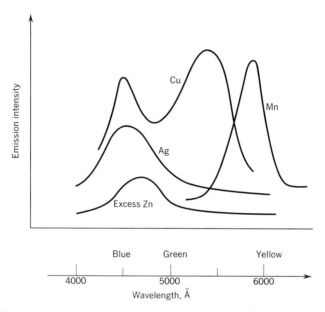

Fig. 11. Luminescence spectra of zinc sulfide phosphors activated with Cu, Ag, Mn, and excess zinc, stimulated by ultraviolet light. The vertical scales of each curve are expressed in arbitrary units to facilitate their comparison.

manganese ions substitute for divalent zinc atoms and emit yellow light, following irradiation with ultraviolet light. The blue light emitted when Cu^{1+}, Ag^{1+}, or excess zinc is incorporated in the crystal is believed to be caused by transitions between energy states introduced in the forbidden-energy region by vacancies present in the crystal. On the other hand, the pronounced green emission, when copper is incorporated, is attributed to divalent copper atoms substituting for zinc. The exact processes leading to such characteristic luminescence are not clearly established, although sufficient evidence is available to permit the formulation of a phenomenological theory. The activator ions introduce localized states in the forbidden gap. Similar states can be introduced by any coactivators or vacancies present; however, luminescence in ZnS seems to be independent of the type of coactivators present. Two such possible states are depicted in Fig. 12. Suppose the energy level at A represents an activator state, say, introduced by Ag^{1+} in ZnS. The level at B then represents a vacancy, or coactivator anion state. When an incident photon produces an electron-hole pair, the conduction electron can fall into either level. If it returns to the activator state at A via transition 1, a visible-light photon of corresponding energy is released. If, instead, it undergoes transition 2 into the level at B, the energy is dissipated thermally and no radiation is emitted. Thus the vacancy acts as an electron trap. (The associated hole produced in the valence band can become trapped by A, so that it is again able to accept a photoelectron via transition 1.) Similarly, an electron trapped at B can be thermally excited to a state in the conduction

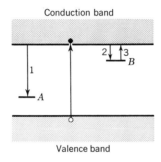

Conduction band

Valence band

Fig. 12

band, transition 3, and subsequently undergo the radiative transition 1. Similar band models can be constructed for other luminescence processes.

The intensity of luminescence after the excitation radiation has been turned off is called the *afterglow*. It obviously depends on the rate at which the luminescence decays, so that it is a function of the density of conduction electrons at any instant of time $N(t)$ and their apparent lifetime τ. Suppose that the probability for a conduction electron to undergo transition 1 in Fig. 12 is $1/\tau$ per second. If a center at A represents the ground state of an activator atom, then the lifetime is virtually independent of temperature and the number of other centers present. The intensity of luminescence at any instant I is equal to the number of transitions taking place (photons emitted) and is given by the time rate of change of N, or by the product on N times the transition probability, so that

$$I = -\frac{dN(t)}{dt} = \frac{N(t)}{\tau} \tag{14}$$

where the minus sign in (14) denotes a decreasing density of conduction electrons.

Integrating the right side of (14),

$$\ln N(t) = \ln N_0 - \frac{t}{\tau}$$

and
$$N(t) = N_0 e^{-t/\tau} \tag{15}$$

where N_0 is the conduction electron density when the excitation light is turned off. The rate of decay of the intensity is then given, according to (14), by

$$I(t) = \frac{N_0}{\tau} e^{-t/\tau} = I_0 e^{-t/\tau} \tag{16}$$

where I_0 is the initial intensity.

The effective lifetime in actual crystals ranges from 10^{-8} sec to several minutes. Such longer lifetimes are observed in certain alkali halides activated by thallium-ion inclusions. The reason for this longer afterglow is explained by assuming that other centers, such as the trapping center B in Fig. 12, are present in the crystal. Suppose the lifetime of a trapped electron is τ_B and the energy difference between the bottom of the conduction band and state B is denoted by E; then the probability per unit time of transition 3 is given by

$$\frac{1}{\tau} = \frac{1}{\tau_B} e^{-E/kT}. \tag{17}$$

An electron trapped at a center like B is unable to take part in the luminescent transition 1 until it is returned to a state in the conduction band. Assuming that transitions of type 1 (in Fig. 12) are more probable than those denoted by 2, the instantaneous intensity of luminescence is determined by the rate at which the thermally activated transition 3 takes place. Direct substitution of (17) into (16) then gives the temperature dependence of the afterglow:

$$I(t) = \left(\frac{N_0}{\tau_B} e^{-E/kT}\right) \exp\left[-\left(\frac{t}{\tau_B}\right) e^{-E/kT}\right]. \tag{18}$$

It is clear from (18) that as $T \to 0$ deg K, the intensity decreases very rapidly, because the rate at which the traps B can be emptied decreases. When the crystal is subsequently heated, the thermal energy necessary for transition 3 can be measured, so that thermoluminescence can be used to study the energy levels of trapping centers.

Stimulated emission. As discussed above, when an electron occupies a quantum state in the conduction band, it can undergo a transition to an empty lower-energy state spontaneously. When such transitions are accompanied by the emission of light photons, the process is called *spontaneous emission*. Reverse transitions, in which photons are absorbed, are also possible and constitute the well-known absorption process. In general, a permanent transition between two quantum states can be described by a relation like

$$E_1 - E_2 = \pm h\nu \tag{19}$$

where the plus sign denotes absorption of a quantum of energy and the minus sign denotes emission. It is reasonable to expect (and can be proved rigorously from quantum mechanics) that the probabilities for a transition in either direction are equal under identical conditions. (This is sometimes called the microscopic reversibility of a quantum process.)

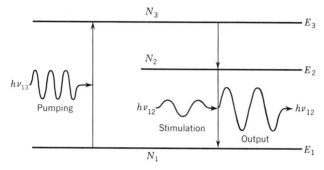

FIG. 13. Energy-level diagram of three-level laser. Electrons excited from E_1 to E_3 subsequently decay to E_2 and are then stimulated to E_1.

Suppose that the two energies in (19) correspond, respectively, to an excited state and the ground state of an atom. Next suppose that the atom is placed in a radiation field of frequency ν determined by (19). It follows from the above that the atom can now undergo a transition in either direction (absorptive or emissive), depending only on its initial state. If the transition is accompanied by the emission of a quantum of energy $h\nu$, the process is called *stimulated emission*. Like the more familiar reverse process of absorption, stimulated emission is proportional to the intensity of the incident radiation. Note also that the frequency of the emitted radiation is the same as that of the stimulating radiation, in view of the reversibility implied by (19).

Consider next the energy states labeled E_1 and E_2 in Fig. 13. The number of absorbed photons of energy, $h\nu_{12}$ (equal to the energy difference $E_2 - E_1$), is proportional to the number of electrons in the lower state, N_1. Similarly, the number of emitted photons of energy $h\nu_{12}$, caused by electronic transitions from state E_2 to state E_1, is proportional to the number of electrons in the upper state, N_2. The ratio of electrons in the two states at equilibrium is determined by the Boltzmann relation, so that

$$\frac{N_2}{N_1} = e^{-(E_2-E_1)/kT}. \tag{20}$$

Since $E_2 > E_1$, Eq. (20) shows that $N_1 > N_2$, which means that more photons are absorbed than are emitted. This is, of course, the usual state of affairs, as discussed in previous sections of this chapter. On the other hand, if by some means N_2 can be maintained larger than N_1, more photons are emitted than are absorbed, and this is called *light amplification by stimulated emission of radiation*, commonly shortened to *laser*.

A possible way to obtain the condition $N_2 > N_1$ is to introduce a third allowed energy level, E_3 in Fig. 13. Incident photons of energy, $h\nu_{13} = E_3 - E_1$, are then absorbed by electronic transitions to the

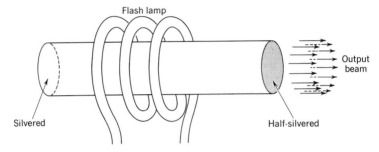

Fig. 14. Model of laser.

quantum states in E_3, and electrons in such states spontaneously drop to the state E_2. If the probability of a subsequent transition from E_2 to E_1 is less than that of the nonradiative transition $E_3 \rightarrow E_2$, N_2 soon becomes larger than N_1, provided that the incident intensity of photons (of energy $h\nu_{13}$) is sufficiently large. The process of making $N_2 > N_1$ is called *pumping*, and the population density of E_1 and E_2 is said to become *inverted*.

The first successful laser made use of the energy levels introduced by substituting Cr^{3+} ions for aluminum in a pale-pink ruby (Al_2O_3) crystal. The energy-band model of a $(Cr,Al)_2O_3$ crystal is not unlike that shown in Fig. 13. The photon energies of green and yellow light are appropriate for exciting the electrons to level E_3, and the wavelength of the radiation emitted in the transition $E_2 \rightarrow E_1$ is 6,943 Å and accounts for the characteristic red fluorescences of ruby. An actual laser (Fig. 14) has the shape of a cylindrical rod whose opposite ends are ground flat and parallel and are partially silvered so as to reflect part of the light incident at right angles to them and transmit part of it. An electronic flash lamp is used to pump the electrons, that is, to raise most of the chromium atoms to the excited level E_3. As some of them return to the ground state via the states E_2, they emit red light, which passes out through one half-silvered end of the crystal. Part of this light is reflected back into the crystal, however, and serves to stimulate further emission. This laser therefore acts as a generator of radiation rather than an amplifier, although it is made to amplify a self-generated signal by virtue of the reflecting surfaces placed at its ends. Since only photons which move strictly parallel to the cylinder axis are reflected back and forth sufficiently to cause appreciable stimulation, the emergent light beam is also strictly parallel. Moreover, because the emission has phase coherence with the stimulation, the beam has both space and time coherence, exactly analogous to a radio-frequency signal but in contrast to a conventional light source in which photons are generated at random.

A relatively large amount of power is required to attain inversion in ruby because the population density N_1 in the ground state E_1 is large. The existence of energy-band models in which E_1 is not the ground state has been investigated, therefore, in several materials. For example, the energy states introduced by neodymium ions in calcium tungstate are of this type. By reducing the temperature and decreasing electron-phonon interactions, the inversion can be increased still further, so that a continuously operating laser is possible instead of one in which pumping is limited to very intense bursts by heating effects generated in the crystal. It also has been demonstrated that a continuously operating laser can be constructed by using a mixture of helium and neon gases in place of a crystal. Pumping is achieved in this system by collisions between neon atoms and excited helium atoms, produced by an electric discharge in the gas. These collisions supply the energy needed to raise the neon atoms to an excited state E_2, from which they can be stimulated to a state E_1 lying above the ground state of the gas. The spectral purity of the stimulated radiation is so remarkable that it has considerable practical importance. The spectral width of the emission from a gas laser, for example, is less than 10^3 cps, at a frequency of 10^{11} cps, and the power output per unit bandwidth is 10^5 times that of a square centimeter of the sun's surface. Also, because the beam is so highly monochromatic, the beam emitted by a laser can be focused to a spot size of the order of the wavelength of the radiation. Power densities of 10^{12} W/m^2 are easily attainable and are capable of melting or vaporizing even the most refractory materials. The potentialities of laser beams in communication applications are best illustrated by the fact that 80 million television channels can be contained in the spectral region corresponding to that of visible light. The strict parallelism of the beam means that the final cross section of a beam transmitted from a ruby laser on earth is less than 2 miles when it has reached the moon.

It is possible to invert the population of the allowed energy states by processes other than photon absorption. As discussed in Chap. 10, minority-carrier injection at a *p-n* junction increases the concentration of carriers above that existing under equilibrium conditions. For example, with sufficiently strong injection currents, the population of the conduction band may become inverted relative to that of deep-lying impurity states. Following this inversion, stimulated emission may accompany electronic transitions from states in the conduction band to the impurity states. Such laser action has been produced in GaAs *p-n* junctions. Not only does the efficiency of converting electric power to optical radiation in such devices approach 100 per cent, but it is possible to modulate the coherent beam by simple electrical modulation of the injection current.

It should be noted that similar transitions also can be stimulated at non-optical frequencies. In fact, this phenomenon was first observed at microwave frequencies in a device that is called a *maser*. The energy states in a maser can arise from the magnetic-energy levels of a paramagnetic ion in a crystal, like those discussed in Chap. 13 in connection with electron paramagnetic resonance. The efficiency of masers is also improved by operation at liquid-helium temperatures. This has an added advantage of reducing the internal noise to extremely low levels, so that masers are the most sensitive detectors and amplifiers of microwave signals yet devised.

Transmission of light

Refraction. As discussed in Chap. 12, when an electric field traverses any medium, it induces electric dipoles in that medium that oppose the field. The velocity of the moving field is decreased thereby, and its propagation direction becomes altered by a corresponding amount. This phenomenon is called *refraction* and occurs for all kinds of electromagnetic radiation, although the magnitude of the index of refraction n depends on the frequency of the radiation. Suppose a light ray passes from air into an insulator crystal that is transparent to the light as shown in Fig. 15. The angle of refraction r is related to the angle of incidence i according to *Snell's law*,

$$\frac{\sin i}{\sin r} = \frac{n_2}{n_1} \tag{21}$$

where both angles are measured from the normal to the interface, and n_1 is the refractive index of air and n_2 of the crystal. Since no refraction can occur in vacuum, its index of refraction is set equal to unity and the other indices are determined accordingly with the aid of (21). The index of refraction of air is 1.0003, so that air is used as the reference standard whenever a precision of less than 3 parts in 10^4 is adequate. As already discussed in this chapter, some of the incident light may be reflected at the surface, in which case the angle of reflection equals the angle of incidence. The amount of light reflected at normal incidence is determined by (7) and depends on the refractive index. Note from (21) that when the refractive index of the crystal n_2 is less than the index of the surrounding medium n_1, it is possible that, for some angle of incidence, $(n_1/n_2) \sin i > 1$. Since $\sin r$ cannot exceed unity, none of the light can be refracted in this case, so that all of it must be reflected. Consequently, when a light beam passes from one medium to a less refractive medium, total reflection occurs for angles of incidence greater than some critical value.

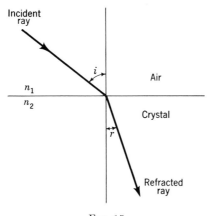

F$_{\text{IG}}$. 15

The velocity of light in a dielectric medium is determined by the index of refraction, so that $v = c/n$, where c is the velocity of light in vacuum ($c = 2.998 \times 10^8$ m/sec). The velocity of light is reduced in the medium primarily by electronic polarization, because the much heavier ions cannot respond to the rapidly varying field at these frequencies. This is the reason why $n^2 = \kappa_f$, the high-frequency dielectric constant. Accordingly, it becomes possible to rewrite the Clausius-Mosotti equation (Chap. 12),

$$\frac{n^2 - 1}{n^2 + 2} = \frac{1}{3\epsilon_0} \sum_i n_i \alpha_i \tag{22}$$

which is called the *Lorenz-Lorentz equation*. When n is very nearly equal to unity, so that $n^2 + 2 \simeq 3$, Eq. (22) becomes

$$n^2 = \kappa_e = 1 + \frac{1}{\epsilon_0} \sum_i n_i \alpha_i. \tag{22a}$$

Multiplying both sides of (21) by M/ρ, where M is the molecular weight and ρ is the density,

$$\frac{M}{\rho}\left(\frac{n^2 - 1}{n^2 + 2}\right) = \frac{1}{3\epsilon_0} N_0\alpha \tag{23}$$

where N_0 is Avogadro's number, and α is the polarizability of a molecule, so that (23) can be used to calculate the *molar polarizability*, or the *molar refractivity*. This equation allows one to relate the index of refraction of a crystal to its structure.

Birefringence. The index of refraction of glasses and of crystals belonging to the isometric system is independent of the direction of propagation of the light. Such crystals are said to be optically isotropic. It is important for the discussion in this section to realize that, when light propagates along a certain direction, the electric field varies sinusoidally at right angles to the propagation direction. The electronic displacements and the resulting dipoles, therefore, are also orthogonal to the propagation direction. In tetragonal and hexagonal crystals, a light ray traveling parallel to c encounters an isotropic structure, since $a_1 = a_2$ in these crystals. (Note that this is just like a light ray traveling along a_3 in a cubic crystal.) A light ray traveling in any other direction, however, encounters an anisotropic structure, so that it is not surprising that it travels with a different velocity. The c axis is therefore an optically unique axis in such crystals, and it is called the *optic axis*. Since there is only one unique direction in tetragonal and hexagonal crystals, they are called *uniaxial* crystals.

In order to study the properties of anisotropic crystals it is necessary to use plane-polarized light, that is, light whose electric field is constrained to alternate in a single plane containing the direction of propagation. The polarization direction is usually called the *vibration direction* of the light. When a ray of plane-polarized light traverses a uniaxial crystal parallel to the optic axis, then, regardless of the vibration direction of the electric field, the index of refraction is the same and is usually denoted ω. When a ray traverses the crystal along a direction which is normal to the optic axis, the index of refraction depends on the vibration direction of the electric field. If the vibration direction is also normal to the optic axis, then the index is, as before, ω. When the vibration direction is parallel to the optic axis, the index of refraction is different and is called ε. When the vibration direction lies between these two directions, then the light ray is split into two components whose electric fields are constrained to vibrate in directions respectively parallel and normal to the optic axis, as shown in Fig. 16. The component parallel to the optic axis has a vibration amplitude $\varepsilon \sin \phi$, and its propagation velocity is determined by ε. The two indices of refraction ε and ω, therefore, are called the *principal* indices of refraction of a uniaxial crystal. If $\varepsilon > \omega$, the crystal is said to be *positive uniaxial,* and if $\omega > \varepsilon$, the crystal is said to be *negative uniaxial.*

Returning to Fig. 16, it is clear that, since the velocities of propagation of the two components are different, the two components emerge from the crystal with a relative phase difference whose magnitude depends not only on the relative propagation velocities (indices of refraction), but also on the thickness of the crystal. Suppose that the thickness of the crystal is such that one component is exactly one-half wavelength behind the other components; that is, they differ in phase by 180 deg. Then the emerging light has the two electric-field components directed as shown in Fig. 17, and the resultant electric field is rotated by an amount 2ϕ from its original direction shown in Fig. 16. The phase of the component whose

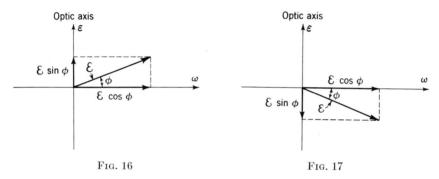

Fig. 16   Fig. 17

vibration direction is parallel to the optic axis is equal to $2\pi\nu t/v_\varepsilon$, where ν is the frequency of the incident light, t is the thickness of the crystal, and v_ε is its propagation velocity. Similarly, the phase of the other component is $2\pi\nu t/v_\omega$. The difference in phase δ, therefore, is

$$\begin{aligned}
\delta &= \frac{2\pi\nu t}{v_\varepsilon} - \frac{2\pi\nu t}{v_\omega} \\
&= \frac{2\pi\nu t}{c}(\epsilon - \omega) \\
&= \frac{2\pi t}{\lambda}(\epsilon - \omega)
\end{aligned} \tag{24}$$

since c/ν is the wavelength of the light in vacuum. Relation (24) can be used to determine the thickness of a so-called *half-wave plate* by setting $\delta = \pi$. Note that the relative phases depend on the frequency (wavelength) of the light.

Suppose that the incident light ray is neither parallel nor normal to the optic axis. For example, consider light incident at point A, in a direction which is normal to one of the natural cleavage faces of a calcite rhombohedron, as shown in Fig. 18. The incident light ray is observed to split up into two rays. This phenomenon is called *double refraction*. One ray, designated the *O* ray in Fig. 18, traverses the crystal without refraction according to Snell's law for normal incidence. (When $i = 0°$, $\sin i = 0$, and $\sin r = 0$.) The other ray is refracted, however, and violates Snell's law. For this reason it is called the *extraordinary ray E* to distinguish it from the *ordinary ray O*. Regardless of the vibration direction of the incident light, the extraordinary ray is constrained to vibrate in a plane containing the optic axis and the incident ray called the *principal plane* (shaded plane in Fig. 18). The ordinary ray, on the other hand, is constrained to vibrate in a direction at right angles to the principal plane. The relative velocities of the two rays are determined by

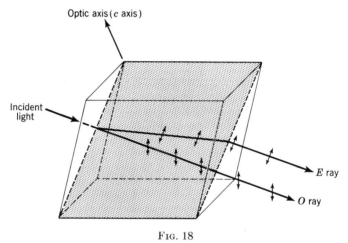

FIG. 18

the two principal indices of refraction, and the difference between these indices is used to measure the birefringence of a crystal.

Calcite, $CaCO_3$, is a negative uniaxial crystal (for the *D* line of sodium, $\lambda = 5,893$ Å, $\varepsilon = 1.486$, and $\omega = 1.658$) whose relatively large birefringence (0.172 for Na *D*) can be easily understood when its crystal structure is considered. The structure of calcite resembles that of a NaCl cube tipped so that its [111] direction is parallel to the *c* axis of calcite (Fig. 19). Note that the CO_3 groups form triangles whose planes are oriented normal to the *c* axis, which is parallel to the optic axis of the crystal. The large difference between the indices of refraction, for light vibrating parallel to and light vibrating normal to the optic axis, is due to this orientation of the carbonate groups. Consider a carbonate group consisting of three oxygen atoms lying at the corners of an equilateral triangle and the carbon atom at its center, as shown in Fig. 20. When the vibration direction of the electric field is normal to the plane of the triangle

(Fig. 20A), the oxygen atoms are polarized, so that the induced dipoles are also normal to the plane of the triangle. Their direction is such that it opposes the direction of the external field. Each dipole, therefore, tends to depolarize its neighboring oxygen dipoles. When the vibration direction of the field is parallel to the plane of the triangle, then two of the

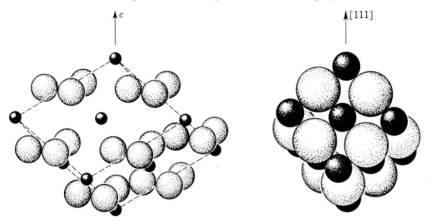

FIG. 19. Crystal structure of calcite, CaCO$_3$, compared with that of NaCl. Only the Ca atoms and CO$_3$ groups situated on the front faces of the rhombohedral unit cell are shown. (The carbon atoms are hidden from sight by the much larger oxygen atoms.)

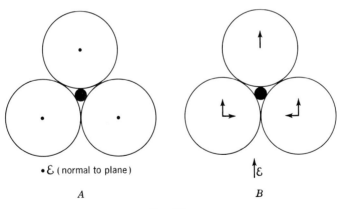

FIG. 20

induced dipoles tend to reinforce the third and, simultaneously, the third dipole tends to reinforce the other two. This can be seen from the induced field directions indicated by small arrows in Fig. 20B. Although the fields in two adjacent oxygen atoms oppose each other, it can be shown that the average increase in the induced-electric-field intensity exceeds the average depolarizing field intensity. Accordingly, a light ray vibrating in the plane of the CO_3 groups in calcite is retarded more than a light ray vibrating in a direction normal to the plane. Since the optic axis of calcite is normal to the plane of the CO_3 groups and since the index of refraction is inversely proportional to the propagation velocity, $\varepsilon < \omega$ and calcite is optically negative.

Crystals belonging to the orthorhombic, monoclinic, and triclinic systems have two optic axes and are called *biaxial* crystals. There are three mutually perpendicular vibration directions and three indices of refraction called α, β, and γ. These directions of vibration bear definite relations to the crystallographic symmetry of a biaxial crystal, and the magnitudes of the indices of refraction are intimately related to its structure. This relationship can be seen in Table 1 by comparing the refractive indices of quartz, which has a three-dimensional network structure, of hornblende, which has a double-chain structure, and of muscovite, which has a sheet structure. As expected from their structures, the birefringence of quartz is much smaller than that of the more anisotropic structures of hornblende and muscovite. Note that the birefringence of biaxial crystals is determined by the numerical difference between the largest and the smallest index.

Table 1
Birefringence of silicates
(Na *D* line $\lambda = 5,893$ Å)

Mineral	Indices of refraction	Birefringence
Quartz	$\varepsilon = 1.553$, $\omega = 1.554$	$\varepsilon - \omega = 0.009$
Hornblende	$\alpha = 1.629$, $\beta = 1.642$, $\gamma = 1.653$	$\gamma - \alpha = 0.024$
Muscovite	$\alpha = 1.561$, $\beta = 1.590$, $\gamma = 1.594$	$\gamma - \alpha = 0.033$

It should be obvious from the above discussion that the optical properties of materials are intimately related to their crystal structure. Thus the indices of refraction of insulators can be used for identification purposes similarly to the powder method of x-ray crystallography, discussed in Chap. 2. In fact, mineralogists have compiled extensive reference tables which enable them to identify the mineral constituents of rocks by examining thin sections under a polarized-light microscope.

Suggestions for supplementary reading

Paul F. Kerr, *Optical mineralogy*, 3d ed. (McGraw-Hill Book Company, Inc., New York, 1959).

Humboldt W. Leverenz, *An introduction to luminescence in solids* (John Wiley & Sons, Inc., New York, 1950).

James W. Meyers, Molecular generators and amplifiers, in *Molecular science and molecular engineering*, edited by Arthur R. von Hippel (The Technology Press of the Massachusetts Institute of Technology and John Wiley & Sons, Inc., New York, 1959), pp. 342–348.

N. F. Mott and R. W. Gurney, *Electronic processes in ionic crystals*, 2d ed. (Oxford University Press, London, 1957).

Suggestions for further reading

Adrianus J. Dekker, *Solid state physics* (Prentice-Hall, Inc., Englewood Cliffs, N.J., 1957).

Charles Kittel, *Introduction to solid state physics*, 2d ed. (John Wiley & Sons, Inc., New York, 1956).

Exercises

1. In the visible region of the electromagnetic spectrum $(\omega_0^2 - \omega^2) \simeq 10^{30}$ sec^{-2}. Assuming $N \simeq 10^{27}$ atoms per cubic meter, calculate κ_e and n. How do these values compare with tabulated values for the alkali halides? HINT: The damping constant γ is usually quite small when compared with ω.

2. Consider the metallic-reflection region in Fig. 1. On the short-wavelength side it is bounded by $\kappa_e = 0$. Show that this corresponds to $\chi' = -1$ and occurs when $\omega = \omega_0 + \gamma/2$, provided γ is so small that γ^4 is negligible by comparison with γ. Also show that $n = \kappa$ at this point and that $n \to 0$ as $\omega \to \omega_0$. Show that total reflection occurs when $n = 0$.

3. According to the discussion of the zone theory in Chap. 7, it can be shown that the energy of a quantum state lying on the Brillouin-zone boundary $E_{hkl} = (h^2/8ma^2)(h^2 + k^2 + l^2)$, where a is the length of the unit-cell edge. Calculate λ_{min} for sodium. ($a = 4.29$ Å in the body-centered cubic structure of Na.) HINT: The first Brillouin zone for a *bcc* lattice is bounded by $((110))$ planes, and the value of k_0 can be determined as shown in Chap. 7. (ANSWER: $\lambda_{min} = 6{,}200$ Å.)

4. Assuming that the effective masses of a hole and an electron in an exciton are both equal to the free-electron mass, calculate the first three exciton levels according to Eq. (11) for NaCl and KCl. (Use the κ_f values listed in Chap. 12.) Compare your values with those determined from the absorption curves in Fig. 5. HINT: Compare the energy differences between the observed exciton levels.

5. The lifetime τ of an exciton is about 10^{-8} sec, and it undergoes a Brownian-like movement for which the average mean square displacement $\langle r^2 \rangle = \frac{2}{3}l v\tau$, where $l \simeq 10$ Å is the mean free path for scattering by the thermally vibrating ions. If its propagation velocity in a crystal is comparable with that of a thermal electron, what is the average displacement of an exciton? (ANSWER: $\simeq 10^{-6}$ m.)

6. As a simple example of an electron-hole pair interaction energy consider a sodium atom having a first ionization potential of 5.12 eV and a second ionization potential of 47 eV. Construct an energy-level diagram for a sodium ion based on these values and determine the energy required to excite a $2p$ electron in sodium to a $3s$ state. The experimental value turns out to be 33 eV. What is the binding energy for such a $3s$ electron-$2p$ hole pair?

7. Silicon exhibits a characteristic metallic luster when viewed with visible light, yet it becomes transparent when viewed with infrared. Why is this not true of iron? Quantitatively, what can be inferred about a crystal that is transparent to visible light but is opaque to ultraviolet?

8. In describing trapping centers with the aid of the band model (Chap. 8), the term *deep* state is used to denote a state far removed from the appropriate band edge, and *shallow* state for one close to this edge. Draw a band model of a crystal containing F, F', and V centers and indicate which are the deep and which are the shallow traps. Indicate the radiative transitions by drawing arrows in the diagram.

9. Sometimes the time dependence of the afterglow of a phosphor follows a *power-law decay*. If there are N free electrons and N empty centers at any instant of time, the intensity $I = \alpha N^2$. Derive the expression for the time dependence of the afterglow by analogy to Eq. (16) in the text. [ANSWER: $I(t) = \alpha N_0^2/(N_0 \alpha t + 1)^2$.]

10. Using the Lorenz-Lorentz equation, calculate the two refractive indices of calcite by assuming that the molar refractivity in calcite is equal to the sum of the ionic refractivity of $Ca^{2+}(R = 1.99)$ and the carbonate groups ($R_\varepsilon = 8.38$, $R_\omega = 11.32$). Why do you think that the calculated ratio ε/ω is the same as the measured ratio 1.486/1.658 but the actual magnitudes are different?

11. It is often convenient to represent the optical properties of a crystal by drawing a prolate or oblate spheroid whose major and minor axes are the indices of refraction. Consider the principal planes of such an *indicatrix* for:

(a) A uniaxial positive crystal whose indices of refraction are $\varepsilon = 2.5$ and $\omega = 1.5$.

(b) A uniaxial negative crystal for which $\varepsilon = 1.8$ and $\omega = 2.0$.

12. Suppose a plane-polarized light beam is incident on the two crystals in Exercise 11. Show the vibration directions and relative wavefronts of the ordinary and extraordinary rays if the angle of incidence on the crystal in (a) is 30 deg and on the crystal in (b) is 40 deg.

Appendixes

Appendix 1. physical
constants

The Commission on Symbols, Units and Nomenclature of the International Union of Pure and Applied Physics has recently issued a report recommending the adoption of certain changes in the use of symbols for various physical quantities and concepts. These recommendations have been issued in the interest of international uniformity, and their rapid adoption will greatly aid in removing confusion occasioned by the use of different symbols to represent the same quantities. This recommendation has received the endorsement of the American Institute of Physics, who will henceforth adopt these symbols in all their publications. In furtherance of this goal, these symbols have been used exclusively in this book. To help acquaint the reader with the new symbols, a listing of the most common symbols is given below.

Unit	Abbreviation	Unit	Abbreviation
ampere	A	lux	lx
milliampere	mA	oersted	Oe
angstrom	Å	ohm	Ω
coulomb	C	kilohm	kΩ
microcoulomb	μC	megohm	MΩ
curie	Ci	poise	P
day	day	second	sec
debye	D	microsecond	μsec
decibel	dB	nanosecond	nsec
dyne	dyn	volt	V
electron volt	eV	watt	W
farad	F	year	yr
picofarad ($= 10^{-12}$ F)	pF	biot	Bi
gauss	G	candela	cd
hertz	Hz	fermi	F
kilogauss	kG	franklin	Fr
henry	H	gilbert	Gi
hour	h	maxwell	Mx
joule	J	neper	Np
kilojoule	kJ	newton	N
kilowatt-hour	kWh	nit	nt
liter	liter	tesla ($=$ weber/m²)	T
lumen	lm	weber	Wb

Physical constants

Symbol	Constant	Magnitude
m	mass of an electron	9.108×10^{-31} kg
AMU	$\frac{1}{16}$ of mass of oxygen atom	1.660×10^{-27} kg
...	mass of hydrogen atom	1.008 AMU
		1.673×10^{-27} kg
...	1 kg equivalent	8.987×10^{16} J
...	1 AMU equivalent	9.311×10^{2} MeV
...	1 MeV equivalent	1.783×10^{-30} kg
eV	1 electron volt	1.602×10^{-19} J
		23.0 kg-cal/mole
MeV	1 million electron volts	1.602×10^{-13} J
h	Planck's constant	6.625×10^{-34} J-sec
N_0	Avogadro's number	6.025×10^{26} AMU/kg (or molecules/kg-mole)
k	Boltzmann's constant	1.380×10^{-23} J/deg K
		8.617×10^{-5} eV/deg K
R	Rydberg (gas) constant	8.317 J/deg K-mole
		1.98 cal/deg K-mole
e	charge of one electron	1.60×10^{-19} C
e/m	charge/mass of one electron	1.759×10^{11} C/kg
c	velocity of light	2.998×10^{8} m/sec
Å	angstrom unit	10^{-10} m
μ_0	permeability of vacuum	1.257×10^{-6} Wb/A-m
ϵ_0	permittivity of vacuum	8.85×10^{-12} C²/N-m²
hc/e	wavelength of photon for $h\nu = 1$ eV	12,398 Å
k/e	energy per deg K	8.616×10^{-5} eV/deg K
kT/e	energy at 300 deg K	0.0258 eV

displayed in Fig. 34*B*. This cross section can be used to determine the radius of a sphere that just fits inside the octahedral void. It is left to Exercise 12 to prove that $r = 0.414R$.

Voids in body-centered cubic packing. Although not immediately apparent in Fig. 29, there are also two types of voids present in a body-

Appendix 2. systems
of units

The choice of units for measuring physical quantities such as force, mass, charge, and so forth, is quite arbitrary, provided only that the units selected are self-consistent. For example, Newton's second law has the general form

$$F = kma \tag{1}$$

where k is a proportionality constant whose magnitude and dimensions are determined by the units chosen to specify F, m, and a. For convenience, $k \equiv 1$ in the *centimeter-gram-second* (cgs) and *meter-kilogram-second* (mks) systems of units, in which F is expressed in dynes (g-cm/sec²) and newtons (kg-m/sec²), respectively.

Other units of measurement have also been proposed, and two other systems, cgs-esu (*electrostatic units*) and cgs-emu (*electromagnetic units*), are of interest in connection with the subject matter of this book. Because the magnitudes of electric current, magnetic-field intensity, etc., are quite large when expressed in these units, a third set, called *practical units*, is often used, particularly in engineering studies. In order to simplify matters, it has been generaly agreed that mks units best represent mechanical, electrical, magnetic, and other magnitudes, so that they have been adopted for use in this book.

Electrical and magnetic units can be defined in terms of Ampère's law, which relates a magnetic field $d\mathbf{B}$ to the current element $i\,d\mathbf{l}$, producing it a distance r away:

$$d\mathbf{B} = k_1 i \frac{d\mathbf{l} \times \mathbf{r}}{r^3}. \tag{2}$$

The scale constant

$$k_1 \equiv \frac{\mu_0}{4\pi} = 10^{-7} \text{ Wb/A-m} \tag{3}$$

in the mks system in which $\mu_0 = 1.26 \times 10^{-6}$ Wb/A-m ($= 4\pi \times 10^7$ H/m).

Similarly, writing Coulomb's law for the force between two charges q_1 and q_2,

$$F = k_2 \frac{q_1 q_2}{r^2} \tag{4}$$

the scale constant, in mks,

$$k_2 \equiv \frac{1}{4\pi\epsilon_0} = 9 \times 10^9 \text{ N-m}^2/\text{C}^2 \tag{5}$$

and $\epsilon_0 = 8.85 \times 10^{-12}$ C^2/N-m^2. With the above definitions, the mks system is internally consistent.

By comparison, k_1 is set equal to unity in the emu system and Ampère's law is used to define the unit of current called an *abampere*. Similarly, $k_2 = 1$ in the esu system, and Coulomb's law defines the unit of charge called a *statcoulomb*. In fact, in the cgs system, magnetic quantities are defined only in emu, and electric quantities in esu, so that it is necessary to convert from one to the other whenever both electric and magnetic quantities appear in the same equation. These relations are indicated in Table 1.

Finally, it should be noted that the factor 4π has been explicitly introduced into Ampère's law (3) and Coulomb's law (5) by the form of the scale constants adopted in the mks system. By these definitions, the quantity 4π is removed from many other relations, for example, Maxwell's equations for electromagnetic fields. This particular choice of scale constants is called the mks *rationalized* system of units and has been adopted in this book because of its internal consistency and general acceptance. The relations between the most important quantities in mks and cgs systems are listed in Table 1.

Table 1
Conversion units

mks	cgs		
$B = \kappa_m \mu_0 H$ (1 Wb $= 10^8$ Mx) $\quad$ $I = \sigma\mathcal{E}$ (σ Ω^{-1}-m^{-1}) $\quad$ $D = \kappa_e\epsilon_0\mathcal{E}$ $\quad$ $c = 2.998 \times 10^8$ m/sec $\quad$ $\mu_0 = 4\pi \times 10^{-7}$ H/m $\quad$ $\epsilon_0 = 8.85 \times 10^{-12}$ F/m	$B = \kappa_m H$ (1 Mx $= 1$ G-cm^2) $\quad$ $I = \sigma\mathcal{E}$ (σ Ω^{-1}-cm^{-1}) $\quad$ $D = \kappa_e\mathcal{E}$ $\quad$ $c = 2.998 \times 10^{10}$ cm/sec		
	esu		emu
1 C $\quad$ 1 A (1 C/sec) $\quad$ 1 V $\quad$ 1 F = 1 C/V	$= 3 \times 10^9$ escoulombs $\quad$ $= 3 \times 10^9$ esamperes $\quad$ $= \frac{1}{300}$ esvolt $\quad$ $= 9 \times 10^{-11}$ cm		$= \frac{1}{10}$ abcoulomb $\quad$ $= \frac{1}{10}$ abamperes $\quad$ $= 10^8$ abvolts
1 J = 1 V $\times$ 1 C $\quad$ $= 10^7$ ergs	1 erg = 1 esvolt $\times$ 1 escoulomb $\quad$ $= 10^{-7}$ J	1 erg = 1 abvolt $\times$ 1 abcoulomb $\quad$ $= 10^{-7}$ J	

κ_m and κ_e are dimensionless and have the same value in both systems.
1 eV $\simeq 4.803 \times 10^{-10}$ escoulomb $\times \frac{1}{300}$ esvolt $\simeq 1.601 \times 10^{-12}$ erg.

Appendix 3. atomic radii

Ionic radii. As discussed in Chapter 1, ionic radii are usually determined by subdividing the cation-anion separation in crystals by some means. Because the bonding in, say, alkali halides is not purely ionic, however, different values are obtained for different compounds. For example, the radius of Li^{1+} calculates as follows:

$$0.68 \text{ Å from Li-F}$$
$$0.76 \text{ Å from Li-Cl}$$
$$0.78 \text{ Å from Li-Br}$$
$$0.82 \text{ Å from Li-I}$$

Pauling derived a set of *crystal radii* by assuming that the radius of O^{2-} is 1.40 Å. He further deduced certain crystal radii from the so-called univalent radii according to

$$R_{crystal} = R_{univalent} \, z^{-2/(n-1)}. \tag{1}$$

A list of Pauling's crystal radii is given in Table 1.

Ahrens used the radius of F^{1-} ($= 1.33$ Å) instead of oxygen to determine the radii of a number of cations. Using these as standards, he deduced the radii of the other cations by employing their ionization potentials I in the relation

$$R_{cation} = \frac{K}{I^n} \tag{2}$$

where K has a constant value for each element considered, and n is assumed to have the same value for all the elements. He then used the pairs of ions Mn^{2+} and Mn^{7+} and Tl^{1+} and Tl^{3+} as calibration ions. The cation radii deduced by Ahrens are listed in Table 2. Because of the procedure used in deriving their values, these radii are correct, strictly speaking, only for cations having six-fold coordination.

Covalent radii. When atoms form electron-pair bonds with each other, the interatomic separation depends on the number of electron-pair bonds formed. In

446

the case of hybrid-bond formation, it also depends on the types of orbitals used. For example, the interatomic separation for two carbon atoms is as follows:

> 1.542 Å when single bonds are formed
> 1.330 Å when double bonds are formed
> 1.204 Å when triple bonds are formed.

Similarly, the radius of covalently bonded nickel depends on the orbitals used, as follows:

> 1.21 Å if octahedral d^2sp^3 orbitals are used
> 1.22 Å if square dsp^2 orbitals are used
> 1.23 Å if tetrahedral sp^3 orbitals are used.

As can be seen from the above, it is not possible to assign a single radius to cover all cases of covalent bonding. Moreover, it should be remembered that most atoms do not form purely covalent bonds in crystals. Nevertheless, a number of elements appear to have very nearly the same radii in a number of crystals in which they are believed to be covalently bonded. Table 3 lists such radii for a number of elements. These radii are appropriate, strictly speaking, only when the atoms have tetrahedral coordination.

Table 1
Crystal radii of Pauling,† in angstrom units

4−	3−	2−	1−	0	1+	2+	3+	4+	5+	6+	7+
			H 2.08	He‡ 0.93	Li 0.60	Be 0.31	B 0.20	C 0.15	N 0.11	O 0.09	F 0.07
C 2.60	N 1.71	O 1.40	F 1.36	Ne‡ 1.12	Na 0.95	Mg 0.65	Al 0.50	Si 0.41	P 0.34	S 0.29	Cl 0.26
Si 2.71	P 2.12	S 1.84	Cl 1.81	Ar‡ 1.54	K 1.33	Ca 0.99	Sc 0.81	Ti 0.68	V 0.59	Co 0.52	Mn 0.46
					Cu 0.96	Zn 0.74	Ga 0.62	Ge 0.53	As 0.47	Se 0.42	Br 0.39
Ge 2.72	As 2.22	Se 1.98	Br 1.95	Kr‡ 1.69	Rb 1.48	Sr 1.13	Y 0.93	Zr 0.80	Nb 0.70	Mo 0.62	
					Ag 1.26	Cd 0.97	In 0.81	Sn 0.71	Sb 0.62	Te 0.56	I 0.50
Sn 2.94	Sb 2.45	Te 2.21	I 2.16	Xe‡ 1.90	Cs 1.69	Ba 1.35	La 1.15	Ce 1.01			
					Au 1.37	Hg 1.10	Tl 0.95	Pb 0.84	Bi 0.74		

† Linus Pauling, *The nature of the chemical bond*, 2d ed. (Cornell University Press, Ithaca, N.Y., 1948), p. 346.

‡ Univalent radii.

Table 2
Ionic radii of cations,† in angstrom units

Ion	Radius	Ion	Radius	Ion	Radius
Ac^{3+}	1.18	Cs^{1+}	1.67	Mn^{2+}	0.80
				Mn^{3+}	0.66
Ag^{1+}	1.26	Cu^{1+}	0.96	Mn^{4+}	0.60
Ag^{2+}	0.89	Cu^{2+}	0.72	Mn^{7+}	0.46
Al^{3+}	0.51	Dy^{3+}	0.92	Mo^{4+}	0.70
				Mo^{6+}	0.62
Am^{3+}	1.07	Er^{3+}	0.89		
Am^{4+}	0.92			N^{3+}	0.16
		Eu^{3+}	0.98	N^{5+}	0.13
As^{3+}	0.58				
As^{3+}	0.46	F^{7+}	0.08	Na^{1+}	0.97
At^{7+}	0.62	Fe^{2+}	0.74	Nb^{4+}	0.74
		Fe^{3+}	0.64	Nb^{5+}	0.69
Au^{1+}	1.37			Nd^{3+}	1.04
Au^{3+}	0.85	Fr^{1+}	1.80		
B^{3+}	0.23	Ga^{3+}	0.62	Ni^{2+}	0.69
Ba^{2+}	1.34	Gd^{3+}	0.97	Np^{3+}	1.10
				Np^{4+}	0.95
Be^{2+}	0.35	Ge^{2+}	0.73	Np^{7+}	0.71
		Ge^{4+}	0.53		
Bi^{3+}	0.96			O^{6+}	0.10
Bi^{5+}	0.74	Hf^{4+}	0.78		
				Os^{6+}	0.69
Br^{5+}	0.47	Hg^{2+}	1.10		
Br^{7+}	0.39			P^{3+}	0.44
		Ho^{3+}	0.91	P^{5+}	0.35
C^{4+}	0.16				
		I^{5+}	0.62	Pa^{1+}	1.13
Ca^{2+}	0.99	I^{7+}	0.50	Pa^{3+}	0.98
				Pa^{5+}	0.89
Cd^{2+}	0.97	In^{3+}	0.81		
				Pb^{2+}	1.20
Ce^{3+}	1.07	Ir^{4+}	0.68	Pb^{4+}	0.84
Ce^{4+}	0.94				
		K^{1+}	1.33	Pd^{2+}	0.80
Cl^{5+}	0.34			Pd^{4+}	0.65
Cl^{7+}	0.27	La^{3+}	1.14		
				Pm^{3+}	1.06
Co^{2+}	0.72	Li^{1+}	0.68		
Co^{3+}	0.63			Po^{6+}	0.67
		Lu^{2+}	0.85		
Cr^{3+}	0.63			Pr^{3+}	1.06
Cr^{6+}	0.52	Mg^{2+}	0.66	Pr^{4+}	0.92

Table 2 (*Continued*)

Pt^{2+}	0.80	Se^{4+}	0.50	Ti^{3+}	0.76
Pt^{4+}	0.65	Se^{7+}	0.42	Ti^{4+}	0.68
Pu^{3+}	1.08	Si^{4+}	0.42	Tl^{1+}	1.47
Pu^{4+}	0.93			Tl^{3+}	0.95
Ra^{2+}	1.43	Sm^{3+}	1.00	Tm^{2+}	0.87
Rb^{1+}	1.47	Sn^{2+}	0.93	U^{4+}	0.97
		Sn^{4+}	0.71	U^{6+}	0.80
Re^{4+}	0.72	Sr^{2+}	1.12	V^{2+}	0.88
Re^{7+}	0.56			V^{3+}	0.74
				V^{4+}	0.63
Rh^{3+}	0.68	Ta^{5+}	0.68	V^{5+}	0.59
		Tb^{3+}	0.93	W^{4+}	0.70
Ru^{4+}	0.67	Tb^{4+}	0.81	W^{6+}	0.62
S^{4+}	0.37				
S^{6+}	0.30	Tc^{7+}	0.56	Y^{3+}	0.92
				Yb^{3+}	0.86
Sb^{3+}	0.76	Te^{4+}	0.70		
Sb^{5+}	0.62	Te^{6+}	0.56	Zn^{2+}	0.74
Sc^{3+}	0.81	Th^{4+}	1.02	Zr^{4+}	0.79

† L. H. Ahrens, The use of ionization potentials, Part 1, Ionic radii of the elements, *Geochim. et Cosmochim. Acta*, vol. 2 (1952), pp. 155–169.

Table 3
Tetrahedral covalent radii,† in angstrom units

	Be	B	C	N	O	F
	1.06	0.88	0.77	0.70	0.66	0.64
	Mg	Al	Si	P	S	Cl
	1.40	1.26	1.17	1.10	1.04	0.99
Cu	Zn	Ga	Ge	As	Se	Br
1.35	1.31	1.26	1.22	1.18	1.14	1.11
Ag	Cd	In	Sn	Sb	Te	I
1.53	1.48	1.44	1.40	1.36	1.32	1.28
Au	Hg	Tl	Pb	Bi		
1.50	1.48	1.47	1.46	1.46		

† A. F. Wells, *Structural inorganic chemistry*, 2d ed. (Oxford University Press, London, 1950), p. 50.

Metallic radii. Probably the simplest radii to determine are those of metal elements. These can be determined by halving the known $Me\text{-}Me$ distances in the metal crystals. Nevertheless, two problems arise in this connection. First of all, the interatomic separations are usually not the same for different allotropic modifications. Secondarily, in the case of noncubic metals, there are two or more interatomic distances that occur in the same crystal. Moreover, these metals do not necessarily have the same radii in alloys that they have in the elemental crystals. The radii listed in Table 4, therefore, represent a range of values that are most commonly observed. As in the case of ionic and covalent radii discussed above, the calculation of interatomic separations by simple addition of the values in Table 4 gives only approximately accurate results.

Table 4
Metallic radii, in angstrom units

Li 1.52	Be 1.1-1.14	B ~1.0	C 0.71-0.77						
Na 1.85	Mg 1.60	Al 1.43	Si 1.17	P 1.09					
K 2.31	Ca 1.97	Sc 1.60-1.65	Ti 1.44-1.47	V 1.31	Cr 1.25	Mn 1.23-1.48	Fe 1.24	Co 1.25	Ni 1.24
Cu 1.28	Zn 1.33-1.45	Ga 1.22-1.40	Ge 1.22	As 1.25-1.57	Se 1.16-1.73		Ru 1.32-1.35	Rh 1.34	Pd 1.37
Rb 2.46	Sr 2.15	Y 1.80-1.83	Zr 1.58-1.61	Nb 1.43	Mo 1.36	Tc 1.35-1.36	Os 1.34-1.36	Ir 1.35	Pt 1.38
Ag 1.44	Cd 1.49-1.64	In 1.62-1.69	Sn 1.40-1.59	Sb 1.45-1.68	Te 1.43-1.73				
Cs 2.63	Ba 2.17	La 1.36-1.87	Hf 1.57-1.60	Ta 1.43	W 1.37	Re 1.37-1.38			
Au 1.44	Mg 1.50	Tl 1.70-1.73	Pb 1.75	Bi 1.55-1.74	Po 1.64-1.67				
			Th 1.80	Pa 1.60-1.62	U 1.50				

Appendix 4. periodic chart of the elements

Group / Period	I	II	III	IV	V	VI	VII	VIII	Zero
1	1 H Hydrogen 1.008								2 He Helium 4.003
2	3 Li Lithium 6.940	4 Be Beryllium 9.013	5 B Boron 10.82	6 C Carbon 12.01	7 N Nitrogen 14.01	8 O Oxygen 16.000	9 F Fluorine 19.00		10 Ne Neon 20.18
3	11 Na Sodium 22.99	12 Mg Magnesium 24.32	13 Al Aluminum 26.98	14 Si Silicon 28.09	15 P Phosphorus 30.97	16 S Sulfur 32.06	17 Cl Chlorine 35.45		18 Ar Argon 39.94
4	19 K Potassium 39.10	20 Ca Calcium 40.08	21 Sc Scandium 44.96	22 Ti Titanium 47.90	23 V Vanadium 50.95	24 Cr Chromium 52.01	25 Mn Manganese 54.94	26 Fe Iron 55.85 · 27 Co Cobalt 58.94 · 28 Ni Nickel 58.71	
	29 Cu Copper 63.54	30 Zn Zinc 65.38	31 Ga Gallium 69.72	32 Ge Germanium 72.60	33 As Arsenic 74.91	34 Se Selenium 78.96	35 Br Bromine 79.91		36 Kr Krypton 83.80
5	37 Rb Rubidium 85.48	38 Sr Strontium 87.63	39 Y Yttrium 88.92	40 Zr Zirconium 91.22	41 Nb Niobium 92.91	42 Mo Molybdenum 95.95	43 Tc Technetium 99	44 Ru Ruthenium 101.1 · 45 Rd Rhodium 102.9 · 46 Pd Palladium 106.4	
	47 Ag Silver 107.9	48 Cd Cadmium 112.4	49 In Indium 114.8	50 Sn Tin 118.7	51 Sb Antimony 121.8	52 Te Tellurium 127.6	53 I Iodine 126.9		54 Xe Xenon 131.3
6	55 Cs Cesium 132.9	56 Ba Barium 137.4	57 La Lanthanum A 138.9	72 Hf Hafnium 178.5	73 Ta Tantalum 180.9	74 W Tungsten 183.9	75 Re Rhenium 186.2	76 Os Osmium 190.2 · 77 Ir Iridium 192.2 · 78 Pt Platinum 195.1	
	79 Au Gold 197.0	80 Hg Mercury 200.6	81 Tl Thallium 204.4	82 Pb Lead 207.2	83 Bi Bismuth 209.0	84 Po Polonium 210	85 At Astatine 210		86 Rn Radon 222
7	87 Fr Francium 223	88 Ra Radium 226	89 Ac Actinium B 227						

Lanthanide series (A)	Actinide series (B)
58 Ce Cerium	90 Th Thorium
59 Pr Praseodymium	91 Pa Protactinium
60 Nd Neodymium	92 U Uranium
61 Pm Promethium	93 Np Neptunium
62 Sm Samarium	94 Pu Plutonium
63 Eu Europium	95 Am Americium
64 Gd Gadolinium	96 Cm Curium
65 Tb Terbium	97 Bk Berkelium
66 Dy Dysprosium	98 Cf Californium
67 Ho Holmium	99 Es Einsteinium
68 Er Erbium	100 Fm Fermium
69 Tm Thulium	101 Md Mendelevium
70 Yb Ytterbium	102 No Nobelium
71 Lu Lutetium	103 Lw Lawrencium

INDEX